Reagents for Organic Synthesis

Fieser and Fieser's

Reagents for Organic Synthesis

VOLUME THIRTEEN

Mary Fieser
Harvard University

Janice G. Smith
Mount Holyoke College

WILEY

A WILEY-INTERSCIENCE PUBLICATION
JOHN WILEY & SONS
NEW YORK • CHICHESTER • BRISBANE •
TORONTO • SINGAPORE

ISBN 0-471-63007-1
ISSN 0271-616X

Printed in the United States of America

10 9 8 7 6 5 4 3 2

PREFACE

This volume of Reagents discusses research published for the most part during 1985 and the first half of 1986. We are particularly grateful to Richard T. Peterson and Marcello DiMare, who provided valuable advice and help during the preparation of the manuscript and who also supervised the proofreading. The difficulties of proofreading have been greatly alleviated by the careful work of many colleagues including John C. Kappos, Wei-guo Su, Philip A. Carpino, Stephen W. Wright, Robert W. Hahl, Seiichi P. T. Matsuda, Francis S. Hannon, Paul Da Silva Jardine, Stephen W. Kaldor, Jonathan A. Ellman, Kristin M. Lundy, Greg C. Fu, Guy V. Lamoureux, and Erick M. Carreira.

Special thanks are extended to Dr. Daniel C. Smith, who helped prepare the indexes for this volume. The indexes now conform to the style used for the Collective Index for Volumes 1 to 12, which will be available shortly. Dr. Smith also provided the photograph of some of the proofreaders.

MARY FIESER
JANICE G. SMITH

October 20, 1987

CONTENTS

A

Acetaldehyde triphenylmethylsulfenimine, $CH_3CH=NSC(C_6H_5)_3$ **(1),** m.p. 129–135°. Tritylsulfenimines can be prepared by reaction of aldehydes or ketones with triphenylmethanesulfenamide, $(C_6H_5)_3CSNH_2$, m.p. 119–122°, in CH_2Cl_2 in the presence of pyridinium *p*-toluenesulfonate.[1]

β-*Lactams*.[2] Lithium enolates of esters react with S-tritylsulfenimines to form N-tritylsulfenyl β-lactams in which the *cis*-isomer predominates. Several reagents can be used for reductive removal of the STr group: Bu_3P (115°), $ISi(CH_3)_3$ (25°), Li–NH_3.

Example:

[1] B. P. Branchaud, *J. Org.*, **48**, 3531 (1983).
[2] D. A. Burnett, D. J. Hart, and J. Liu, *J. Org.*, **51**, 1929 (1986).

Acetic-formic anhydride.

Deoxygenation of amine oxides. Trialkylamine N-oxides and dialkylarylamine N-oxides are converted to the tertiary amines on reaction with this anhydride in CH_2Cl_2 at 25°.[1]

[1] N. Tokitoh and R. Okazaki, *Chem. Letters*, 1517 (1985).

Example:

Acetohydroxamic acid, $CH_3CONHOH$ (**1**), m.p. 89–92°. Preparation.[1] Supplier: Aldrich.

Amidation of alkyl aryl ethers.[2]   Either anisole or phenetole undergoes *para*-amidation exclusively on reaction with acetohydroxamic acid in polyphosphoric acid (yields 50–60%). Thus phenetole is converted into the drug phenacetin in ~50% yield (equation I).

[1] W. N. Fishbein, J. Daly, and C. Streeter, *Anal. Biochem.*, **28**, 13 (1969).
[2] J. March, and J. S. Engenito, Jr., *J. Org.*, **46**, 4304 (1981); *idem, Org. Syn.* submitted (1985).

1-Acetoxy-2,4-hexadiene, $CH_3CH{=}CHCH{=}CHCH_2OCOCH_3$ (**1**). The alcohol is available from Aldrich.

Diels-Alder reactions.   Although 2,4-hexadiene shows only slight regioselec-tivity in Diels-Alder reactions, the 1-acetoxy derivative (**1**) undergoes regioselective cycloaddition with a variety of dienophiles.[1]

Example:

[1] G. A. Kraus, S. Yue, and J. Sy, *J. Org.*, **50**, 283 (1985).

3-Acetoxy-2-tributylstannylmethyl-1-propene, $CH_2{=}C{\Large\diagup^{CH_2SnBu_3}_{\diagdown CH_2OAc}}$ (1).

The reagent is prepared in three steps from methallyl alcohol (38% yield).

4-Methylenetetrahydrofurans and 4-methylenepyrrolidines.[1]   The adducts formed by Lewis acid-catalyzed reaction of **1** with an aldehyde or ketone undergo cyclization to 4-methylenetetrahydrofurans in the presence of a Pd(0) catalyst prepared from $Pd(OAc)_2$, $P(C_6H_5)_3$, and BuLi, and a base, DBU or $N(C_2H_5)_3$ (1.5 equiv.).

Example:

$$C_6H_5CHO + 1 \xrightarrow[92\%]{\substack{BF_3{\cdot}O(C_2H_5)_2 \\ CH_2Cl_2, -78°}} \underset{C_6H_5}{\overset{OH\quad CH_2}{\diagup}}{\diagdown}OAc \xrightarrow[70\%]{Pd(0), DBU} \underset{C_6H_5}{O}{\diagup}CH_2$$

A similar reaction of **1** with an imine followed by cyclization results in a 4-methylenepyrrolidine.

Example:

$$C_6H_5CH{=}N(CH_2)_2CH_3 + 1 \xrightarrow[88\%]{} \underset{C_6H_5}{\overset{C_3H_7\diagdown NH\quad CH_2}{}}OAc \xrightarrow[72\%]{} \underset{C_6H_5}{\overset{C_3H_7\diagdown N}{}}{=}CH_2$$

[1] B. M. Trost and P. J. Bonk, *Am. Soc.*, **107**, 1778 (1985).

3-Acetoxy-2-trimethylsilylmethyl-1-propene, (1).

4-Methylene-1-cyclopentenes.[1]   All attempts to effect cycloaddition of the acetate **1** (**9**, 454; **11**, 578) to alkynes are unsuccessful, but **1** does add to norbornadienes (prepared from alkynes) in the presence of a palladium(0) catalyst to form adducts that afford 4-methylene-1-cyclopentenes on flash vacuum thermolysis (equation I).

$$(I)\quad \underset{R^2}{\overset{R^1}{\underset{|}{\overset{|}{\underset{C}{\overset{C}{\parallel}}}}}} \xrightarrow{C_2H_6} \underset{R^2}{\overset{R^1}{}} + \underset{(CH_3)_3SiCH_2}{\overset{AcOCH_2}{\underset{|}{\overset{|}{C}}{=}CH_2}} \xrightarrow[25-90\%]{\substack{1)\ Pd(0),\ THF,\ \Delta \\ 2)\ 530-755°}} \underset{R^2}{\overset{R^1}{}}{=}CH_2$$

$$\mathbf{1}$$

[1] B. M. Trost, J. M. Balkovec, and S. R. Angle, *Tetrahedron Letters*, **27**, 1445 (1986).

3-Acyl-1,3-oxazolidine-2-thiones, chiral (1). Nagao and co-workers[1] have prepared the chiral 3-acetyl-1,3-oxazolidine-2-thiones (**1a** and **1b**) and used them to effect diastereoselective aldol reactions. The two chiral auxiliaries show, as expected, opposite diastereoselectivities, but contrast with the diastereoselectivities observed with chiral 4-alkyl-2-oxazolidones (**11**, 379–381). This aldol reaction has been used to prepare the chiral azetidinone **4** (equation I) and (+)-Prelog-Djerassi lactone.

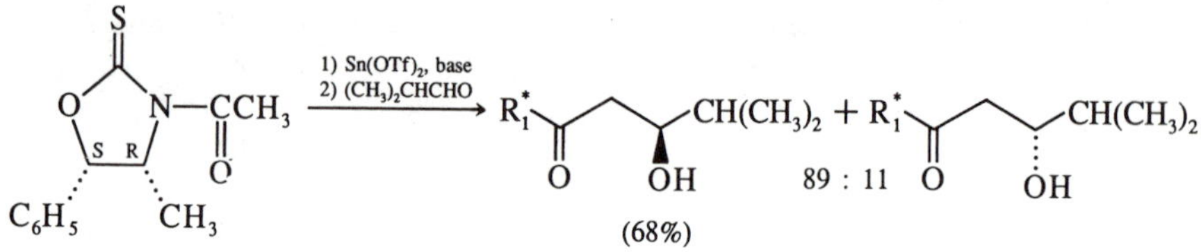

(R, S)-**1a** = R₁*COCH₃

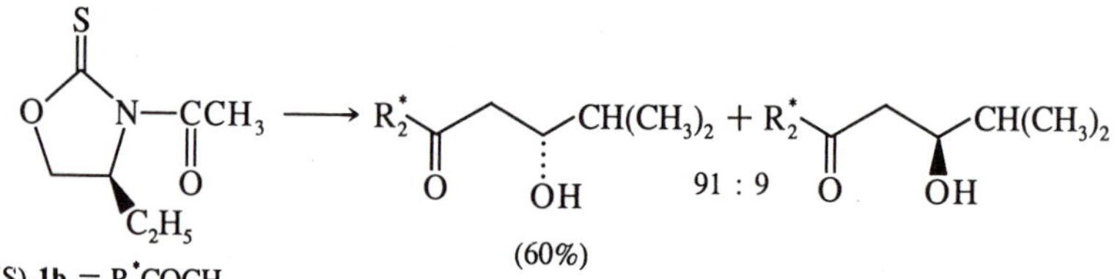

(S)-**1b** = R₂*COCH₃

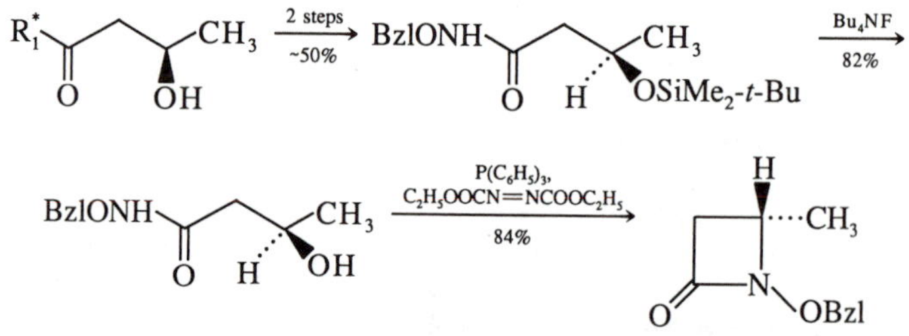

4

[1] Y. Nagao, S. Yamada, T. Kumagai, M. Ochiai, and E. Fujita, *J.C.S. Chem. Comm.*, 1418 (1985); Y. Nagao, T. Inoue, K. Hashimoto, Y. Hagiwara, M. Ochiai, and E. Fujita, *ibid.*, 1419 (1985).

(*o*-Acylphenoxy)acetyl chlorides,   (**1**).

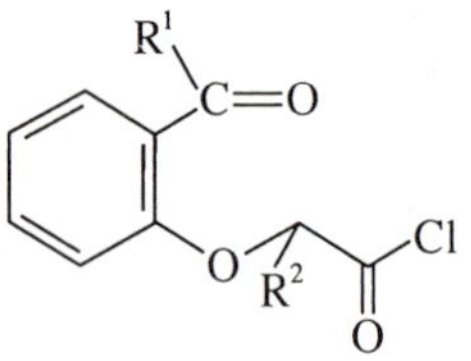

(*o*-Acylphenoxy)acetic acids are prepared by reaction of *o*-acylphenols and αbromoacetic acids in THF with NaH (51–78% yield). The acid chlorides are formed on reaction with oxalyl chloride in C_6H_6.

Intramolecular [2+2]cycloadditions; benzofurans.   On dehydrochlorination with triethylamine in refluxing C_6H_6, these acid chlorides are converted into (*o*-acylphenoxy)ketenes, which undergo a [2+2]cycloaddition reaction to form β-lactones, which are converted to benzofurans by spontaneous decarboxylation in 53–82% yield.[1]

Example:

[1] W. T. Brady and Y. F. Giang, *J. Org.*, **51**, 2145 (1986).

Alkylaluminum halides.

Intramolecular Sakurai reaction.   Allylic and propargylic silanes can undergo a Lewis acid catalyzed intramolecular Sakurai reaction.[1] In cyclization to hydrindanones, the stereochemical outcome can differ from that obtained by fluoride ion catalysis (presumably kinetically controlled cyclization), equation (I).[2]

	77%	1:1
F⁻	77%	1:1
$C_2H_5AlCl_2$	62%	1:3

Intramolecular addition to conjugated dienones is particularly useful because it permits entry to fused cyclooctane or cyclohexane rings (Sakurai product).[3]
Example:

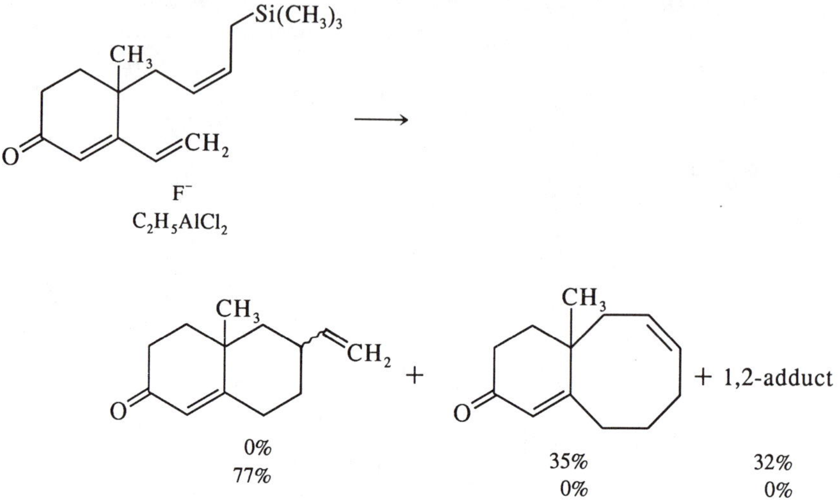

This Sakurai reaction has been used for a stereoselective synthesis of nootkatone (2),[4] an eremophilane sesquiterpene. Thus the trienone 1 cyclizes to 2 at 0° in the presence of $C_2H_5AlCl_2$ in 50% yield. $TiCl_4$ and $BF_3 \cdot O(C_2H_5)_2$ are less effective catalysts (28% and 23% yields of 2).

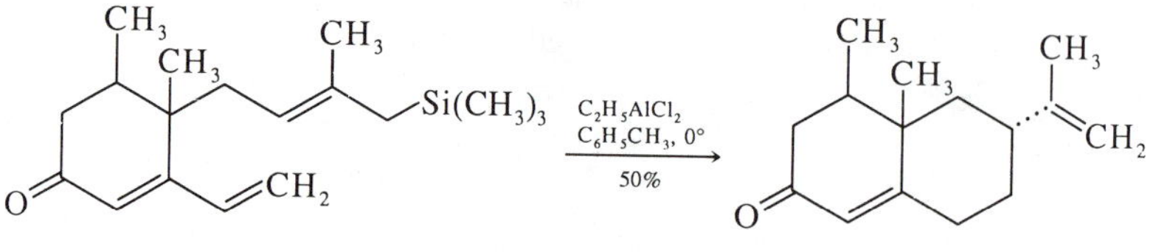

Diels-Alder cyclization of 2,8,10-undecatrienals.[5]   These unsaturated alde-hydes undergo intramolecular Diels-Alder cyclization, particularly under Lewis acid catalysis. The reaction is highly *endo*-selective. Silyl-protected alcohol groups at C_4 and C_7 can be present, and *t*-butyldimethylsilyl ethers show a strong axial pref-erence.

Example:

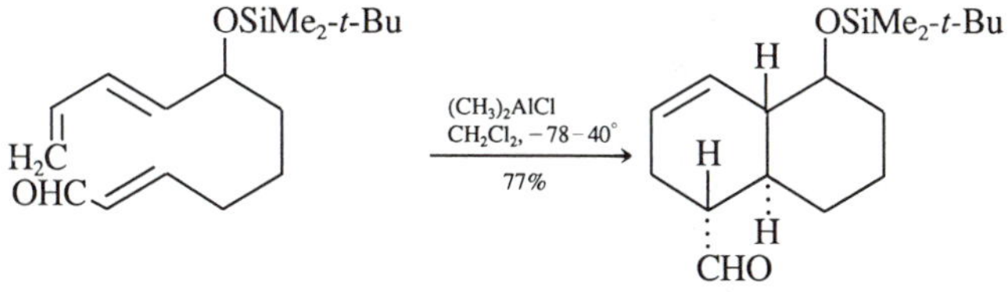

This reaction was used for a synthesis of the hydronaphthalenecarboxylic acid **1**, a subunit of the macrocyclic antitumor antibiotic chlorothricolide.

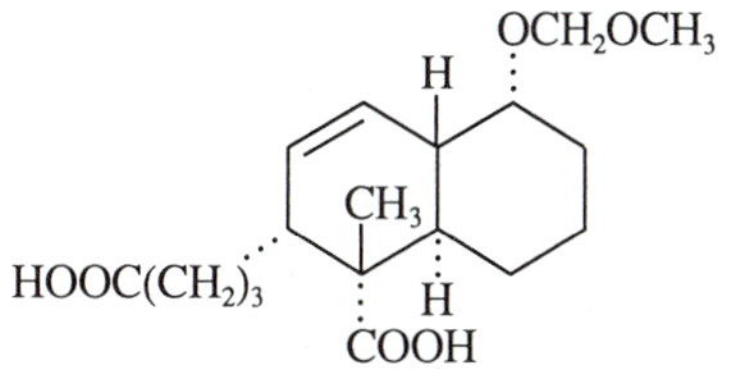

1

***Bicyclic alcohols with angular methyl groups* (11, 10–11).** The reaction of 2,5-dimethylmethylenecyclopentane (**1**) with acrolein (1 equiv.) and $(CH_3)_2AlCl$ involves two sequential ene reactions to give the expected bicyclic alcohol **2** (69%) and in addition the corresponding ketone **3** (12% yield), even in the absence of an oxidant.[6] In the presence of 2 equiv. of acrolein, the reaction results in **3** in 60% yield and only 1% of **2**. Apparently, the alkoxide precursor to **2** undergoes

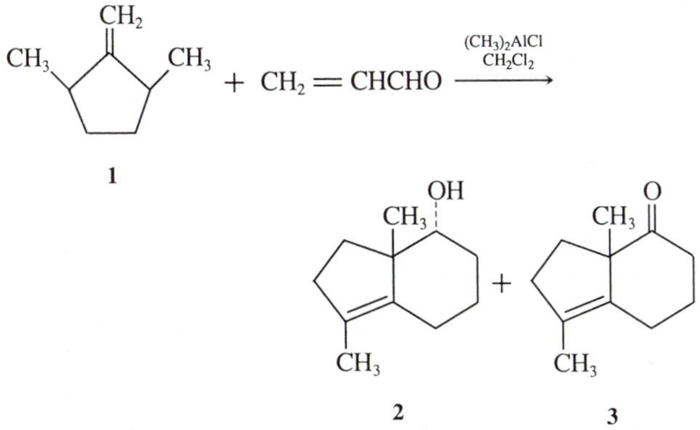

2   **3**

in situ Oppenauer oxidation to give **3**. The oxidation can be prevented by use of excess $(CH_3)_2AlCl$.

[1] D. Schinzer, S. Sólyom, and M. Becker, *Tetrahedron Letters*, **26**, 1831 (1985).
[2] G. Majetich, J. Defauw, K. Hull, and T. Shawe, *ibid.*, **26**, 4711 (1985).
[3] G. Majetich, K. Hull, J. Defauw, and R. Desmond, *ibid.*, **26**, 2747 (1985); G. Majetich, K. Hull, and R. Desmond, *ibid.*, **26**, 2751 (1985).
[4] G. Majetich, M. Behnke, and K. Hull, *J. Org.*, **50**, 3615 (1985).
[5] J. A. Marshall, J. E. Audia, J. Grote, and B. G. Shearer, *Tetrahedron*, **42**, 2893 (1986).
[6] B. B. Snider and B. E. Goldman, *ibid.*, **42**, 2951 (1986).

Alkyldimesitylboranes, (Mes)₂BR.

syn-**1,2-Diols.** Oxidation of the intermediate obtained by reaction of aromatic aldehydes with the anion of ethyldimesitylborane results selectively in *syn*-1,2-diols.[1]

Example:

$C_6H_5CHO \xrightarrow[\substack{2)\ H_2O_2,\ ^-OH \\ 84\%}]{1)\ (Mes)_2BCHCH_3}$

(*syn/anti* = 92:8)

1,3-Diols. The anions of (Mes)₂BC₂H₅ and (Mes)₂BCH₃ react with substituted oxiranes to give products that are oxidized to 1,3-diols. The reaction is regioselective in that the least substituted carbon atom is the site of attack. The reaction shows modest to high *syn*-selectivity.[2]

Example:

(*syn/anti* = 7:1)

[1] A. Pelter, D. Buss, and A. Pitchford, *Tetrahedron Letters*, **26**, 5093 (1985).
[2] A. Pelter, G. Bugden and R. Rosser, *ibid.*, **26**, 5097 (1985).

N-Alkyl-N-fluoroarenesulfonamides.

Alkenyl fluorides.[1] Alkenyllithium compounds, obtained from reaction of alkenyl iodides with *t*-butyllithium, are converted into alkenyl fluorides by reaction with an N-*t*-butyl-N-fluorobenzenesulfonamide (**1**) in THF/ether/pentane at −120 → 20°; yields 71–88%.

1) t-BuLi
2) **1**, −120 → 20°

74%

S. H. Lee and J. Schwartz, *Am. Soc.*, **108**, 2445 (1986).

Allyl chloroformate, $ClCO_2CH_2CH=CH_2$ (AocCl,**1**), b.p. 109–110°. Supplier: Aldrich.

β-*Glycosidation of* D-*glucosamine.*   N-Allyloxycarbonyl derivatives of amines are obtained by reaction with **1** and $N(C_2H_5)_3$. This derivative (**2**) of triacetyl-D-glucosamine undergoes selective β-glycosidation by virtue of anchimeric assistance.[1]

The protective group can be removed by reaction of a catalytic amount of $Pd[P(C_6H_5)_3]_4$ and excess dimethyl malonate. This transfer reaction does not affect usual protective groups. On the other hand, the acetyl groups can be removed selectively by hydrazinolysis.

Protection of nucleosides.[2]   The allyloxycarbonyl (Aoc) group is recommended for protection of the amino and imide groups of nucleoside bases as well as the sugar hydroxyl groups. It is stable to fluoride ion and to dichloroacetic acid, used to deblock usual protecting groups of nucleoside bases, but is easily removed by brief treatment with several palladium catalysts in the presence of nucleophiles. Ammonium formates are particularly effective in the presence of Pd(0), $Pd_2(dba)_3 \cdot$ $CHCl_3$, $PdCl_2$.

[1] P. Boullanger and G. Descotes, *Tetrahedron Letters*, **27**, 2599 (1986).
[2] Y. Hayakawa, H. Kato, M. Uchiyama, H. Kajino, and R. Noyori, *J. Org.*, **51**, 2400 (1986).

Allyltributyltin.

Homoallylamines.[1] In the presence of BF_3 etherate or $TiCl_4$ allyltributyltin (**1**) reacts with aldimines to give homoallylamines in moderate to high yield.
Example:

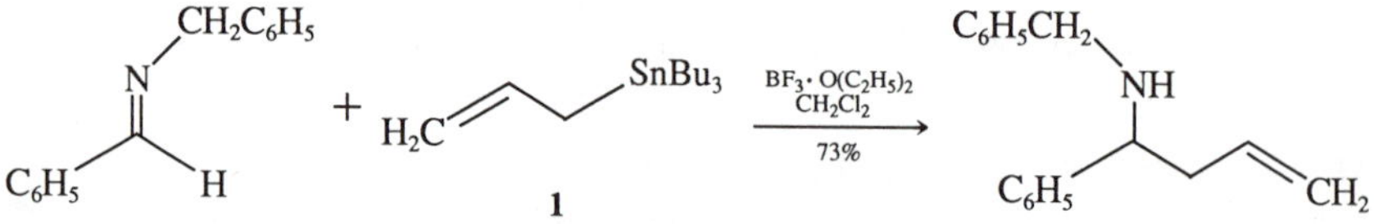

A similar reaction occurs with crotyltributyltin to furnish mainly *syn*-β-methyl homoallylamines under controlled conditions in which the aldimine and the Lewis acid catalyst are allowed to complex for some time before addition of the reagent.
Example:

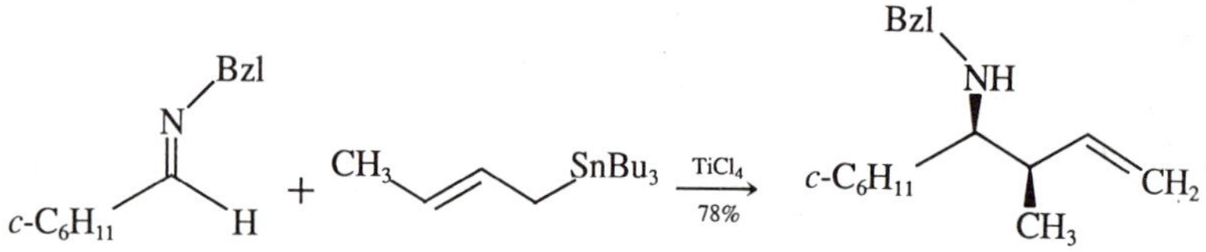

(*syn/anti* = 23:1)

α-Allylation of pyridinium salts. The reagent (**1**) reacts with N-alkoxycarbonylpyridinium salts (**2**) regioselectively to afford α-allyl-1,2-dihydropyridines.[2]
Example:

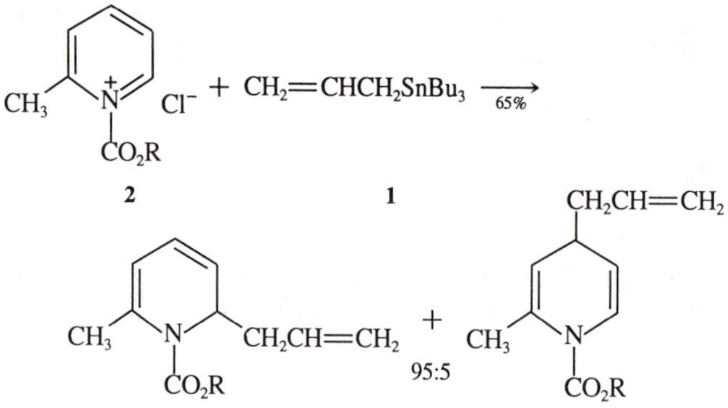

[1] G. E. Keck and E. J. Enholm, *J. Org.*, **50**, 146 (1985).
[2] R. Yamaguchi, M. Moriyasu, M. Yoshioka, and M. Kawanisi, *ibid.*, **50**, 287 (1985).

Allyltriisopropylsilane, $CH_2=CHCH_2Si(i\text{-}Pr)_3$ (**1**), b.p. $130°/16$ mm. The silane is prepared by reaction of $CH_2=CHCH_2MgBr$ with triisopropylsilyl triflate (96% yield).

α-Silyl aldehydes. The anion of this silane is alkylated by primary alkyl iodides or bromides with much higher γ-selectivity than that observed with the anion of allyltrimethylsilane (equation I). The products can be converted via the epoxides into α-silyl aldehydes.[1]

$$6\text{-}35{:}1$$

[1] J. M. Muchowski, R. Naef, and M. L. Maddox, *Tetrahedron Letters*, **26**, 5375 (1985).

Allyltrimethylsilane.

Conjugate allylation. Majetich *et al.*[1] have compared this silane, in combination with Bu_4NF or $TiCl_4$, and lithium diallylcuprate as reagents for conjugate addition to various Michael acceptors, and conclude that the first reagent is generally superior. It is particularly useful for allylation of α,β-unsaturated esters and nitriles. The Sakurai reaction (allyltrimethylsilane + $TiCl_4$) is only useful for allylation of α,β-enones; use of the silane and fluoride ion in this case leads to both 1,2- and 1,4-adducts. Organocuprate allylations are highly substrate-dependent.

The report includes details for drying commercial Bu_4NF under high vacuum (<0.1 mm) at 25° for 30 minutes. This material is dissolved in DMF and treated with flame-dried 4 Å molecular sieves. Freshly prepared solutions should be used, but stock solutions are effective for about one day.

Intramolecular conjugate allylation (**12**, **25**).[2] Fluoride ion catalyzes intramolecular Michael additions of allyltrimethylsilane to α,β-enones as well as α,β-unsaturated esters, nitriles, and amides; Lewis acid catalysis is not effective. The method is particularly suited to cyclopentane annelations.[2]

Homoallylic ethers.[3]   Reaction of aldehydes with allyltrimethylsilane and $Cl_2Ti(OR)_2$ (prepared from LiOR and $TiCl_4$) results in homoallylic ethers in 50–95% yield.
   Example:

$$C_6H_5CHO + Cl_2Ti[OCH(CH_3)_2]_2 + CH_2{=}CHCH_2Si(CH_3)_3 \xrightarrow[-75 \to 20°]{CH_2Cl_2,}$$

$$\underset{90\%}{}$$

$$\overset{\underset{\displaystyle C_6H_5}{\quad}}{\underset{}{}}\ \text{OCH(CH}_3)_2 \quad CH_2$$

This reaction provides a method for preparation of optically active homoallylic alcohols **(3)** by use of $Cl_2Ti(OR)_2$ prepared from (S)-(—)-1-phenylethanol **(1)**.

Allylation of arenes and alcohols.   Allyltrimethylsilane and some related allyltrialkylmetal reagents of Sn and Ge in combination with iodosylbenzene (1 equiv.) activated by BF_3 etherate (0.25–1 equiv.) allylate aromatics in CH_2Cl_2 at $-30°$ to $25°$. Under these conditions alcohols are converted into allyl ethers, even though iodosylbenzene is a known oxidant for alcohols.[4]
   Example:

α-C-Allylglycopyranosides.[5]   Reaction of protected glycopyranoses with allyl-trimethylsilane in the presence of a Lewis acid (usually BF_3 etherate) in 1,2-dichloroethane results in about equimolar amounts of the α- and β-allyl-C-glycopyranosides. However, in acetonitrile the α-anomer is formed with high selectivity (up to 99:1).

Example:

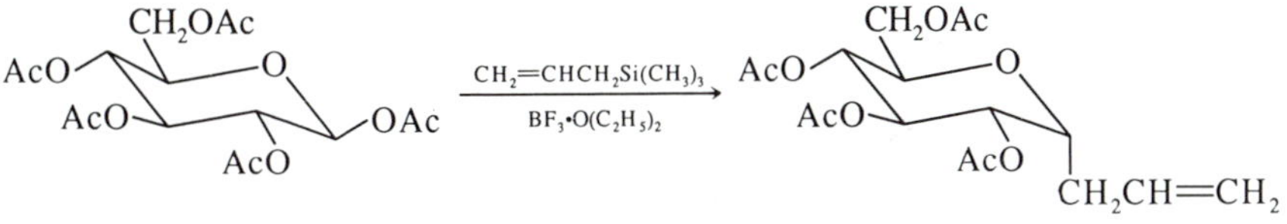

$ClCH_2CH_2Cl$	72%	α/β = 1:1
CH_3CN	81%	α/β = 95:5

[1] G. Majetich, A. Casares, D. Chapman, and M. Behnke, *J. Org.*, **51**, 1745 (1986).
[2] G. Majetich, R. W. Desmond, Jr. and J. J. Soria, *ibid.*, **51**, 1753 (1986).
[3] R. Imwinkelried and D. Seebach, *Angew. Chem. Int. Ed.*, **24**, 765 (1985).
[4] M. Ochiai, E. Fujita, M. Arimoto, and H. Yamaguchi, *Chem. Pharm. Bull.*, **33**, 41 (1985).
[5] A. Giannis and K. Sandhoff, *Tetrahedron Letters*, **26**, 1479 (1985).

Allyltriphenylsilane, $(C_6H_5)_3SiCH_2CH{=}CH_2$ (**1**, m.p. 92–94°). Preparation.[1]

1,3-Dienes.[2]   The titanium reagent **2**, obtained from **1** by metallation and reaction with $Ti(O\text{-}i\text{-}Pr)_4$, reacts with aldehydes to give β-triphenylsilyl alcohols (**3**). These can be converted into either (E)- or (Z)-1,3-dienes by treatment with acid or base.

Example:

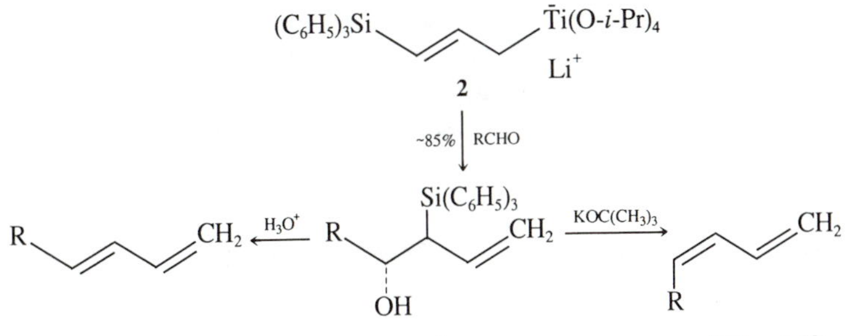

[1] H. Gilman and D. Aoki, *J. Organomet. Chem.*, **2**, 44 (1964).
[2] Y. Ikeda and H. Yamamoto, *Bull. Chem. Soc. Japan*, **59**, 657 (1986).

Allyltriphenyltin.

Allylation of α-methylthio aldehydes.[1]   This reaction, when catalyzed by a Lewis acid, particularly $SnCl_4$, proceeds with high *anti*-selectivity. In contrast, the reaction with α-benzyloxy aldehydes is highly *syn*-selective.

Example:

$$R = SCH_3 \qquad 80\%$$
$$R = OCH_2C_6H_5 \qquad 64\%$$

94:6
4:96

[1] M. Shimagaki, H. Takubo, and T. Oishi, *Tetrahedron Letters*, **26**, 6235 (1985).

Alumina.

Mono esters of dicarboxylic acids.[1]   Aliphatic straight-chain dicarboxylic acids when adsorbed on alumina react with diazomethane to form monomethyl esters in quantitative yield. Terephthalic acid, isophthalic acid, and 1,4-cyclohexanedicarboxylic acid are also converted selectively to the monomethyl esters. However, phthalic acid does not show any enhanced selectivity under these conditions. Evidently selectivity for monoesterification results from adsorption of one of the acid groups on alumina.

Enamines.   In the presence of alumina (Woelm 200 neutral) *cis*-α,β-epoxysilanes can be opened by pyrrolidine or morpholine to give α-amino-β-hydroxysilanes, from which the pure *cis*-enamine can be obtained by KH-induced *syn* β-elimination (equation I). The isomeric *trans*-α,β-epoxysilane is inert to this ring

cleavage; hence the pure *cis*-enamine can be prepared from a mixture of *cis*- and *trans*-epoxysilanes. The *cis*-enamine is readily isomerized to the pure *trans*-enamine by traces of acids.[2]

Aldol condensation.[3]   Benzaldehydes condense with alkyl aryl ketones in the presence of basic alumina to form chalcones in 70–85% yield. Benzaldehydes

$$ArCHO + CH_3\overset{\overset{\displaystyle O}{\displaystyle \|}}{C}Ar' \xrightarrow[70-85\%]{Al_2O_3} ArCH{=}CHCOAr'$$

also undergo a similar condensation with cyclic ketones to form dienones, but the reaction with acyclic ketones is very slow.

Wittig, Wittig-Horner, and Knoevenagel reactions.[4] These reactions can be carried out with Al_2O_3 or KF supported on Al_2O_3 without solvent, but addition of water catalyzes both Wittig and Wittig-Horner reactions. Under these conditions trimethylsulfonium iodide undergoes reaction to form epoxides (equation I).

$$(I) \quad (CH_3)_3\overset{+}{S}\overset{-}{I} + C_6H_5CHO \xrightarrow[75\%]{KF,\ Al_2O_3} C_6H_5CH\overset{\displaystyle O}{\overset{\diagdown}{\underset{}{}}}CH_2$$

The reaction of diethyl cyanomethylphosphonate in the presence of Al_2O_3 with aliphatic or aromatic aldehydes in the absence of solvent results mainly in the product of Knoevenagel reaction (equation II).

$$(II) \quad NCCH_2PO(OC_2H_5)_2 + RCHO \xrightarrow[70-80\%]{} \underset{H}{\overset{R}{\diagdown}}C{=}C\underset{PO(OC_2H_5)_2}{\overset{CN}{\diagup}}$$

Conjugate addition of nitroalkanes to enones.[5] This addition can be effected with basic alumina in the absence of a solvent at 25° in 50–88% yield.
Example:

$$+ CH_2{=}CHCOCH_3 \xrightarrow{Al_2O_3}_{88\%}$$

[1] H. Ogawa, T. Chihara, and K. Taya, *Am. Soc.*, **107**, 1365 (1985).
[2] P. F. Hudrlik, A. M. Hudrlik, and A. K. Kulkarni, *Tetrahedron Letters*, **26**, 139 (1985).
[3] R. S. Varma, G. W. Kabalka, L. T. Evans, and R. M. Pagni, *Syn. Comm.*, **15**, 279 (1985).
[4] F. Texier-Boullet, D. Villemin, M. Ricard, H. Moison, and A. Foucaud, *Tetrahedron*, **41**, 1259 (1985).
[5] G. Rosini, E. Marotta, R. Ballini, and M. Petrini, *Synthesis*, 237 (1986).

Aluminum chloride.

1,4-*Addition* to 1,3-*dienes*. In the presence of $AlCl_3$, β-(ethylthio)nitroalkenes react with 1,3-dienes to form *cis*-1,4-disubstituted alkenes.[1]

Example:

Aminolysis of esters.[2]   Reaction of esters with dialkylamines proceeds slowly if at all, but can be catalyzed effectively by $AlCl_3$ or $ZrCl_4$. The reaction of α-chloro carboxylic esters proceeds with only slight racemization (equation I).

(I) $CH_3\overset{\underset{\mid}{Cl}}{C}HCOOCH_3 + HN(C_2H_5)_2 \xrightarrow[77\%]{AlCl_3,\ C_6H_5CH_3} CH_3\overset{\underset{\mid}{Cl}}{C}HCON(C_2H_5)_2$

(S, 98%)   [(S/R) = 82:18]

9,10-Dihydroacridines; carbazoles.[3]   Addition of 2-azidodiphenylmethane (**1**) to a suspension of $AlCl_3$ in CH_2Cl_2 results in liberation of N_2 and formation of 9,10-dihydroacridine (**2**). A similar reaction with 2-azidobiphenyl results in carbazole (83% yield).

1   **2**

Stereoselective Friedel-Crafts alkylation.[4]   Alkylation of benzene with methyl (S)-2-(mesyloxy)propionate, derived from (S)-lactic acid, under Friedel-Crafts conditions (2 equiv. of $AlCl_3$) affords methyl (S)-phenylpropionate in 50–80% chemical yield and as high as 97% optical yield. Unfortunately extension to other aromatics results in mixtures of isomeric products.

[1] K. Fuji, S. P. Khanapure, M. Node, T. Kawabata, and A. Ito, *Tetrahedron Letters*, **26**, 779 (1985).
[2] R. D. Gless, Jr., *Syn. Comm.*, **16**, 633 (1986).

[3] H. Takeuchi, M. Maeda, M. Mitani, and K. Koyama, *J.C.S. Chem. Comm.*, 287 (1985).
[4] O. Piccolo, F. Spreafico, G. Visentin, and E. Valoti, *J. Org.*, **50**, 3945 (1985).

Aluminum iodide, AlI$_3$. AlI$_3$ can be prepared *in situ* by reaction of aluminum foil (Merck) with I$_2$ in CS$_2$. The slightly yellow solution can be used as is for conversion of alkyl chlorides to alkyl iodides at $-78°$ to $20°$ in about 80 to 100% yield.[1]

[1] F. J. Arnaiz and J. M. Bustillo, *Org. Syn.*, submitted (1985).

(1S, 2S)-2-Amino-3-methoxy-1-phenyl-1-propanol (1).
  Chiral-**trans-1,2-*disubstituted*-1,2-*dihydronaphthalenes*** (cf. **12**, 310–311).[1] Addition of an organolithium to chiral (1-naphthalyl)oxazolines (**2**, derived from **1**) followed by quenching with CF$_3$COOH results in the salt **3**. This product is reduced by LiAlH$_4$ to a single *trans*-1,2-diaxial diastereomer **4**.

4 (88% ee)

4-Aryl-1,4-dihydropyridines.[2]   These compounds are important calcium channel blockers or enhancers depending on the stereochemistry at C$_4$. An enantioselective synthesis of the (S)-isomers (blockers) is based on addition of an aryllithium to a chiral 3-dihydrooxazolyl-5-methoxycarbonylpyridine (**2**). Thus addition

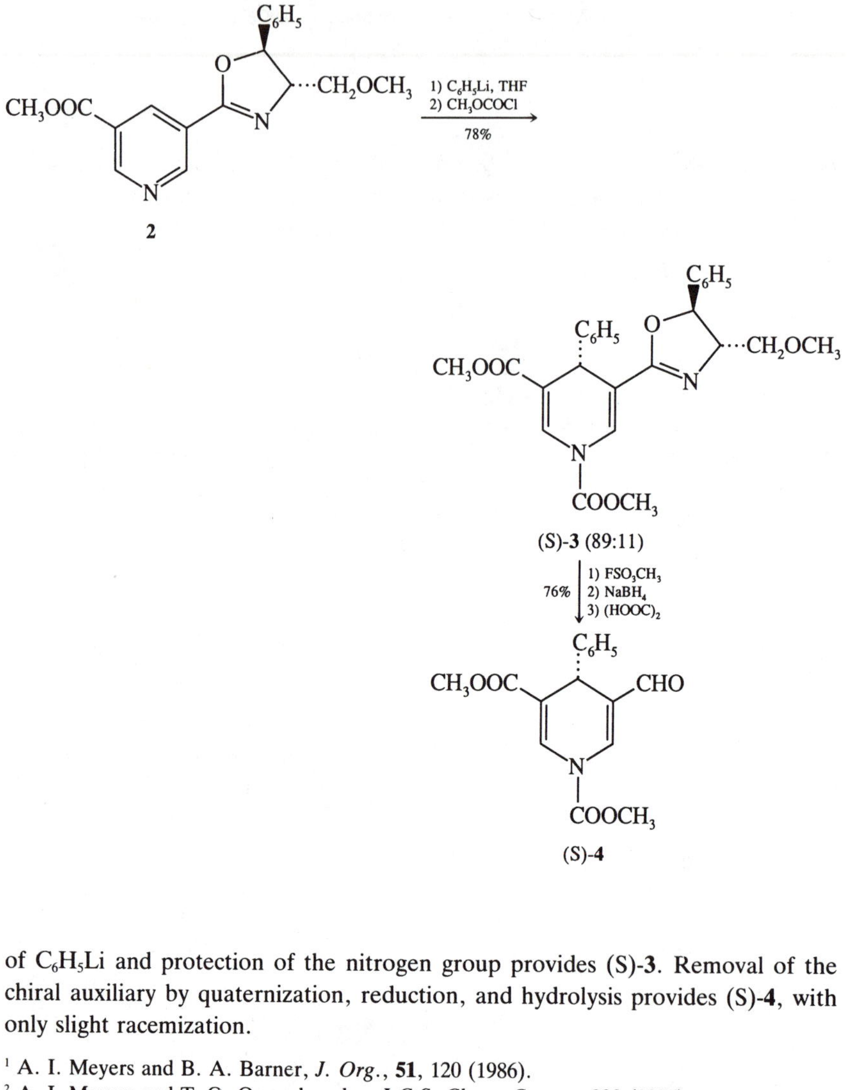

2

1) C₆H₅Li, THF
2) CH₃OCOCl
78%

(S)-3 (89:11)

1) FSO₃CH₃
76%   2) NaBH₄
3) (HOOC)₂

(S)-4

of C₆H₅Li and protection of the nitrogen group provides (S)-3. Removal of the chiral auxiliary by quaternization, reduction, and hydrolysis provides (S)-4, with only slight racemization.

[1] A. I. Meyers and B. A. Barner, *J. Org.*, **51**, 120 (1986).
[2] A. I. Meyers and T. O. Oppenlaender, *J.C.S. Chem. Comm.*, 920 (1986).

(1S, 2S)-2-Amino-1-phenyl-1,3-propanediol (1).

Chiral 4,4-disubstituted cyclohexenones.[1] The chiral bicyclic lactam **2**, obtained by reaction of 4-acetylbutanoic acid with **1**, on dialkylation gives mainly the diastereomer from *endo*-attack, with the highest diastereoselectivity obtained by using the larger electrophile in the second alkylation. Hydride reduction followed by hydrolysis furnished 4,4-dialkylated cyclohexenones (**4**) in >95% ee.

$$CH_3CO(CH_2)_3COOH \xrightarrow[60-65\%]{1} \mathbf{2} \xrightarrow[67\%]{\substack{1)\ LDA,\ CH_3I \\ 2)\ LDA,\ C_6H_5CH_2Br}}$$

2

$$\mathbf{3\ (97:3)} \xrightarrow{53\%} \mathbf{4\ (>95\%\ ee)}$$

[1] A. I. Meyers, B. A. Lefker, K. T. Wanner, and R. A. Aitken, *J. Org.*, **51**, 1936 (1986).

Arenediazonium tetrafluoroborates.

α-Aryl ketones.[1] Arenediazonium tetrafluoroborates react with silyl enol ethers in pyridine at 0° to afford α-aryl ketones in 60–75% yield. This reaction fails or proceeds in low yield in other solvents.

α-Amino esters.[2] Reaction of benzenediazonium tetrafluoroborate with ketene silyl acetals provides either α-hydrazono or α-azo esters, both of which are converted into α-amino esters on catalytic hydrogenation.

Example:

$$C_6H_5CH{=}C \begin{array}{l} OCH_3 \\ OSi(CH_3)_3 \end{array} \xrightarrow[83\%]{\substack{C_6H_5N_2BF_4 \\ py,\ 0°}} \underset{NHC_6H_5}{\overset{C_6H_5}{\underset{}{N}}}{\diagdown}COOCH_3 \xrightarrow[\sim100\%]{\substack{H_2,\ Pd/C, \\ CH_3OH}} \underset{NH_2}{\overset{C_6H_5}{}}{\diagdown}COOCH_3$$

(E/Z = 7:1)

[1] T. Sakakura, M. Hara, and M. Tanaka, *J.C.S. Chem. Comm.*, 1545 (1985).
[2] T. Sakakura and M. Tanaka, *ibid.*, 1309 (1985).

Arene(tricarbonyl)chromium complexes.

Hydrogenation catalysts.[1] The complexes are superior catalysts for stereoselective hydrogenation of alkynes to (Z)-alkenes. Naphthalenetricarbonylchromium[2] is particularly useful because hydrogenations with this catalyst proceed in THF

at 45°. In contrast, use of (methyl benzoate)tricarbonylchromium requires a temperature of 120° and high H_2 pressure (70 kg/cm^2).
    Example:

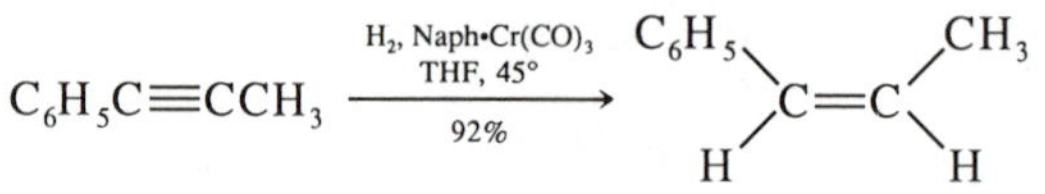

$$C_6H_5C\equiv CCH_3 \xrightarrow[92\%]{\substack{H_2, \text{ Naph·Cr(CO)}_3 \\ \text{THF, 45°}}} \underset{H}{\overset{C_6H_5}{\diagdown}}C=C\underset{H}{\overset{CH_3}{\diagup}}$$

In addition cisoid α,β-enones and imines can be reduced.
    Examples:

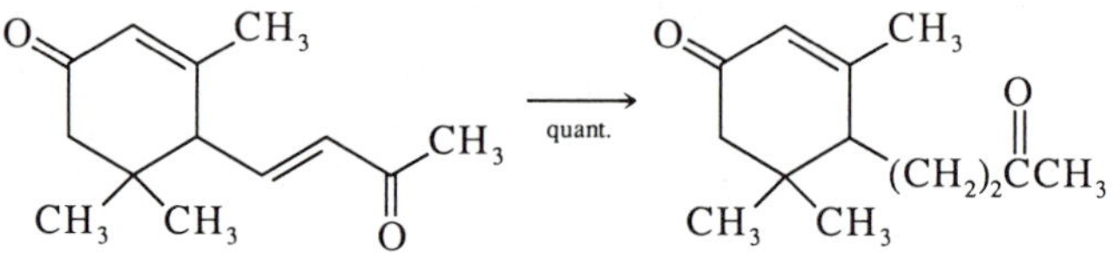

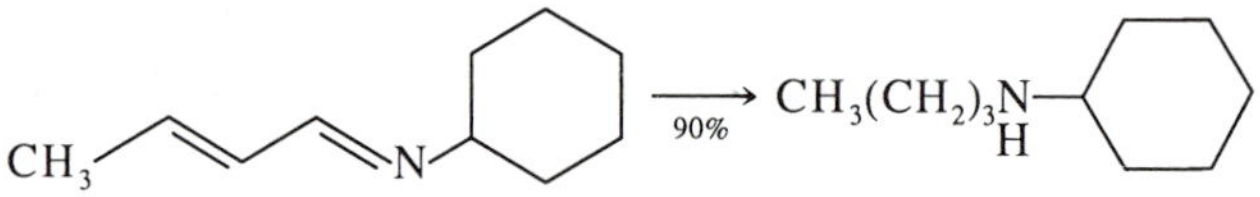

    This hydrogenation provides the key step in a synthesis of cyanocarbacyclin (3) from a 2-cyano-1,3-diene (1).[3] Hydrogenation of (Z)-1 catalyzed by

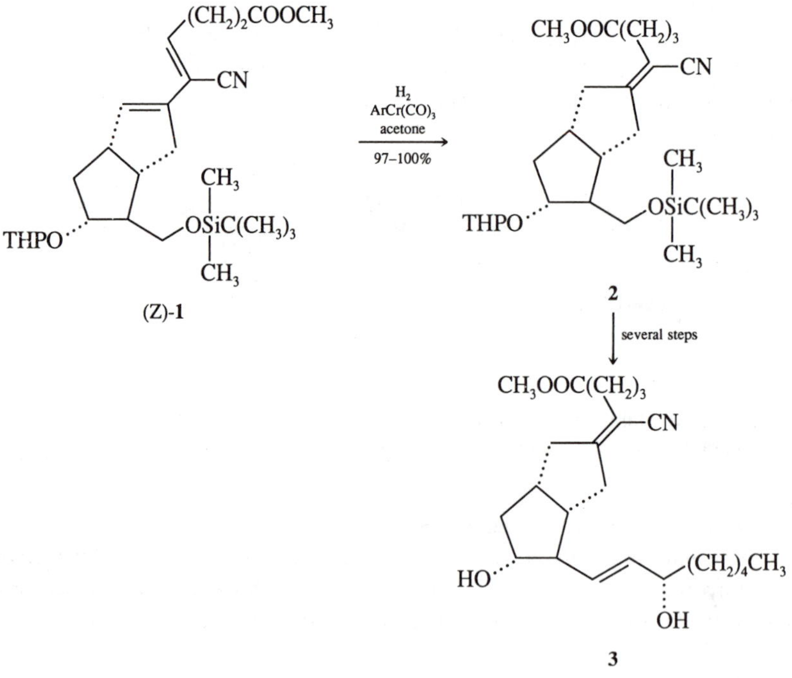

3

$(C_6H_5COOCH_3)Cr(CO)_3$ or (naphthalene)-$Cr(CO)_3$ proceeds by 1,4-addition to provide **2** in essentially quantitative yield. The (E)-isomer of **1** does not undergo this hydrogenation because of steric hindrance.

Asymmetric reduction of α-keto esters.[4] The chromium-complexed α-keto ester **1** is reduced to the carbinol by sodium borohydride with surprisingly high

	1		
$NaBH_4$, $-70°$		70%	76% de
$LiB[CH(CH_3)C_2H_5]_3H$, $-70°$		90%	90% de

asymmetric induction. As expected, lithium tri-*sec*-butylborohydride is even more selective.

Stereospecific benzylic hydroxylation.[5] The anion of the chromium tricarbonyl complex (**1**) of (+)-N,N-dimethylamphetamine is hydroxylated by oxodiperoxymolybdenum(pyridine)(hexamethylphosphoric triamide) to give the optically pure complex (**2**) of (+)-N-methylpseudoephedrine (**3**).

Stereoselective benzylic alkylation.[6] Benzylic acetates of arenechromium tricarbonyls react with R_3Al or $(C_2H_5)_2Zn–TiCl_4$ to form *exo*-alkyl chromium complexes.

Example:

(CH$_3$)$_3$Al   99%   R = CH$_3$
(C$_2$H$_5$)$_2$Zn, TiCl$_4$ 62%   R = C$_2$H$_5$

O-Aryloximes; benzofurans.[7]   Aryl halides complexed by Cr(CO)$_3$ react under phase-transfer catalysis with acetone oxime to give complexed O-aryloximes in 75–98% yield. The free oxime can be obtained by treatment with I$_2$ (55–80% yield). O-Aryloximes are converted into benzofurans when heated in ethanol with H$_2$SO$_4$.
    Example:

(R = CH$_3$, OCH$_3$)

[1] M. Sodeoka and M. Shibasaki, *J. Org.*, **50**, 1147 (1985).
[2] G. Yagupsky and M. Cais, *Inorg. Chim. Acta*, **12**, L27 (1975).
[3] M. Shibasaki and M. Sodeoka, *Tetrahedron Letters*, **26**, 3491 (1985).
[4] A. Solladié-Cavallo and J. Suffert, *ibid.*, **26**, 429 (1985).
[5] J. Blagg and S. G. Davies, *J.C.S. Chem. Comm.*, 653 (1985).
[6] M. Uemura, K. Isobe, and Y. Hayashi, *Tetrahedron Letters*, **26**, 767 (1985).
[7] A. Alemagna, C. Baldoli, P. D. Buttero, E. Licandro, and S. Maiorana, *J.C.S. Chem. Comm.*, 417 (1985).

Arylselenocarboxamides. $ArC\overset{Se}{\underset{NH_2}{\diagdown}}$ (1).

These reagents are easily prepared by reaction of nitriles with Se, CO, and H$_2$O.[1]
    Deoxygenation of epoxides.[2]   Epoxides are converted into alkenes stereo-specifically on reaction with benzeneselenocarboxamide at 0–20°.

Examples:

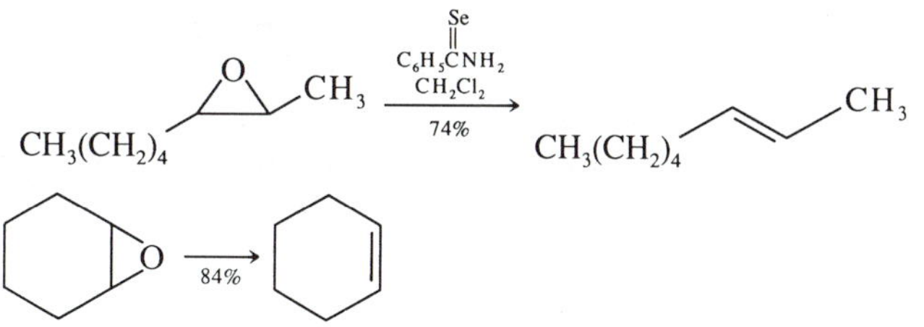
Se
$C_6H_5CNH_2$
CH_2Cl_2
74%
$CH_3(CH_2)_4$
CH_3
84%

[1] A. Ogawa, J. Miyake, Y. Karasaki, S. Murai, and N. Sonoda, *J. Org.*, **50**, 384 (1985).
[2] A. Ogawa, J. Miyake, S. Murai, and N. Sonoda, *Tetrahedron Letters*, **26**, 669 (1985).

2-Arylsulfonyloxaziridines.

Improved preparation:[1]

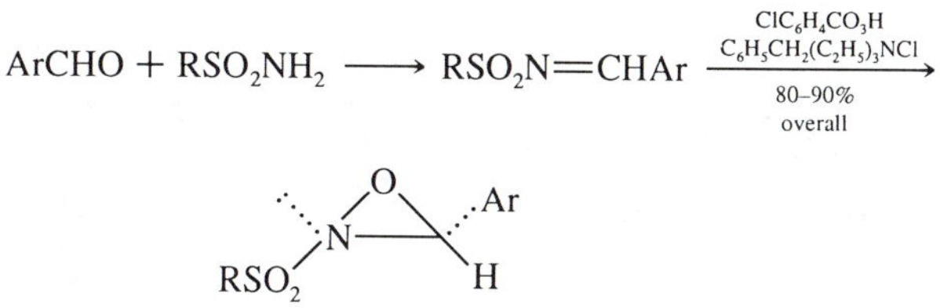
$ArCHO + RSO_2NH_2 \longrightarrow RSO_2N{=}CHAr$
$ClC_6H_4CO_3H$
$C_6H_5CH_2(C_2H_5)_3NCl$
80–90% overall
RSO_2
Ar
H

Optically active α-hydroxy acids.[2] The enolate of the amide (**2**) derived from phenylacetic acid and L-prolinol (**10**, 332) is oxidized by 2-(phenylsulfonyl)-3-phenyloxaziridine (**1**) to give optically active α-hydroxy amides (**3**). Significantly, the configuration of **3** depends upon the base. The lithio enolate (LDA) is converted to the (S)-isomer in >95% de, whereas the sodio enolate, generated with $NaN[Si(CH_3)_3]_2$, is converted into the (R)-isomer in 93% de.

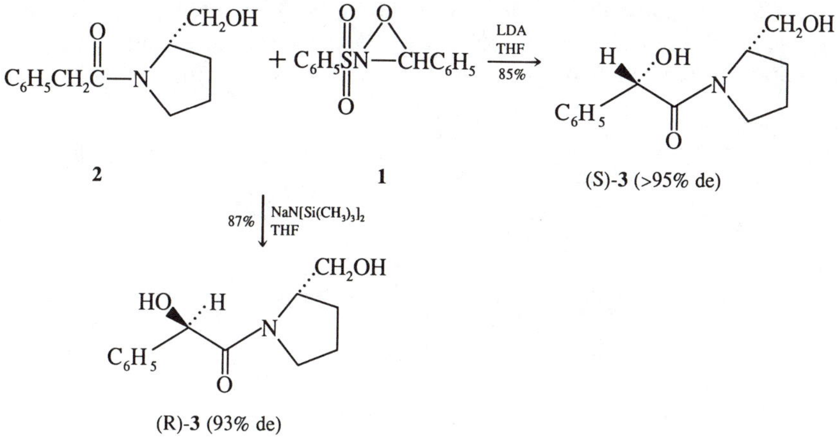
$C_6H_5CH_2C$—N
CH_2OH
+ C_6H_5SN—CHC_6H_5
LDA
THF
85%
H
OH
C_6H_5
2
1
(S)-3 (>95% de)
87%
$NaN[Si(CH_3)_3]_2$
THF
HO
H

(R)-**3** (93% de)

Asymmetric hydroxylation of chiral imides.[3]   The (Z)-sodium enolates of the chiral imides **2** and **3** undergo asymmetric hydroxylation on reaction with 2-(phenylsulfonyl)-3-phenyloxaziridine (**1**). The products [(R)-**4** and (S)-**6**] are solvolyzed to (R)- and (S)-α-hydroxy esters. This hydroxylation can also be effected with MoOPh, which is much less reactive than **1** but slightly more stereoselective. In general **1** is preferred to MoOPh because of the higher yields.
Examples:

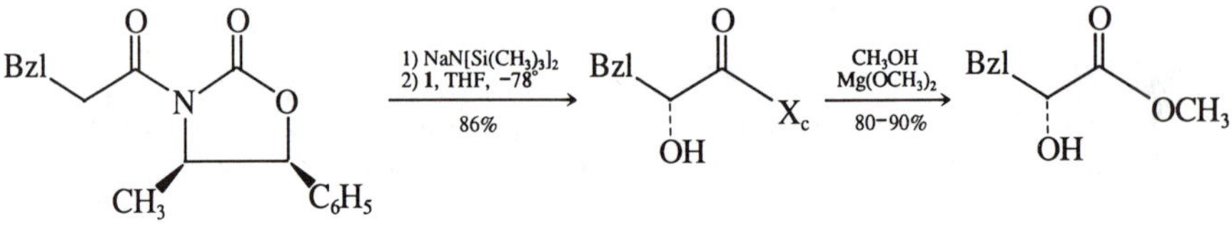

2   (R)-**4** (R/S = 94:6)   (R)-**5**

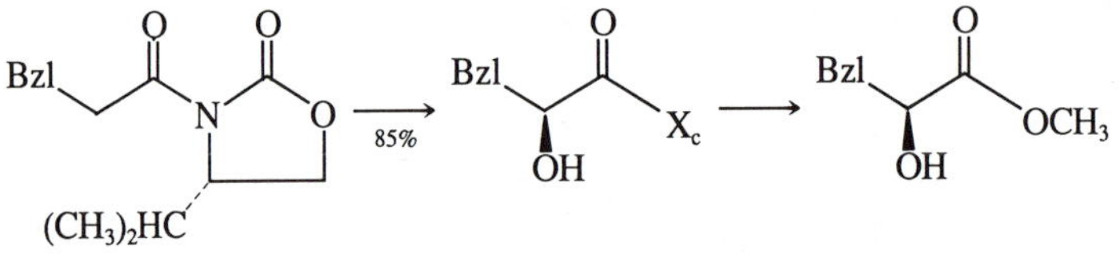

3   (S)-**6** (S/R = 95:5)   (S)-**5**

For another use of related chiral alkylsulfonyloxaziridines, see (Camphorylsulfonyl)oxaziridines, this volume.

[1] F. A. Davis, L. C. Vishwakarma, and O. D. Stringer, *Org. Syn.* submitted (1985).
[2] F. A. Davis and L. C. Vishwakarma, *Tetrahedron Letters*, **26**, 3539 (1985).
[3] D. A. Evans, M. M. Morrissey, and R. L. Dorow, *Am. Soc.*, **107**, 4346 (1985).

Azidotrimethylsilane.

Cleavage of epoxides.[1]   The epoxide **1** is cleaved by $N_3Si(CH_3)_3$ and a catalytic amount of $ZnCl_2$ to the hydroxy azide **2** in 90% yield. The corresponding benzoate (**3**) on hydrogenation undergoes O → N benzoyl transfer to provide **4** in 95% ee.

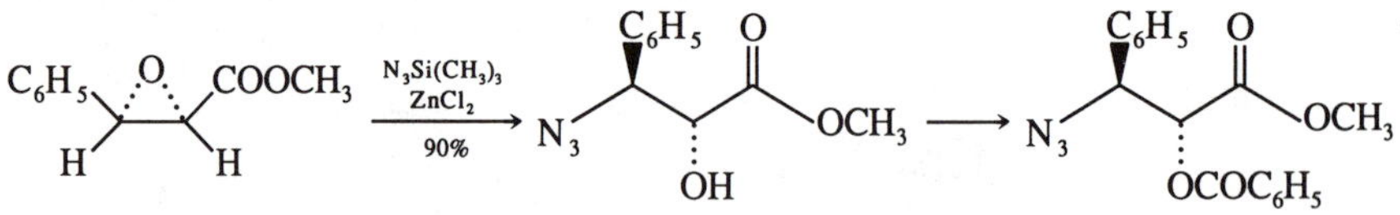

(2S, 3R)-**1**   **2**   **3**

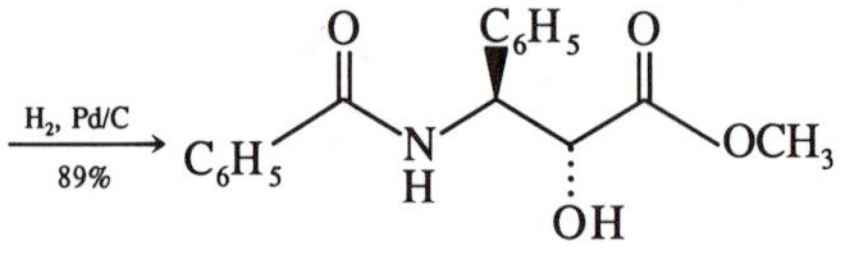

4 (95% ee)

β-*Hydroxy-α-amino acids*.[2] Hydrazoic acid (HN_3) can be generated *in situ* from $N_3Si(CH_3)_3$ and CH_3OH in DMF. Reagent generated in this way reacts with the epoxide **1** to give **2** in high yield. The azide group can be reduced in high yield

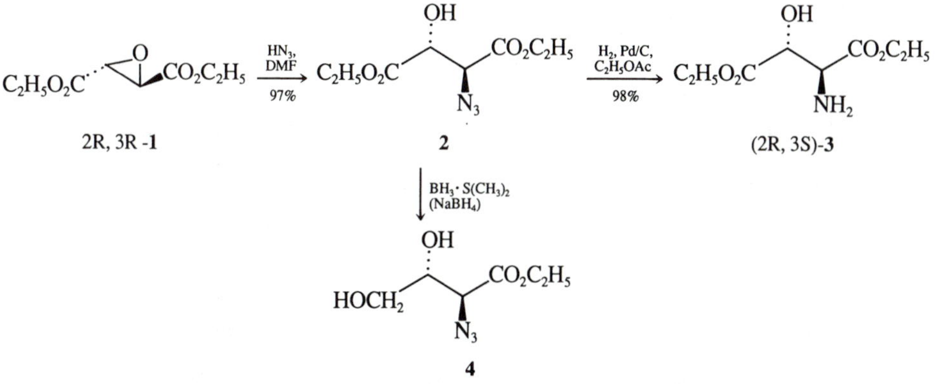

by catalytic hydrogenation to **3**, a derivative of L-aspartic acid. Reaction of **2** with $BH_3 \cdot S(CH_3)_2$ catalyzed by $NaBH_4$ affects only the ester group α to the hydroxy group to provide **4**, which can be converted to a derivative of *erythro*-β-hydroxy-methyl-L-serine.

Cleavage of 2,3-*epoxy alcohols*.[3] The combination of $(C_2H_5)_2AlF$ (1 equiv.) and $N_3Si(CH_3)_3$ cleaves these epoxides regio- and stereoselectively to provide *anti*-3-azido-1,2-diols.

Example:

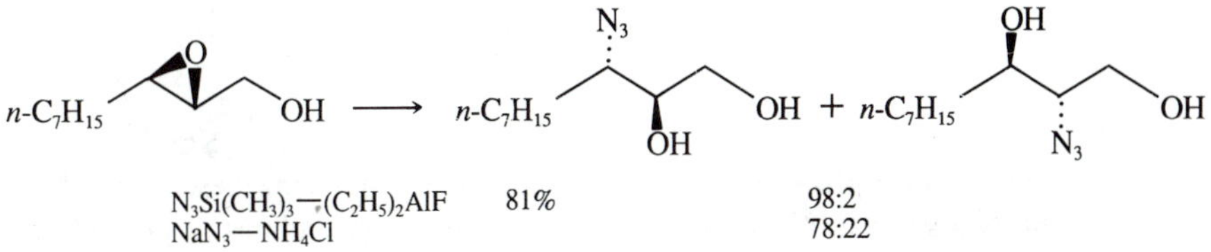

[1] J.-N. Denis, A. E. Greene, A. A. Serra, and M.-J. Luche, *J. Org.*, **51**, 46 (1986).
[2] S. Saito, N. Bunya, M. Inaba, T. Moriwake, and S. Torii, *Tetrahedron Letters*, **26**, 5309 (1985).
[3] K. Maruoka, H. Sano, and H. Yamamoto, *Chem. Letters*, 599 (1985).

B

Benzeneselenenyl halides.

Phenylselenolactonization.[1]   This reaction has been used to prepare 24,20-steroidal lactones (**3a** and **3b**). Thus reaction of **1a** and **1b**, prepared in several steps from 3β-methoxy-5-androstene-17-one, with C_6H_5SeCl (2 equiv.) and $N(C_2H_5)_3$ (1 equiv.) yields directly the unsaturated lactones **2a** and **2b** which are reduced by diimide to **3a** and **3b**.

Intramolecular amidoselenylation.[2]   The reaction of a benzeneselenenyl halide with an N-alkenylacetamide in CH_3CN can afford pyrrolidines or piperidines by an intramolecular amidoselenylation. C_6H_5SeBr is usually superior to C_6H_5SeCl for this reaction. Addition of SiO_2 increases the yield. The C_6H_5Se group can be removed reductively by $(C_6H_5)_3SnH/AIBN$ or by nickel boride at 0°.

Examples:

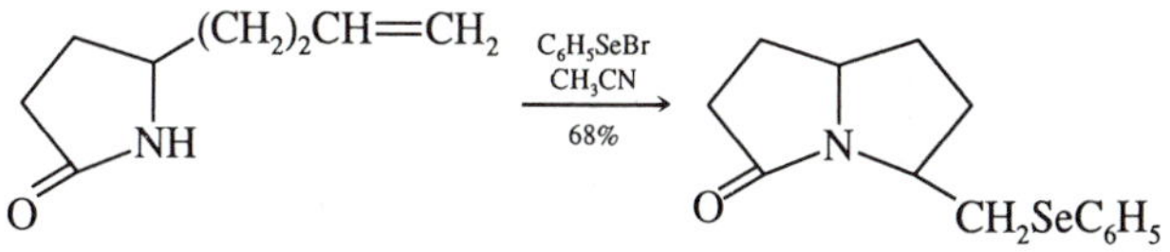

α-Halo-α, β-unsaturated ketones.[3]   Reaction of C_6H_5SeBr or C_6H_5SeCl with a cyclic or an acyclic α-diazo ketone in CH_2Cl_2 at 20° results in the α-halo-α-phenylseleno adduct in generally high yield. The products can undergo dehydro-halogenation to α-phenylseleno-α,β-enones or selenoxide fragmentation to α-halo-α,β-enones. Reaction of the α-diazo ketone with C_6H_5SeCl and AgOAc results in an α-acetoxy-α-phenylseleno adduct.

Example:

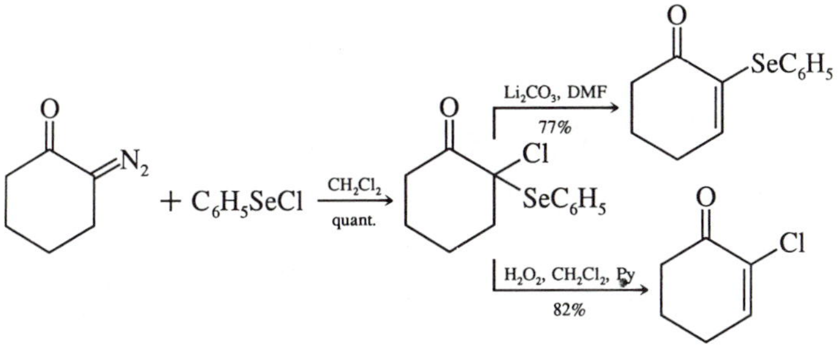

[1] M. Kocor and B. Bersz, *Tetrahedron*, **41**, 197 (1985).
[2] A. Toshimitsu, K. Terao, and S. Uemura, *J. Org.*, **51**, 1724 (1986).
[3] D. J. Buckley and M. A. McKervey, *J.C.S. Perkin I*, 2193 (1985).

Benzeneselenenyl trichloride. $C_6H_5SeCl_3$. The reagent can be prepared by reaction of $(C_6H_5Se)_2$ with SO_2Cl_2 in $CHCl_3$. It can be stored at $-20°$ for at least one month.

Reaction with ketones.[1]   The reagent reacts with ketones in ether to introduce a $C_6H_5SeCl_2$ group into the α-position. The products undergo selenoxide elimination to give α,β-enones; they are reduced by thiourea to α-phenylseleno ketones.

Example:

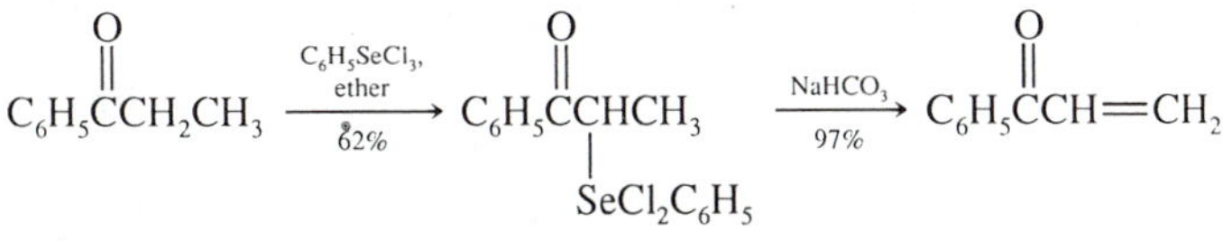

[1] L. Engman, *Tetrahedron Letters*, **26**, 6385 (1985).

Benzeneseleninic anhydride–Cyanotrimethylsilane.

α-Cyano amines.[1]   Dehydrogenation of a secondary amine with $(C_6H_5SeO)_2O$ in the presence of $NCSi(CH_3)_3$ (or NaCN) gives α-cyano amines.

Example:

[1] D. H. R. Barton, A. Billion, and J. Boivin, *Tetrahedron Letters*, **26**, 1229 (1985).

1-Benzotriazolyl diethyl phosphate,

(1).

The reagent is prepared by reaction of 1-hydroxybenzotriazole with diethyl chlorophosphate and $N(C_2H_5)_3$.

Amides and peptides.[1]   Amides are obtained in 93–96% yield by reaction of **1** and $N(C_2H_5)_3$ with carboxylic acids and amines in DMF (preferred solvent). A similar condensation of amino acid esters and Boc- or Cbo-amino acids proceeds in 91–96% yield and with practically no racemization.

[1] S. Kim, H. Chang, and Y. K. Ko, *Tetrahedron Letters*, **26**, 1341 (1985).

Benzoyl *t*-butyl nitroxide,

(1).

Preparation[1]:

1 (green oil)

Phenols → *o-Quinones.*[2] This nitroxide is comparable to Fremy's salt for oxidation of monohydric phenols to quinones, but yields are low with simple phenols. The final step in a recent synthesis of methoxatin (**4**, a cofactor for a bacterial methanol dehydrogenose) required oxidation of the phenolic ring of **2** to an *o*-quinone (**3**) and ester hydrolysis. The usual oxidant, Fremy's salt, is not useful for

                    2   **3** R = CH₃
                                                                 4 R = H

this oxidation because **2** is insoluble in aqueous solutions. The desired oxidation can be effected in CH_2Cl_2/CH_3OH (9:1) with **1** in 93% yield.[3]

[1] P. F. Alewood, S. A. Hussain, T. C. Jenkins, M. J. Perkins, A. H. Sharma, N. P. Y. Siew, and P. Ward, *J.C.S. Perkin I*, 1066 (1978).
[2] S. A. Hussain, T. C. Jenkins, and M. J. Perkins, *ibid.*, 2809 (1979).
[3] A. R. MacKenzie, C. J. Moody, and C. W. Rees, *Tetrahedron*, **42**, 3259 (1986).

(Benzylamino)acetonitrile, $C_6H_5CH_2NHCH_2CN$ (**1**). The nitrile is obtained as the hydrochloride, m.p. 171°, by reaction of the hydrochloride of benzylamine with KCN and aqueous formaldehyde (65% yield).[1]

  β-*Lactams.*[2] This N-(cyanomethyl)amine reacts with lithium ester enolates to afford 4-unsubstituted β-lactams in 60–75% yield (equation I). This reaction is generally applicable to N-(cyanomethyl)amines. 3-Amino-β-lactams can be ob-

tained by reaction of the enolate of a silyl-protected glycine ester (**2**, equation II) with **1** or a chiral cyanoamine such as **4**.

(II)

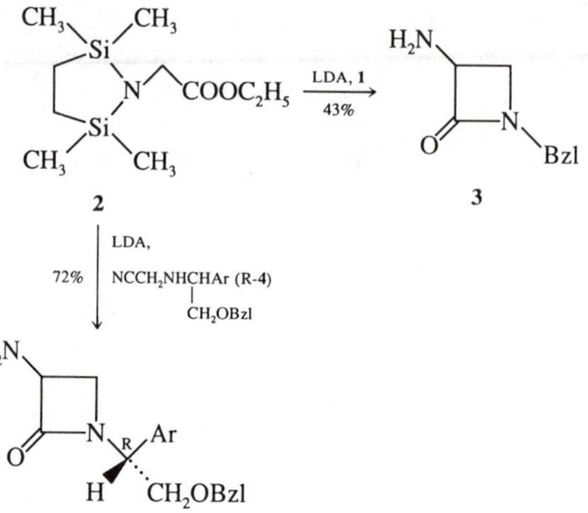

$$\xrightarrow[43\%]{\text{LDA, 1}}$$

2   **3**

72% | LDA,
NCCH$_2$NHCHAr (R-4)
CH$_2$OBzl

(R)-**5** (>95% ee)

[1] W. Baker, W. D. Ollis, and V. D. Poole, *J. Chem. Soc.*, 307 (1949).
[2] L. E. Overman and T. Osawa, *Am. Soc.*, **107**, 1698 (1985).

Benzylchlorobis(triphenylphosphine)palladium.

Coupling of RCOCl with tetraorganotins   (**9**, 41–42; **12**, 44). Complete details are available for a typical Pd-catalyzed coupling of an acid chloride with an organotin reagent (equation I). Usually trimethyltin reagents are preferred because

(I)

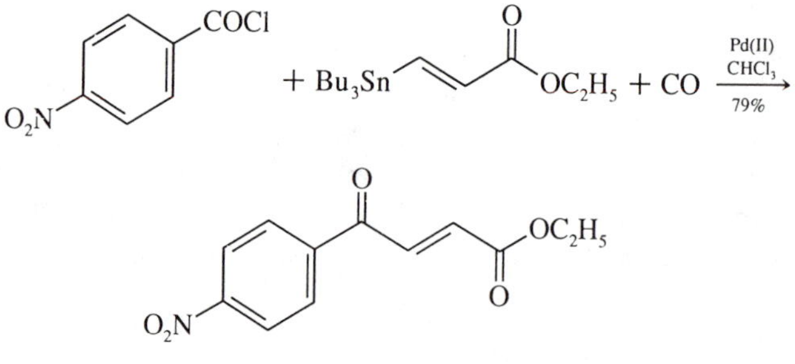

the by-product, $(CH_3)_3SnCl$, can be removed by aqueous extraction. In the present case the by-product, Bu_3SnCl, is converted by KF into insoluble Bu_3SnF, which is removed by filtration.[1]

[1] A. F. Renaldo, J. W. Labadie, and J. K. Stille, *Org. Syn.* submitted (1985).

Benzyl isocyanate, $C_6H_5CH_2N=C=O$ (**1**).

2-Amino-1,3-diols.[1]   This reagent can be used as an equivalent of NH_3 in a synthesis of 2-amino-1,3-diols from chiral 2,3-epoxy alcohols. Thus reaction of the epoxy alcohol **2** with NaH and **1** affords **a,** which can be isolated or hydrolyzed to the protected 2-amino-1,3-diol. In all cases, the nitrogen nucleophile adds selectively to C_2. This transformation was used in a synthesis of (+)-*erythro*-dihydrosphingosine (**4**).

(+)-D-**4**

[1] W. R. Roush and M. A. Adam, *J. Org.*, **50**, 3752 (1985).

N-Benzyl-N-methoxymethyl-N-(trimethylsilyl)methylamine (1). Preparation:

1, b.p. 78–80°/0.5 mm.

1,3-Dipolar cycloadditions; pyrrolidines.[1]   In the presence of LiF, **1** is converted into the azomethine ylide **a,** which undergoes 1,3-dipolar cycloaddition with alkenes to form pyrrolidines. Sonication increases the rate and the yield of this cycloaddition.

Example:

[1] A. Padwa and W. Dent, *Org. Syn.* submitted (1985); A. Padwa, Y.-Y. Chen, W. Dent, and H. Nimmesgern, *J. Org.*, **50**, 4006 (1985).

Benzyl trichloroacetimidate, $HN{=}C\overset{\displaystyle CCl_3}{\underset{\displaystyle OCH_2C_6H_5}{\Big\backslash}}$ (1). The reagent is prepared

by reaction of benzyl alcohol in ether with NaH and then with trichloroacetonitrile. Allyl trichloroacetimidate (**2**) is prepared in the same way.

Protection of hydroxyl groups.[1] Benzyl or allyl ethers of carbohydrates are obtained in satisfactory yield by reaction with **1** or **2** in cyclohexane in the presence of triflic acid (20°). Trichloroacetamide precipitates and the desired ethers are obtained from the filtrate after neutralization.

[1] H.-P. Wessel, T. Iversen, and D. R. Bundle, *J.C.S. Perkin I*, 2247 (1985).

Birch reduction.

Enantioselective Birch reduction-alkylation.[1] The chiral benzoic acid derivative **1**, prepared by condensation of *o*-hydroxybenzoic acid with L-prolinol followed by cyclization (Mitsunobu reaction), undergoes Birch reduction (K, NH_3, THF, *t*-butyl alcohol) followed by alkylation with C_2H_5I to give essentially only **2**. Acid hydrolysis returns the chiral auxiliary and provides the 2-alkylated cyclohexenone **3**.

The anthranilic acid derivative **4**, prepared from isatoic anhydride and L-proline, on Birch reduction and alkylation affords pure **5**, which is hydrolyzed by acid to the aminolactone **6**, with the absolute configuration opposite to that of **3**.

[1] A. G. Schultz and P. Sundararaman, *Tetrahedron Letters*, **25**, 4591 (1984); A. G. Schultz, P. J. McCloskey, P. Sundararaman, and J. P. Springer, *ibid.*, **26**, 1619 (1985).

Bis(acetonitrile)chloronitropalladium(II)–Copper(II) chloride. $(CH_3CN)_2PdClNO_2$ and $CuCl_2$ (1:4) when heated in *t*-butyl alcohol at 55° form a brown solid (1) of uncertain constitution. This Pd(II) complex catalyzes air oxidation of terminal alkenes to give aldehydes as the major product.[1]

Examples:

$$C_6H_5CH{=}CH_2 \xrightarrow{\ 1\ } C_6H_5CH_2CHO$$

$$C_6H_{13}CH{=}CH_2 \longrightarrow C_6H_{13}CH_2CHO\ +\ C_6H_{13}COCH_3$$
$$55{:}45$$

$$CH_2{=}CHCH_2OH \longrightarrow (CH_3)_3COCH_2CH_2CHO\ +\ (CH_3)_3COCH{=}CHCH_2OH$$
$$70{:}30$$

[1] B. L. Feringa, *J.C.S. Chem. Comm.*, 909 (1986).

Bis(acetonitrile)dichloropalladium(II).

Alkylidenelactones.[1] The alkenylpalladium(II) intermediate obtained in the reaction of lithium alkynoates with $PdCl_2(CH_3CN)_2$ (**12**, 50) can be trapped with allyl halides to give (E)-alkylidenelactones.

Example:

[1] N. Yanagihara, C. Lambert, K. Iritani, K. Utimoto, and H. Nozaki, *Am. Soc.*, **108**, 2753 (1986).

Bis(acetylacetonate)zinc(II), $Zn(acac)_2$.

Coupling of organoboranes.[1] Organoboranes do not undergo palladium catalyzed cross-coupling with aryl or benzyl halides, but use of $Cl_2Pd[P(C_6H_5)_3]_2$ in combination with $Zn(acac)_2$ effects carbonylative coupling to give unsymmetrical ketones in 60–80% yield. Presumably $Zn(acac)_2$ acts as a base and undergoes transmetallation to give an acetylacetonatopalladium(II) intermediate.

Example:

$$B(C_2H_5)_3 + CO + C_6H_5I \xrightarrow[\substack{82\%}]{\substack{Pd(II),\ Zn(II) \\ THF,\ HMPT}} C_6H_5COC_2H_5$$

[1] Y. Wakita, T. Yasunaga, M. Akita, and M. Kojima, *J. Organomet. Chem.*, **301**, C17 (1986).

Bis(benzonitrile)dichloropalladium(II).

Claisen rearrangement of allyl imidates.[1]   This Pd(II) catalyst effects exclusive [3, 3] rearrangement of allylic imidates at 25° to allylic amides. Rearrangement of the chiral (E)-allylic imidate **1** results in a mixture of two chiral allyl amides **2** and **3**. Thermal rearrangement of **1** results only in **2**.

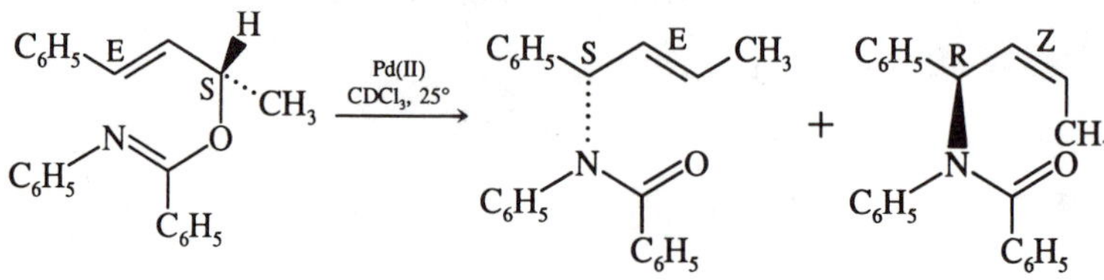

| **1** | **2 (78%)** | **3 (22%)** |

Rearrangement of these substrates with $Pd[P(C_6H_5)_3]_4$ can result in [3, 3] and/ or [1, 3] rearrangement depending on the structure.
Example:

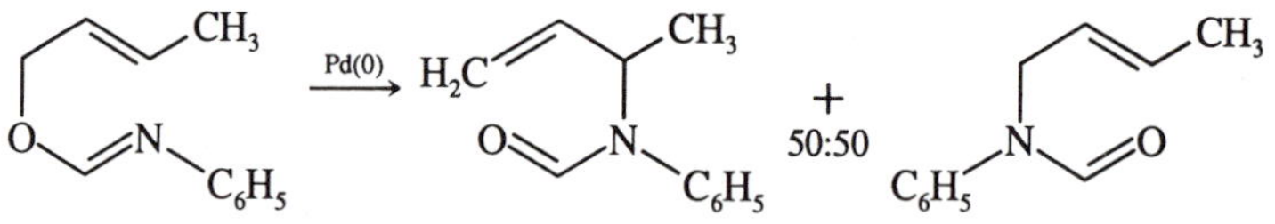

[1] T. G. Schenck and B. Bosnich, *Am. Soc.*, **107**, 2058 (1985).

Bis[1,2-bis(diphenylphosphine)ethane]palladium(0), Pd(dppe)$_2$.

Displacement of allylic gem-diacetates.[1]   In the presence of Pd(dppe)$_2$[2] as catalyst and O,N-bis(trimethylsilyl)acetamide as base, one acetoxy group of an allylic *gem*-diacetate can be displaced by a stabilized nucleophile.
Example:

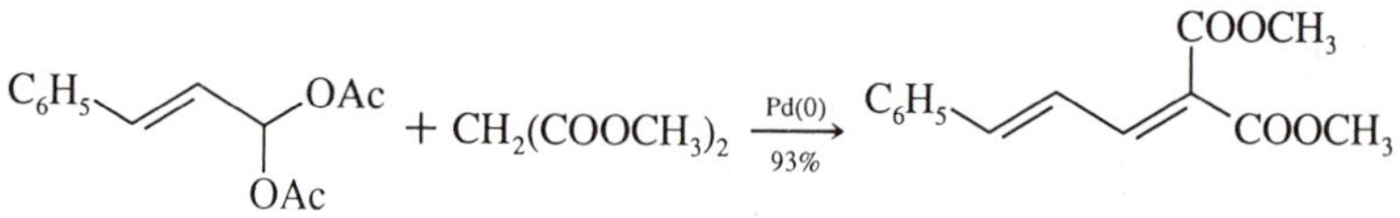

[1] B. M. Trost and J. Vercauteren, *Tetrahedron Letters*, **26**, 131 (1985).
[2] dppe = $(C_6H_5)_2PCH_2CH_2P(C_6H_5)_2$.

Bis(1,5-cyclooctadiene)nickel(0), Ni(COD)$_2$.

[3 + 2]Cycloaddition of methylenecyclopropane with alkenes.[1]   This reaction when catalyzed by Ni(COD)$_2$ and triphenylphosphine can result in either 2,3- or 3,4-disubstituted methylenecyclopentanes depending on the electronic properties of the alkene.

Examples:

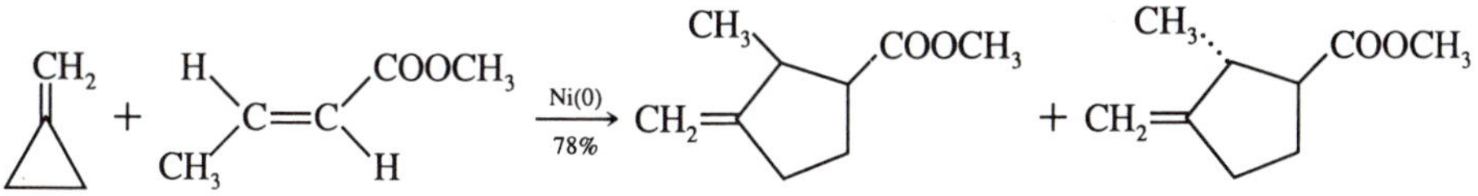

22:78

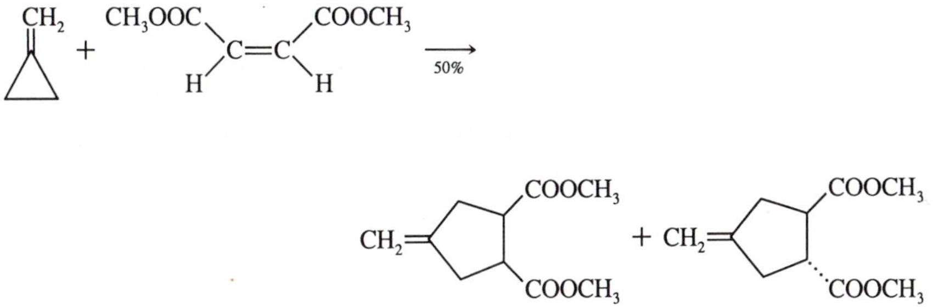

25:75

[1] P. Binger and P. Wedemann, *Tetrahedron Letters*, **26**, 1045 (1985).

Bis(η^5-cyclooctadienyl)ruthenium.

Enol esters.[1]   In the presence of (C$_8$H$_{11}$)$_2$Ru and P(Bu)$_3$ in the molar ratio 1:2, α,β-unsaturated acids add to terminal alkynes to form enol esters as the major product.

Example:

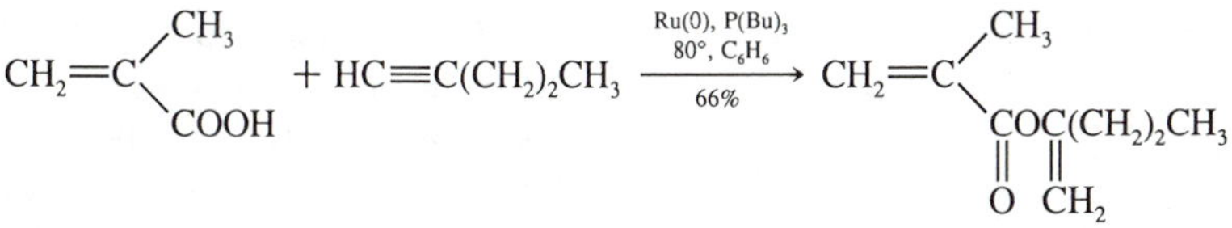

This reaction can be extended to saturated aliphatic carboxylic acids by addition of maleic anhydride to the catalytic system.[2]

Example:

$$CH_3COOH + HC\equiv CPr \xrightarrow[99\%]{\text{, Ru(0), PBu}_3} CH_3\underset{\underset{O}{\|}}{C}\underset{\underset{CH_2}{\|}}{C}Pr$$

[1] T. Mitsudo, Y. Hori, and Y. Watanabe, *J. Org.*, **50**, 1566 (1985).
[2] T. Mitsudo, Y. Hori, Y. Yamakawa, and Y. Watanabe, *Tetrahedron Letters*, **27**, 2125 (1986).

Bis(2,4-dimethyl-3-pentyl) tartrate (1).

Chiral allenylboronic esters.[1] The enantioselectivity in synthesis of homopropargylic esters by the reaction of aldehydes with chiral allenylboronic esters (**11**, 181) is markedly increased by use of bis-2,4-dimethyl-3-pentyl esters of D- or L-tartaric acid rather than the diethyl ester. Yields in the reaction of various saturated aldehydes are 70–90%, and optical yields are consistently greater than 90% and even higher (97–99%) when the aldehyde is present in excess. However, yields are poor in reactions with aryl and α,β-unsaturated aldehydes. This modified procedure was used in a synthesis of (S)-(—)-ipsenol (**2**) from D-(—)-bis(2,4-dimethyl-3-pentyl) tartrate (**1**) (equation I).

(I) $(CH_3)_2CHCH_2CHO + CH_2{=}C{=}CHB(OH)_2 \xrightarrow[78\%]{\text{D-1, C}_6\text{H}_5\text{CH}_3}$ (CH₃)₂CH⎯⎯⎯C≡CH

(>99% ee)

2, >99% ee

[1] N. Ikeda, I. Arai, and H. Yamamoto, *Am. Soc.*, **108**, 483 (1986).

(R)-(+)- and (S)-(−)-2,2′-Bis(diphenylphosphine)-1,1′-binaphthyl (BINAP).

Asymmetric hydrogenation.[1] The chiral ruthenium complex $Ru_2Cl_4[(S)-(—)-BINAP]_2[N(C_2H_5)_3]$ (**1**) has been prepared by reaction of $RuCl_2(COD)$ [COD = 1,5-cyclooctadiene] with (S)-BINAP in the presence of $N(C_2H_5)_3$.[1] This complex in the presence of $N(C_2H_5)_3$ catalyzes the hydrogenation of (E)- or (Z)-α-acylaminoacrylic acids, $R^1CH{=}C(COOH)NHCOR^2$, to derivatives of (S)-alanine in 65–86% ee. The induced chirality is comparable to but opposite to that induced with $Rh[(S)-(−)-BINAP]ClO_4$ or $Rh[(S)-(−)-BINAP](CH_3OH)_2ClO_4$.[2] The complex **1** also catalyzes hydrogenation of prochiral glutaric anhydrides to optically active δ-valerolactones.

Diastereoselective hydrogenation.[3]   The δ-hydroxy-α,β-unsaturated ester (**3**) undergoes hydrogenation in the presence of rhodium [(+)-BINAP]-norbornadiene tetrafluoroborate (**1**) (**10**, 36) almost exclusively to give **4**. This catalyst (**1**) is more effective than rhodium(DIPHOS)norbornadiene tetrafluoroborate (**2**) (**12**, 426).

3		**4**
1	90%	98:2
2		85:15

[1] T. Ikariya, Y. Ishii, H. Kawano, T. Arai, M. Saburi, S. Yoshikawa, and S. Akutagawa, *J.C.S. Chem. Comm.*, 922 (1985).
[2] A. Miyashita, H. Takaya, T. Souchi, and R. Noyori, *Tetrahedron*, **40**, 1245 (1984).
[3] D. A. Evans and M. DiMare, *Am. Soc.,* **108**, 2476 (1986).

[1,4-Bis(diphenylphosphine)butane](norbornadiene)rhodium tetrafluoroborate (1).

Stereospecific hydrogenation of a methylene cyclohexane (2).[1]   Hydrogenation of **2** with a heterogeneous catalyst results in preferential addition of H_2 from the less hindered side to give mainly **4** (R = H). Hydrogenation with the homogeneous

2		**3**	**4**
Pd/C or Pd/BaSO₄	70–80%	2:8	(R = H)
1	99%	10:0	(R = Bzl)

rhodium catalyst **1** results in quantitative formation of **3** (R = Bzl). Hydrogenation with $[(C_6H_5)_3P]_3RhCl$ is almost as stereospecific, and results in **3** and **4** in the ratio 9.5:0.5. The axial hydroxyl group of **2** evidently has a marked effect on the stereochemical course of hydrogenation.

[1] A. S. Machado, A. Olesker, S. Castillon, and G. Lukacs, *J.C.S. Chem. Comm.*, 330 (1985).

trans-2,5-Bis(methoxymethoxymethyl)pyrrolidine (1).

α-Hydroxy carboxylic acids.[1]   N-(Benzyloxyacetyl)-trans-2,5-bis(methoxy-methoxymethyl)pyrrolidine (2) can be prepared from 1 by reaction with benzyloxy-acetyl pivalic anhydride (equation I). The anion of 2 (BuLi or LDA) is alkylated

2, α_D —48.9°

with >96% de by primary halides or by isopropyl triflate. The products are hy-drolyzed without racemization to (S)-α-alkoxy carboxylic acids.

Example:

(96% de)   (S)

Chiral α-amino acids.[2]   The lithium anion of the N-protected glycine amides 3, prepared by reaction of the pyrrolidine with [bis(methylthio)methylene]glycyl pivalic anhydride (DMAP), is alkylated with high diastereoselectivity. The (S)-amino acid (5) is obtained on acid hydrolysis.

[1] M. Enomoto, Y. Ito, T. Katsuki, and M. Yamaguchi, *Tetrahedron Letters*, **26**, 1343 (1985).
[2] S. Ikegami, T. Hayama, T. Katsuki, and M. Yamaguchi, *ibid.*, **27**, 3403 (1986).

2,4-Bis(4-methoxyphenyl)-1,3-dithia-2,4-diphosphetane-2,4-disulfide (Lawesson's reagent) (1).

Thiophene synthesis.[1]   A key step in a synthesis of α-terthienyl (3) from 2-

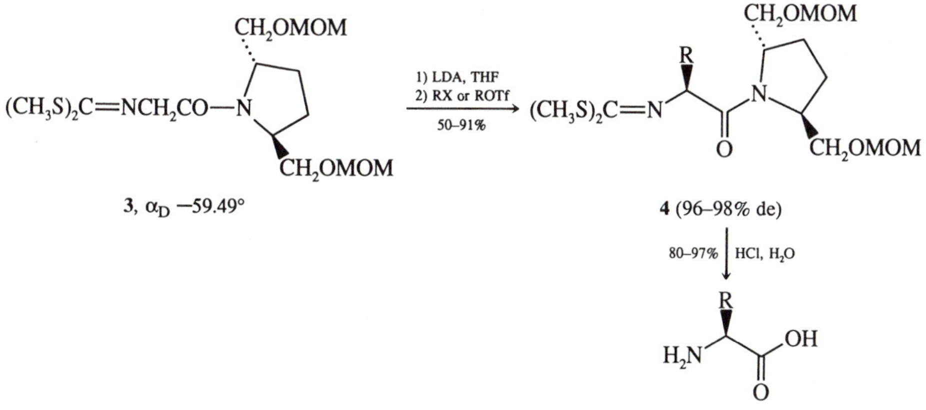

3, α_D −59.49°

4 (96–98% de)

(S)-5 (95–97% ee)

acetylthiophene is the reaction of 1,4-di-2-thienyl-1,4-butanedione (2) with Lawesson's reagent (1) in toluene at 80–90°.

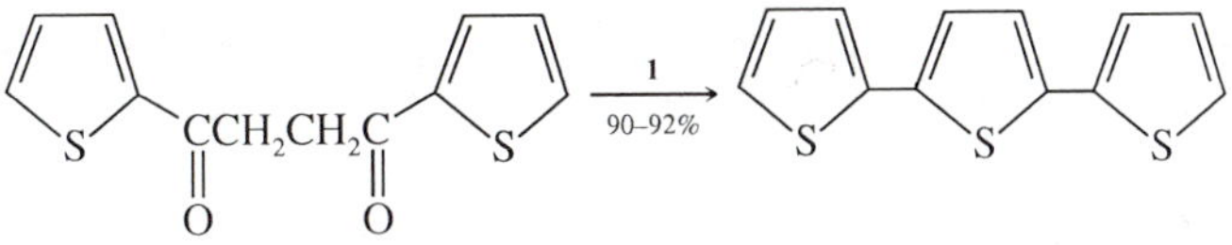

2   3

Review.[2]   Thionation of various substrates with this reagent has been reviewed (100 references).

[1] H. Wynberg and P. Camper, *Org. Syn.*, submitted (1985).
[2] M. P. Cava and M. I. Levinson, *Tetrahedron*, **41**, 5061 (1985).

Bismuth.
    Allylation of aldehydes.[1]   Metallic Bi (1.2 equiv.) effects reaction of allyl bromides or iodides with aldehydes to form homoallylic alcohols in 70–98% yield.
    Example:

[1] M. Wada and K. Akiba, *Tetrahedron Letters*, **26**, 4211 (1985).

N,N-Bis(2-oxo-3-oxazolidinyl)phosphordiamidic chloride (1).
    Coupling of N-alkylamino acids.[1]   Coupling of these amino acids by the usual

reagents suffers from low yields and significant racemization, but can be effected by **1** in combination with $N(C_2H_5)_3$ or diisopropylethylamine at 0–5°.

Example:

$$\underset{\substack{|\\ Bu\text{-}i}}{BocN-CHCOOH} + \underset{\substack{|\\ i\text{-}Bu}}{CH_3NHCHCOOCH_2C_6H_5} \xrightarrow[84\%]{1,\ N(C_2H_5)_3,\ CH_2Cl_2}$$

$$\underset{\substack{|\\ Bu\text{-}i}}{Boc\text{-}N-CHCO}-\underset{\substack{|\\ Bu\text{-}i}}{N-CHCOOCH_2C_6H_5}$$

$$\alpha_D\ -107.3°$$

[1] R. D. Tung and D. H. Rich, *Am. Soc.*, **107**, 4342 (1985).

Bis(2-pyridinethiolato)tin(II), $Sn\left(S \overset{\displaystyle\frown}{} N\right)_2$ (**1**). This stannylene is prepared *in situ* by reaction of 1,1'-dimethylstannocene and 2-mercaptopyridine in CH_2Cl_2 at 20°.

α,β-Dihydroxy ketones.[1]   The tin(IV) enediolates generated from α-diketones with **1** react with aldehydes to form α,β-dihydroxy ketones with moderate diastereoselectivity.

Example:

$$\underset{O\ \ \ \ O}{CH_3\overset{\|}{C}-\overset{\|}{C}CH_3} \xrightarrow[82\%]{\substack{1)\ 1,\ CH_2Cl_2\\ 2)\ C_6H_5(CH_2)_2CHO}} \underset{\substack{|\ \ \ \ |\\ CH_3\ \ OH}}{CH_3\overset{O}{\overset{\|}{}}\overset{OH}{}(CH_2)_2C_6H_5} + \underset{\substack{|\ \ \ \ |\\ CH_3\ \ OH}}{CH_3\overset{O}{\overset{\|}{}}\overset{OH}{}(CH_2)_2C_6H_5}$$

$$(syn) \qquad\qquad 77:23 \qquad (anti)$$

2-Hydroxyalkyl-1,4-diketones.[2]   The bisenolate of 1,4-enediones obtained by reaction with $Sn(SPy)_2$ reacts with an aldehyde to give 2-hydroxyalkyl-1,4-diketones.

Example:

$$\underset{O\ \ \ \ \ \ \ \ \ O}{R^1\overset{\|}{C}CH=CH\overset{\|}{C}R^2} \xrightarrow[66\text{–}92\%]{\substack{Sn(SPy)_2\\ RCHO}} \underset{\substack{|\\ CH_2CR^2\\ \overset{\|}{O}}}{R^1\overset{O}{\overset{\|}{}}\overset{OH}{}R}$$

$$anti/syn = 65\text{–}90{:}35\text{–}10$$

[1] J. Ichikawa and T. Mukaiyama, *Chem. Letters*, 1009 (1985).
[2] T. Mukaiyama, J. Ichikawa, M. Toba, and M. Hayashi, *ibid.*, 1539 (1985).

[Bis(salicylidene-γ-iminopropyl)methylamine]cobalt(II) (1, CoSMDPT).

Cleavage of —CH=CH—.   After observing the formation of vanillin from the oxygenation of lignin catalyzed by **1**, Drago *et al.*[1] examined the cobalt-catalyzed oxygenation of isoeugenol (**2**) as a model substrate. Again vanillin (**3**) was obtained as the major product in addition to some dehydroisoeugenol (**4**).

4 (20%)

[1] R. S. Drago, B. B. Corden, and C. W. Barnes, *Am. Soc.*, **108**, 2453 (1986).

Bis(tributyl)tin oxide, $[(C_4H_9)_3Sn]_2O$.

Selective oxidation of sec-alcohols.[1]   The final steps in a synthesis of spectinomycin (**3**), an aminocyclitol antibiotic, involved a selective oxidation of one of three alcohol groups of **1** to provide **2**, which furnishes **3** on deprotection (H_2/Pd/ C, 90% yield). This oxidation can be carried out by conversion to a tributyltin ether followed by NBS oxidation to an α-hydroxy ketone in 80% yield (crude). Alternatively, reaction with Bu_2SnO provides a stannylidene acetal, which is also oxidized by NBS to **2** in high yield.

[1] S. Hanessian and R. Roy, *Can. J. Chem.*, **63**, 163 (1985).

Borane–Dimethylamine, $(CH_3)_2NH \cdot BH_3$ **(1)**, m.p. 36°. Supplier: Aldrich.

Reduction and N-formylation of folic acid.   Folic acid (**2**) can be converted in one step in 64% yield into folinic acid (leucovorin, **3**) with borane–dimethylamine (15 equiv.) in formic acid at 4°.[1] Folinic acid, a metabolite of folic acid, is used

2

3

clinically as an antidote to folic acid antagonists. The most efficient previous synthesis[2] involves a two-step reduction ($Na_2S_2O_4$, then $NaBH_4$) followed by reaction with methyl formate in DMSO–pyridine (51% overall, crude). The actual reactant may be boron triformate.

[1] R. A. Forsch and A. Rosowsky, *J. Org.*, **50**, 2582 (1985).
[2] E. Khalifa, A. N. Ganguly, J. H. Bieri, and M. Viscontini, *Helv.*, **63**, 2554 (1980).

Borane–Tetrahydrofuran.

Reduction of nitroalkenes. α,β-Unsaturated nitro compounds can be reduced to hydroxylamines by $BH_3 \cdot THF$ in the presence of a catalytic amount of sodium borohydride (equation I).[1]

This reduction can also be conducted in comparable yield with $NaBH_4$ and BF_3 etherate (ratio 1:1.5 equiv.) in THF.[2] The hydroxylamines can be reduced completely to alkylamines on extended reaction with reagent generated *in situ*.[3]

[1] M. S. Mourad, R. S. Varma, and G. W. Kabalka, *J. Org.*, **50**, 133 (1985).
[2] R. S. Varma and G. W. Kabalka, *Org. Prep Proc. Intl.*, **17**, 254 (1985).
[3] *Idem, Syn. Comm.*, **15**, 843 (1985).

Boron tribromide.

Bromoboration; (Z)-1,2-dihalo-1-alkenes.[1]   Reaction of a terminal alkyne (1-octyne) with BBr_3 in CH_2Cl_2 results in a β-bromoalkenylborane, which undergoes brominolysis or iodinolysis with retention to provide (Z)-1,2-dihalo-1-alkenes (equation I).

(I) $CH_3(CH_2)_5C\equiv CH \xrightarrow{BBr_3}$ $\xrightarrow[76\%]{\text{NaI, NaOAc,} \atop \text{chloramine-T}}$

[1] S. Hara, T. Kato, H. Shimizu, and A. Suzuki, *Tetrahedron Letters*, **26**, 1065 (1985).

Boron trichloride.

Aldol reactions.[1]   Several exotic boron derivatives have been used to prepare boron enolates, of particular interest because of their use for selective *syn*-aldol reactions. Actually boron enolates can be generated using BCl_3 and Hünig's base. Dichloroboron enolates are unusually reactive even at $-95°$, and show *syn*-selectivity of 80–95%. Aldol reactions are carried out in CH_2Cl_2 by mixing the ketone and BCl_3 (1:2 equiv.) followed by addition of the base (2 equiv.) and the aldehyde (1 equiv.). Yields are 80–95%.

[1] H.-F. Chow and D. Seebach, *Helv.*, **69**, 604 (1986).

Boron trifluoride etherate.

1,2-Acyl rearrangement of α,β-epoxy ketones.   This rearrangement can be used for synthesis of cyclic spiro-1,3-ketones. Thus the 2-cycloheptylidenecyclopentanone oxide **1** rearranges in the presence of BF_3 etherate at $25°$ within one minute to the spiro-1,3-diketone **2** in 91% yield.[1]

1   2

Glycosylation.   Two laboratories have reported that protected glycosyl fluorides undergo stereoselective glycosylation in CH_2Cl_2 with alcohols or silyl ethers in the presence of BF_3^2 or BF_3·etherate.[3] The liberated HF is trapped with $N(C_2H_5)_3$. Glycosylation of 2,3,4,6-tetrapivaloyl-α-D-glucopyranosyl fluoride occurs stereoselectively to give β-glycosides in high yield.

The Swiss report[3] includes a variation of the Mitsunobu reaction for preparation of acid-sensitive glycosyl fluorides (equation I).

Alkynyl ketones.[4]   Aliphatic esters react with lithium acetylides in the presence of this Lewis acid in THF at $-78°$ to form alkynyl ketones in 40–80% yield.

$$R^1COOC_2H_5 + R^2C{\equiv}CLi \xrightarrow[40-80\%]{\substack{BF_3 \cdot O(C_2H_5)_2, \\ THF, -78°}} R^1\overset{\overset{O}{\|}}{C}C{\equiv}CR^2$$

$$R^2 = H, C_6H_5,$$
$$C_5H_{11}\text{-}n$$

(E)- to (Z)-Isomerization.[5] BF_3 etherate or $C_2H_5AlCl_2$ increases the efficiency of photoisomerization of (E)- to (Z)-cinnamic esters. In the absence of the Lewis acid, optimum (E)- to (Z)-conversion is in the range of 20–55%, whereas conversion of $\geqslant 85\%$ is possible in the presence of the Lewis acids.

Lewis acids inhibit the photochemical deconjugation of α,β- to β,γ-unsaturated butenoic acids and promote rearrangement of (E)- to (Z)-isomers. Lewis acids also promote selective (E)- to (Z)-isomerization of the α,β-double bond of methyl 2,4-hexadienoate.[6]

Diels-Alder reactions of α-ethenylidenecyclanones.[7]   These dienophiles (1) are readily obtained by reaction of lithium acetylide with epoxides followed by oxidation, but tend to polymerize when heated. Fortunately catalysts, such as BF_3 etherate or $ZnCl_2$, permit Diels-Alder reactions to proceed at low temperatures. This cycloaddition provides a regio- and stereoselective route to spirocyclic dienones (2) in fair to good yield.

Example:

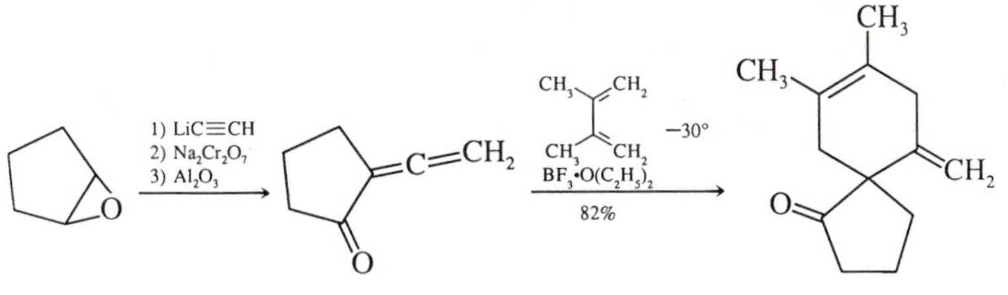

1   2

Cyclocondensation of aldehydes with silyloxydienes (**12**, 312). The stereoselectivity of this reaction depends not only on the catalyst but also on the solvent
as summarized in equation (I) for reactions with 1-methoxy-2-methyl-3-trialkylsi-
lyloxy-1,3-pentadienes **1a** and **1b**.[8]

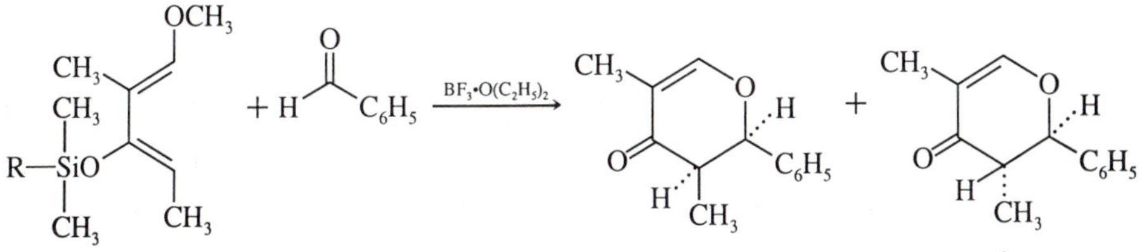

1a, R = CH$_3$	CH$_2$Cl$_2$	1:4.6
	C$_6$H$_5$CH$_3$	2.2:1
1b, R = t-Bu	CH$_2$Cl$_2$	1:2.3
	C$_6$H$_5$CH$_3$	7:1

Retro-Claisen rearrangement.[9]   The formyl bicyclo[2.2.2]octane **1** when heated
with a catalytic amount of HOAc at 110° rearranges to the vinyl ether **2**, probably
because of relief of the strain associated with the *vic*-quaternary carbon centers.
This retro-Claisen rearrangement occurs rapidly at 0° in the presence of BF$_3$·O(C$_2$H$_5$)$_2$
(0.1 equiv.).[10]

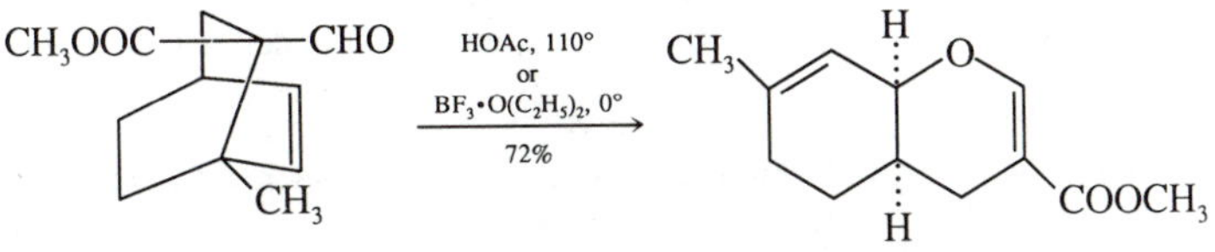

1   2

syn-**Selective addition of crotylmetals to aldehydes** (**10**, 411).   The ability of BF_3 etherate to reverse the usual *anti*-selectivity in the reaction of tributylcrotyltin with aldehydes to provide mainly *syn*-γ-adducts in addition to the α-adduct is observed with the crotylmetal reagents based on Cu, Cd, Hg, Tl, Ti, Zr, and V, and is particularly marked in reactions of Cu, Tl, and Zr. The α/γ ratio is highly dependent on the structure of the aldehyde.[11]

Example:

anti/syn = 81:19

syn/anti = 96:4   34:66   Z/E = 50:50

In contrast, BF_3 etherate does not influence the *syn/anti* ratio in the case of Mg, Zn, or B.

Allylic, benzylic, or tertiary bromides (iodides).   Allylic, benzylic, or tertiary alcohols are converted into bromides or iodides by reaction with BF_3 etherate and $(C_2H_5)_4NBr$ or $(C_2H_5)_4NI$ in CH_2Cl_2 in 60–80% yield. NaBr or NaI can replace $(C_2H_5)_4NX$ in the case of allylic or benzylic alcohols.[12] The combination of BF_3 etherate and $(C_2H_5)_4NX$ is useful for cleavage of ethers.[13]

Carbamates.[14]   A conventional procedure for preparation of carbamates of secondary alcohols involves reaction with an isocyanate and a basic catalyst (pyridine, triethylamine). Actually, BF_3 etherate and $AlCl_3$ are generally superior catalysts for this reaction, effected either in benzene or ether (25°, ½–2 hours).

[1] R. D. Bach and R. C. Klix, *J. Org.*, **50**, 5438 (1985).

[2] K. C. Nicolaou, A. Chucholowski, R. E. Dolle, and J. L. Randall, *J.C.S. Chem. Comm.*, 1155 (1984).

[3] H. Kunz and W. Sager, *Helv.*, **68**, 283 (1985).

[4] M. Yamaguchi, K. Shibato, S. Fujiwara, and I. Hirao, *Synthesis*, 421 (1986).

[5] F. D. Lewis, J. D. Oxman, L. L. Gibson, H. L. Hampsch, and S. L. Quillen, *Am. Soc.*, **108**, 3005 (1986).

[6] F. D. Lewis, D. K. Howard, S. V. Barancyk, and J. D. Oxman, *ibid.*, **108**, 3016 (1986).

[7] J.-L. Gras and A. Guerin, *Tetrahedron Letters*, **26**, 1781 (1985).

[8] S. Danishefsky, K.-H. Chao, and G. Schulte, *J. Org.*, **50**, 4650 (1985).
[9] R. K. Boeckman, Jr., C. J. Flann, and K. M. Poss, *Am. Soc.*, **107**, 4359 (1985).
[10] R. P. Lutz has reviewed catalysts for Cope and Claisen rearrangements, *Chem. Rev.* **84**, 205 (1984).
[11] Y. Yamamoto and K. Maruyama, *J. Organometal. Chem.*, **284**, C45 (1985).
[12] A. K. Mandal and S. W. Mahajan, *Tetrahedron Letters*, **26**, 3863 (1985).
[13] A. K. Mandal, N. R. Soni, and K. R. Ratnam, *Synthesis*, 274 (1985).
[14] T. Ibuka, G.-N. Chu, T. Aoyagi, K. Kitada, T. Tsukida, and F. Yoneda, *Chem. Pharm. Bull.*, **33**, 451 (1985).

Bromine.

Epoxide ring expansion.[1] The reaction of the monoepoxide (**1**) of 1,5-cyclooctadiene with bromine in CH_2Cl_2 results in the two cyclic ethers **2** and **3** with high stereoselectivity.

[1] S. G. Davies, M. E. C. Polywka, and S. E. Thomas, *Tetrahedron Letters*, **26**, 1461 (1985).

B-Bromocatecholborane; B-chlorocatecholborane,

BBr(Cl) (**1**). The reagents can be prepared by reaction of catechol with BBr_3 (or BCl_3); solutions in CH_2Cl_2 are stable for months in the absence of moisture.[1]

Cleavage of ethers[2,3] ***and carbamates.***[2] These reagents are useful for the cleavage of ethers and carbamates. Since the chloride is less reactive than the bromide, it can be more useful for selective cleavages. The tentative order of reactivity for the bromide is Boc > Cbo ≈ *t*-BuOR > $C_6H_5CH_2OR$ > CH_2=$CHCH_2OR$ > R^1R^2CHOR > R^1CH_2OR. Cleavage of carbamates generally requires 2 equiv. of **1**.

[1] W. Gerrard, M. F. Lappert, and B. A. Mountfield, *J. Chem. Soc.*, 1529 (1959).
[2] R. K. Boeckman, Jr., and J. C. Potenza, *Tetrahedron Letters*, **26**, 1411 (1985).
[3] P. F. King and S. G. Stroud, *ibid.*, **26**, 1415 (1985).

Bromodimethylborane.

(Methylthio)methyl ethers (MTM ethers, $ROCH_2SCH_3$).[1] Methoxymethyl (MOM) and (2-methoxyethoxy)methyl (MEM) ethers are converted to MTM ethers by reaction with $(CH_3)_2BBr$ in CH_2Cl_2 at $-78°$ for 1 hour followed by reaction with

(I) $ROCH_2O(CH_2)_2OCH_3 \xrightarrow{(CH_3)_2BBr} [ROCH_2Br] \xrightarrow[72-83\%]{CH_3SH} ROCH_2SCH_3$

(II) $R'CH(OCH_3)_2 \xrightarrow[72-90\%]{\substack{1)\ (CH_3)_2BBr \\ 2)\ RSH}} R'CH\underset{SR}{\overset{OCH_3}{<}}$

CH_3SH and $C_2H_5N[CH(CH_3)_2]_2$ (equation I). Under the same conditions dialkyl acetals can be converted into O,S-acetals (equation II). MEM and MOM ethers can be converted into cyanomethyl ethers ($ROCH_2CN$) by sequential reaction with $(CH_3)_2BBr$ and Bu_4NCN in 80–90% yield.

[1] H. E. Morton and Y. Guindon, *J. Org.*, **50**, 5379 (1985).

9-Bromo-9-phenylfluorene (9-PhFlBr),

(**1**), m.p. 99°. The

reagent is prepared in 80% yield by addition of phenyllithium to fluorenone followed by reaction of the adduct with 48% HBr in toluene.

Protection of amines.[1] The 9-phenylfluorenyl group is introduced by treatment of an amine with **1** in CH_3CN with K_3PO_4 as base and $Pb(NO_3)_2$ as a bromide scavenger. The 9-PhFl group is much more stable than the trityl group to protonolysis, but can be removed with TFA in aqueous CH_3CN.

This protecting group can direct enolate formation of an ester as well as the site of alkylation. Thus the enolate of **2** is alkylated at the benzylic position rather

than the position α to the ester group. This methodology was used in a synthesis of optically pure pipecolates such as **5** from L-asparagine.

[1] B. D. Christie and H. Rapoport, *J. Org.*, **50**, 1239 (1985).

Bromopentacarbonylmanganese, $Mn(CO)_5Br$ (**1**). This yellow metal carbonyl halide is obtained in 73% yield by reaction of manganese carbonyl with Br_2 in CCl_4 at 40°.[1]

γ-*Butyrolactones*.[2]   Reaction of terminal alkynes with CH_3I and $Br(CO)_5Mn$ [or $Mn_2(CO)_{10}$] in equimolar amounts in an atmosphere of carbon monoxide under phase-transfer conditions results in *cis*-and *trans*-2,4-disubstituted γ-butyrolactones in 50–75% yield, with some preference for the *trans*-isomer. Reactions conducted under nitrogen give only traces of the lactones.

Example:

$$C_6H_5C\equiv CH + CH_3I \xrightarrow[R_4NCl,\ NaOH,\ CH_2Cl_2]{Mn(CO)_5Br,}$$

(47%)   (31%)

[1] E. W. Abel and G. Wilkinson, *J. Chem. Soc.*, 1501 (1959).
[2] C. H. Wang and H. Alper, *J. Org.*, **51**, 273 (1986).

N-Bromosuccinimide.

1,4-Benzodioxines.   The reaction of benzo-1,4-dioxanes with NBS catalyzed by dibenzoyl peroxide results in 2,3-dibromo derivatives, which undergo debromination to benzo-1,4-dioxines when treated with sodium iodide in acetone (equation I).[1]

(I)

[1] G. Coudert and G. Guillaumet, *Org. Syn.*, submitted (1985).

***trans*-Bromotetracarbonyl(methylmethylidyne)tungsten,** $CH_3C\equiv W(CO)_4Br$ (**1**), yellow needles.

Preparation[1]:

$$W(CO)_6 + CH_3Li \longrightarrow \left[(CO)_5W\overset{\overset{\displaystyle O}{\|}}{C}CH_3 \right]Li \xrightarrow{(CH_3)_3OBF_4} (CO)_5W[C(OCH_3)CH_3] \xrightarrow{BBr_3} 1$$

Phenol synthesis.[2]   This metal-carbyne (or the corresponding chromium-carbyne) reacts with diynes rapidly at 25° to form phenols in 15–55% yield. Examples:

[1] E. O. Fischer, U. Schubert, and H. Fischer, *Inorg. Syn.*, **19**, 172 (1979).
[2] T. M. Sivavec and T. J. Katz, *Tetrahedron Letters*, **26**, 2159 (1985).

t-**Butyl 4-benzoylperbenzoate, (1)**, m.p. 62–64°.

The perester is obtained in 75% yield by reaction of 4-benzoylbenzoyl chloride with *t*-butyl hydroperoxide and $N(C_2H_5)_3$ in ether.[1]

Photoinitiation of free-radical reactions.[2]   Use of thermal initiators for radical sources, such as AIBN or dibenzoyl peroxide, requires temperatures >50°. This perester, in contrast, decomposes at room temperature or below on irradiation at 360 nm. This mode of initiation can be useful when stereoselectivity is enhanced at lower temperatures.

[1] L. Thijs, S. N. Gupta, and D. C. Neckers, *J. Org.*, **44**, 4123 (1979).
[2] P. Gottschalk and D. C. Neckers, *ibid.*, **50**, 3498 (1985).

t-**Butyldimethylsilyl trifluoromethanesulfonate.**

Preparation[1]: This reagent can be prepared in 66% yield by reaction of triflic acid with isopropenyltrimethylsilane, $CH_2=C(CH_3)Si(CH_3)_3$, which is available from Petrarch or which can be prepared by a Wurtz-Fittig reaction of 2-chloropropene with sodium and chlorotrimethylsilane in ether/HMPT.

—**NHBoc** ⟶ —**NHCbo**.[2]   The removal of Boc groups requires relatively strong acidic conditions. Conversion to a Cbo group, removable under neutral conditions, is useful for acid-sensitive substrates. This reaction can be effected in two steps: reaction with *t*-BuMe₂SiOTf provides an N-*t*-butyldimethylsilyloxycarbonyl group, which is converted to an NCbo group by reaction with benzyl bromide and Bu₄NF (60–90% overall yield).

Example:

[1] P. F. Hudrlik and A. K. Kulkarni, *Tetrahedron Letters*, **26**, 1389 (1985).
[2] M. Sakaitani and Y. Ohfune, *ibid.*, **26**, 5543 (1985).

t-Butyl hydroperoxide–Chromium carbonyl.

Benzylic oxidation.   Tetralins are oxidized to α-tetralones in good yield by (CH₃)₃COOH and a catalytic amount of Cr(CO)₆.[1]
Example:

[1] A. J. Pearson and G. R. Han, *J. Org.*, **50**, 2791 (1985).

t-Butyl hydroperoxide–Dialkyl tartrate–Titanium(IV) isopropoxide.

Catalytic asymmetric epoxidation.[1]   Addition of heat-activated, powdered 3–5 Å molecular sieves to the asymmetric epoxidation of allylic alcohols increases the rate and, more importantly, permits use of catalytic amounts of titanium reagent and the tartrate ester. However, it is still important to use at least a 10% excess of the tartrate ester over Ti(O-*i*-Pr)₄; a 20% excess is usually advisable. In general, 5% of Ti(O-*i*-Pr)₄ and 7.5% of the tartrate ester is used for the catalytic epoxidation.

The main role of the molecular sieves is to protect the catalyst from the deleterious effect of adventitious water. In fact, deliberate addition of only one equivalent of water completely destroys the catalyst. A further advantage of the catalytic reaction is that the chemical yields are higher because the isolation of products is simpler. In general, enantioselectivity is somewhat lower, but usually by only 1–3%.

Perhaps the greatest advantage of the new procedure is that water-soluble products can be converted *in situ* into useful derivatives, after excess hydroperoxide is reduced with trimethyl phosphite. For this purpose, 4-nitrobenzoate (PNB) esters are particularly useful because of high crystallinity. They also can be used as such for further reaction involving epoxide cleavage. Tosylation and silylation *in situ* are also possible.

The advantages of the catalyzed version are particularly apparent in the case of allyl alcohol whose epoxide, glycidol (**2**), is unstable to the catalyst.[2] It can be obtained by catalyzed epoxidation at 0° with cumene hydroperoxide instead of *t*-butyl hydroperoxide in 65% yield and 90% ee. However, for use in a synthesis of the β-adrenergic blocking agent (2S)-propranolol (**5**), the epoxide is not isolated but treated with sodium α-naphthoxide to furnish the diol **3**. The synthesis of **5** is completed by conversion to an epoxide (**4**) and ring opening with isopropylamine.

An alternative one-pot synthesis of **5** from the enantiomer of **2** is outlined in equation (I).

The *in situ* opening is particularly useful for terminal epoxide groups present in compounds such as 2-methylglycidol (**6**), which is opened regioselectively at the terminal C_3-position by a variety of nucleophiles (C_6H_5SH, *sec*-amines, and phenols) in the presence of $Ti(O-i-Pr)_4$.

Example:

6   **7** (92% ee)

Chiral sulfoxides (**12**, 92). Kagan *et al.*[3] have reviewed the asymmetric oxidation of sulfides by a water-modified Sharpless reagent. Optical yields are generally highest in the oxidation of aryl methyl sulfides (~75–90%).

[1] R. M. Hanson, S. Y. Ko, H. Masamune, J. M. Klunder, and K. B. Sharpless, *Am. Soc.*, in press.
[2] J. M. Klunder, S. Y. Ko, and K. B. Sharpless, *J. Org.*, **51**, 3710 (1986); S. Y. Ko and K. B. Sharpless, *ibid.*, **51**, 5413 (1986).
[3] H. B. Kagan, E. Duñach, C. Nemecek, P. Pitchen, O. Samuel, and S.-H. Zhao, *Pure and Appl. Chem.*, **57**, 1911 (1985); E. Duñach and H. B. Kagan, *Nouv. J. Chim.*, **9**, 1 (1985).

***t*-Butyl hydroperoxide–Dibutyltin oxide.** The reagents (1:1) when mixed in benzene (60°) form a soluble dibutyltin oxyperoxide (**1**).

1

Selective epoxidation of allylic alcohols.[1]   This reagent is particularly useful for completely selective epoxidation of a double bond allylic to a hydroxyl group in the presence of another double bond. In this respect it is superior to *t*-BuOOH in combination with $VO(acac)_2$, $Al(O-t-Bu)_3$, or $Ti(O-i-Pr)_4$. The stereoselectivity with **1** is fairly similar to that of *t*-BuOOH–$VO(acac)_2$.

The rate of epoxidation depends on the substitution pattern of the double bond. A monosubstituted double bond is attacked very slowly, whereas a tetrasubstituted double bond reacts rapidly. Thus tri- or tetrasubstituted double bonds can be epoxidized in the presence of a mono- or disubstituted double bond.

[1] S. Kanemoto, T. Nonaka, K. Oshima, K. Utimoto, and H. Nozaki, *Tetrahedron Letters*, **27**, 3387 (1986).

t-Butyl hydroperoxide–Dichlorotris(triphenylphosphine)ruthenium(II).

Oxidation of amines to imines.[1] In the presence of this ruthenium complex secondary amines are oxidized to imines in >70% yield. This reaction is particularly useful for preparation of 1-azadienes.

Example:

$$C_6H_5CH=CHCH_2NHC_6H_5 \xrightarrow[74\%]{\underset{\text{Ru(II), C}_6\text{H}_6}{(CH_3)_3COOH,}} C_6H_5CH=CHCH=NC_6H_5$$

Acyl cyanides.[2] Aromatic and α,β-unsaturated cyanohydrins are oxidized efficiently to acyl cyanides by *t*-butyl hydroperoxide in benzene in the presence of $RuCl_2[P(C_6H_5)_3]_3$.

Examples:

X = H, CH₃, Cl, OCH₃

Benzoyl cyanides are excellent reagents for selective N-benzoylation of amino alcohols.

[1] S.-I. Murahashi, T. Naota, and H. Taki, *J.C.S. Chem. Comm.*, 613 (1985).
[2] S.-I. Murahashi, T. Naota, and N. Nakajima, *Tetrahedron Letters*, **26**, 925 (1985).

t-Butyl hydroperoxide–Vanadyl acetylacetonate.

Epoxidation of allylic and homoallylic alcohols.[1] The diastereomeric diols **1a** and **1b** posess both allylic and homoallylic hydroxyl groups. Oxidation of the

1a, R¹ = CH₃, R² = H
1b, R¹ = H, R² = CH₃

2a
2b

mixture with this reagent followed by reduction provides the triols **2a** and **2b**. Separate experiments show that **1a** is converted stereospecifically into **2a**, and **1b** into **2b**. In this substrate, the epoxidation is controlled by the homoallylic alcohol. The normal effect of the allylic alcohol is observed when the homoallylic group is protected as the MOM ether.

The key step in a total synthesis of cycloeudesmol (**6**) is a highly stereoselective epoxidation of **3** to provide **4**, which can be converted into the cyclopropane **5** in quantitative yield by treatment with 3 equiv. of LDA at $-78 \rightarrow 0°$.[2]

[1] R. W. Irvine, R. A. Russell, and R. N. Warrener, *Tetrahedron Letters*, **26**, 6117 (1985).
[2] M. Ando, K. Wada, and K. Takase, *ibid.*, **26**, 235 (1985).

t-Butyl hypochlorite.

Nitrile oxides.   *t*-Butyl hypochlorite is a convenient reagent for conversion of aldoximes into hydroximinyl chlorides, useful precursors to nitrile oxides (equation I).[1]

[1] C. J. Peake and J. H. Strickland, *Syn. Comm.*, **16**, 763 (1986).

Butyllithium.

Transesterification. Methyl esters of aromatic and α,β-unsaturated acids can undergo transesterification when treated with an alcohol (excess) and BuLi (1 equiv.) in dry THF at 25° (equation I). The yields can be quantitative in reactions

$$\text{(I)} \quad R^1COOCH_3 + R^2OH + n\text{-BuLi} \xrightarrow{\text{THF}} \left[\begin{array}{c} OLi \\ | \\ R^1\!-\!C\!-\!OR^2 \\ | \\ OCH_3 \end{array} \right] \longrightarrow R^1COOR^2$$

with secondary or (particularly) tertiary alcohols, but are low with primary alcohols, except for allylic alcohols. The method is not useful for transesterification of aliphatic esters.[1]

Benzyllithiums. These reagents are generally prepared indirectly from benzyltin or mercury compounds or from benzyl ethers. They can also be prepared by reaction of benzylphenyl or benzylmethyl selenides with *n*-, *sec*- or *t*-BuLi, since cleavage occurs exclusively at the benzyl—selenium bond.[2]

Example:

$$\underset{\underset{C(CH_3)_3}{|}}{\overset{\overset{SeCH_3}{|}}{C_6H_5C}}\!-\!CH_3 \xrightarrow[65\%]{\substack{1)\ \text{BuLi} \\ 2)\ \text{ICH(CH}_3)_2}} \underset{\underset{C(CH_3)_3}{|}}{\overset{\overset{CH(CH_3)_2}{|}}{C_6H_5C}}\!-\!CH_3$$

Propargyl dianion $(C_3H_2Li_2)$. This anion can be prepared by dilithiation of allene with BuLi in 1:1 ether/hexane. Use of THF $(-50°)$ or BuLi/TMEDA results in a mixture of propargylide and allenyl anions. The anion couples readily with alkyl and allyl halides to give terminal alkynes. The intermediate lithium acetylide can also react with various electrophiles.[3]

Example:

$$C_3H_2Li_2 \xrightarrow{C_6H_5CH_2Cl} [C_6H_5CH_2CH_2C\!\equiv\!CLi] \xrightarrow[83\%]{H^+} C_6H_5CH_2CH_2C\!\equiv\!CH$$

$$80\% \downarrow CH_2O$$

$$C_6H_5CH_2CH_2C\!\equiv\!CCH_2OH$$

[1] O. Meth-Cohn, *J.C.S. Chem. Comm.*, 695 (1986).
[2] M. Clarembeau and A. Krief, *Tetrahedron Letters*, **26**, 1093 (1985); A. Krief, M. Clarembeau, and P. Barbeaux, *J.C.S. Chem. Comm.*, 457 (1986).
[3] J. Hooz, J. G. Calzada, and D. McMaster, *Tetrahedron Letters*, **26**, 271 (1985).

Butyllithium–Potassium *t*-butoxide.

Asymmetric Wittig rearrangement (**12**, 96). The Wittig rearrangement has been used to control the configuration of two of the four chiral centers in a synthesis

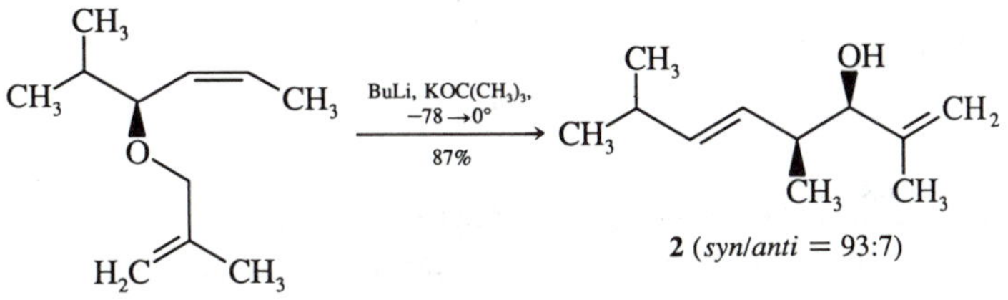

2 (*syn/anti* = 93:7)

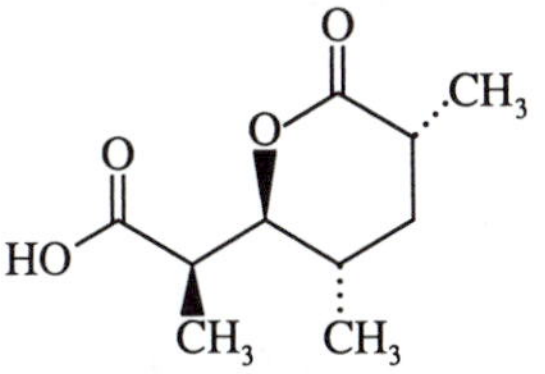

3

of (+)-Prelog-Djerassi lactone (**3**).[1] Thus rearrangement of the ether **1** provides the desired *syn*-isomer **2** (93:7 ratio). A third chiral center was introduced by a stereoselective (15:1) hydroboration of the terminal double bond with dicyclohexylborane.

[1] D. J.-S. Tsai and M. M. Midland, *Am. Soc.*, **107**, 3915 (1985).

Butyllithium–Tetramethylethylenediamine.

Benzanthraquinones.[1]   The skeleton of benz[*a*]anthraquinone is obtained by directed aromatic metallation[2] of **1** with butyllithium–TMEDA (2 equiv.) to provide a dilitho species **a**, which is then coupled with **2** to provide **3**. This product was used for a synthesis of ochromycinone (**4**).

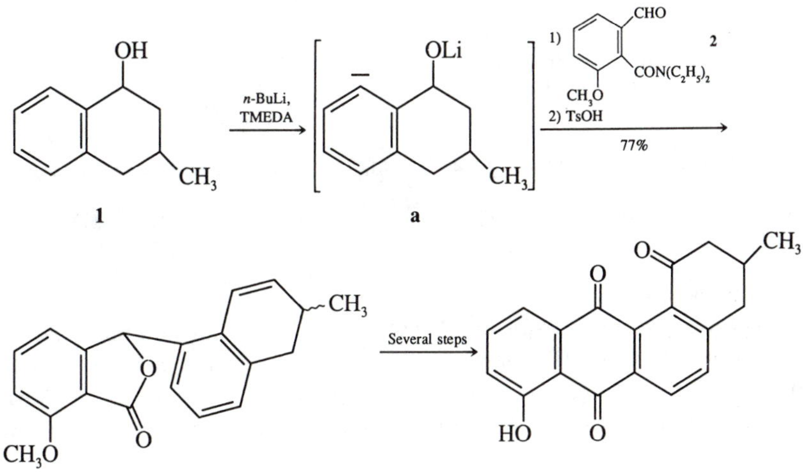

[1] K, Katsuura and V. Snieckus, *Tetrahedron Letters*, **26**, 9 (1985).
[2] P. Beak and V. Snieckus, *Accts. Chem. Research*, **15**, 306 (1982).

t-Butyllithium.

α,β-*Unsaturated acylsilanes* (3).   Allyl trimethylsilyl ethers (**1**) on metalla-tion with *t*-butyllithium (-78 to $-30°$) rearrange to organolithiums that on pro-tonation form (α-hydroxyallyl)silanes (**2**). These silanes are thermally unstable, but can be oxidized by DMSO–oxalyl chloride or by NCS–$S(CH_3)_2$ to α,β-unsat-urated acylsilanes (**3**).[1]
   Example:

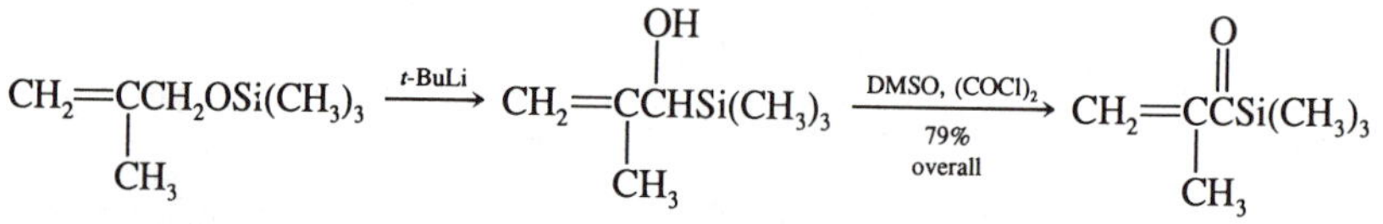

   Cyclization of ω-*iodoepoxides*.   Lithium-halogen exchange (*t*-BuLi) of ω-iodoepoxides results in cyclization to (cycloalkyl)methanols or cycloalkanols. Ad-dition of Lewis acids can affect the regiochemistry.[2]
   Example:

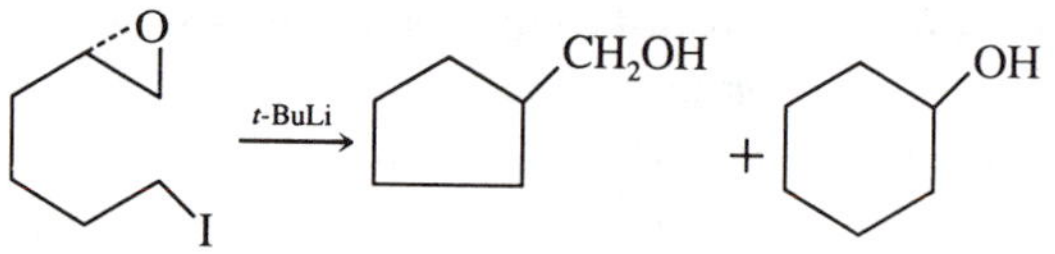

	78%	10:1
BF + $MgBr_2$	65%	>100:1
+ $CuBr \cdot S(CH_3)_2$	59%	1:4.4

[1] R. L. Danheiser, D. M. Fink, K. Okano, Y.-M. Tsai, and S. W. Szczepanski, *J. Org.*, **50**, 5393 (1985).
[2] M. P. Cooke, Jr. and I. N. Houpis, *Tetrahedron Letters*, **26**, 3643 (1985).

t-Butyl perbenzoate.

Benzoyloxylation of β-*lactams*.[1]   Reaction of β-lactams with this reagent in the presence of copper(II) octanoate effects benzoyloxylation at C_4, α to the ni-trogen when the nitrogen is substituted by a phenyl or *t*-butyl group. However, reaction at an exocyclic carbon α to the nitrogen is a competing reaction when possible.

Example:

[1] C. J. Easton and S. G. Love, *Tetrahedron Letters*, **27**, 2315 (1986).

Butylsodium, $CH_3(CH_2)_3Na$. Butylsodium precipitates as a powder when prepared by the reaction of BuLi with $NaOC(CH_3)_3$ in hexane. BuNa is soluble in hexane when complexed with TMEDA.

BuNa/TMEDA/hexane is comparable to BuLi/TMEDA. Thus it deprotonates weak organic acids. Benzylsodium·TMEDA (made from toluene) is obtained as yellow needles. BuK/TMEDA, prepared in the same way, metallates hydrocarbons more readily than BuNa/TMEDA, but also decomposes more readily.[1]

[1] C. Schade, W. Bauer, and P. V. R. Schleyer, *J. Organomet. Chem.*, **295**, C25 (1985).

R-(+)-*t*-Butyl (*p*-tolylsulfinyl)acetate (1).

Chiral propargylic alcohols.[1] The magnesium enolate of **1** undergoes an aldol type condensation with propargylic aldehydes to form adducts (**2**) in 50–75% yield. These are readily reduced to (S)-propargylic alcohols (**3**), obtained in 70–90% ee.

The reaction of **1** with propargylic ketones shows only moderate enantioselectivity (35–50% ee).

[1] G. Solladié, C. Frechou, and G. Demailly, *Nouv. J. Chem.*, **9**, 21 (1985).

C

Cadmium(0). A reactive cadmium powder can be prepared by reaction of $CdCl_2$ with lithium naphthalide in glyme or THF. A more reactive cadmium powder is obtained by reaction of $CdCl_2$ with lithium naphthalide prepared by sonication in TMEDA and toluene. A third form is prepared by treatment of Cd_2Li with 1 equiv. of I_2 to leach the lithium.

The Cd powders can be used in the standard synthesis of ketones from acid chlorides. The advantage is that many functional groups are tolerated.[1]

Example:

$$C_6H_5CH_2Br \xrightarrow[\substack{88\%}]{\substack{1)\ Cd(0),\ C_6H_6 \\ 2)\ C_6H_5COCl}} C_6H_5CH_2\overset{\displaystyle O}{\overset{\|}{C}}C_6H_5$$

The reactive Cd(0) powder reacts with α-halo esters to give a Reformatsky-type reagent.

Example:

$$BrCH_2COOC_2H_5 \xrightarrow[\substack{100\%}]{\substack{1)\ Cd_2Li/I_2 \\ 2)}}$$

[1] E. R. Burkhardt and R. D. Rieke, *J. Org.*, **50**, 416 (1985).

Calcium acetate.

Polyketide synthesis. Polyketides (**2**) are obtained on reaction of the sodium lithium dianion (**1**) of methyl acetoacetate with dimethyl glutarates. When refluxed with $Ca(OAc)_2$ in CH_3OH, **2** cyclizes to 1-oxo-1,2,3,4-tetrahydronaphthalenes in 30–75% yield.[1]

Example:

Since the aromatic products are also glutarates, the sequence can be repeated.

[1] M. Yamaguchi, K. Hasebe, and T. Minami, *Tetrahedron Letters*, **27**, 2401 (1986).

Camphor.

Asymmetric conjugate addition of R₂CuLi.[1]   The chiral crotonyl ester **1**, prepared from (+)-camphor by addition of 1-naphthylmagnesium bromide followed
by esterification with crotonic acid, undergoes conjugate addition with Bu_2CuLi at
−25° to give the adduct **2** in 95% de. Reduction of **2** with $LiAlH_4$ gives (S)-3-

1 (Ar = 1-naphthyl)   2 (95% de)

methyl-1-heptanol (95% ee) with recovery of the chiral auxiliary. The diastereoselectivity is highly dependent on the bulk of the alkyl or aryl group at C_2. No
diastereoselectivity obtains when the aromatic group is replaced by H.

[1] P. Somfai, D. Tanner, and T. Olsson, *Tetrahedron*, **41**, 5973 (1985).

Camphor-10-sulfonic acid (CSA).

Chiral halohydrins; epoxides.[1]   The esters (**2**) of the chiral alcohol **1** derived from camphor-10-sulfonic acid, are converted to α-chloro esters (**3**) by O-silylation and reaction with NCS with high diastereoselectivity. Reduction of **3** with $Ca(BH_4)_2$ results in the recovered auxiliary and the chlorohydrin **4** with clean retention. Cyclization of **4** to the terminal epoxide **5** proceeds with clean inversion.

Asymmetric   hydrogenation   of   camphor-derived   sultamimides (**1**).[2]   Hydrogenation of these imides catalyzed by Pd/C proceeds with >90%

1a, R = *n*-Pr	95%	98:2
1b, R = (CH$_2$)$_2$CH=C(CH$_3$)$_2$	96%	95:5

diastereoselectivity and is markedly higher than that observed with the soluble $Ir(COD)(Py)(PCy_3)PF_6$ catalyst (**10**,116; **12**,151; this volume). Hydrogenation of the (Z)-isomers of **1** shows the same but opposite sense of stereofacial discrimination. Saponification of **2** or **3** with LiOH in aqueous THF affords optically active β-substituted carboxylic acids and the chiral sultam.

Asymmetric acetoxylation of esters.[3]   Reaction of lead tetraacetate with the silyl enolate of the chiral ester **2** derived from camphorsulfonic acid results in α-acetoxylation with high diastereoselectivity. After crystallization **3** is obtained in 95% de. The product can be hydrolyzed to the optically active α-acetoxy carboxylic acid by K_2CO_3 or reduced to the chiral glycol **4** by $LiAlH_4$.

Addition of an organocuprate to the chiral ester (**5**) followed by similar α-acetoxylation results in generation of two contiguous asymmetric centers to give **7** with high optical purity.

Stereoselective spiroketalization.[4] Optimum conditions for thermodynami-

2, R^1 = H, R^2 = CH$_2$OH   **4**, R^1 = H, R^2 = CH$_2$OH
3, R^1 = CH$_2$OH, R^2 = H   **5**, R^1 = CH$_2$OH, R^2 = H

2/3/4/5 = 4.6:1.0:17.2:3.4

cally controlled spiroketalization of **1** to provide **4** in 54% yield employ CSA as catalyst and DMSO as solvent. The product is a precursor to talaromycin A (**6**).

6

Enantioselective lactonization.[5]   Monoprotonation of the disodium salt (**1**) of 4-hydroxypimelic acid in ethanol results in cyclization to a γ-lactone (**2**) in high yield. If 1 equiv. of (lS)-CSA is used optically active **2** is formed in yields as high

(S)-**2** (94% ee)

as 94% ee. The degree of enantioselectivity is highly dependent on the concentration of **1**, being highest at a concentration of 3.5 mmol in 100 ml. of C_2H_5OH containing 0.5% of water. Higher concentrations of water can even result in complete loss of enantioselectivity.

[1] W. Oppolzer and P. Dudfield, *Tetrahedron Letters*, **26**, 5037 (1985).
[2] W. Oppolzer, R. J. Mills, and M. Réglier, *ibid.*, **27**, 183 (1986).
[3] W. Oppolzer and P. Dudfield, *Helv.*, **68**, 216 (1985).
[4] S. L. Schreiber, T. J. Sommer, and K. Satake, *Tetrahedron Letters*, **26**, 17 (1985).
[5] K. Fuji, M. Node, S. Terada, M. Murata, H. Nagasawa, T. Taga, and K. Machida, *Am. Soc.*, **107**, 6404 (1985).

(Camphorylsulfonyl)oxaziridines, (+)-(2R, 8aS)-1 and (−)-(2S, 8aR)-1. The oxaziridines are obtained by oxidation of the sulfonimines (−)-and (+)-**2** derived from *d*- and *l*-10-camphorsulfonic acid, respectively.[1] The oxidation is best conducted with Oxone and 18-crown-6 in benzene/water buffered at pH 8.

(+)-(2R, 8aS)-**1**   (−)-**1**   (−)-**2**

Asymmetric   hydroxylation   of   lithium   enolates   of   esters   and amides.[2]   Hydroxylation of typical enolates of esters with (+)- and (−)-**1** is effected in 75–90% yield and with 55–85% ee. The reaction with amide enolates with (+)- and (−)-**1** results in the opposite configuration to that obtained with ester enolates and with less enantioselectivity. Steric factors appear to predominate over metal chelation.

For other uses of oxaziridines in synthesis, see 2-Arenesulfonyloxaziridines, this volume.

[1] F. A. Davis, J. F. Lemendola, Jr., U. Nadir, E. W. Kluger, T. C. Sedergran, T. W. Panunto, R. Billmers, R. Jenkins, Jr., I. J. Turchi, W. H. Watson, J. S. Chen, and M. Kimura, *Am. Soc.*, **102**, 2000 (1980).

[2] F. A. Davis, M. S. Haque, T. G. Ulatowski, and J. C. Towson, *J. Org.*, **51**, 2402 (1986).

Carbon dioxide.

Cleavage of vinyl epoxides to* cis-*diols.   In the presence of a Pd(0) catalyst, prepared by reaction of BuLi (2 equiv.) with $Pd(OAc)_2$ (1 equiv.) and $P(O-i-Pr)_3$ (excess) in THF, carbon dioxide reacts with vinyl epoxides to give carbonates of vinyl *cis*-1,2-diols.[1]

Examples:

These reactions proceed at room temperature, probably because CO_2 serves as a co-catalyst.

β-Keto acids. Ketones undergo carboxylation when treated with CO_2 at atmospheric pressure and at 25° in acetonitrile in the presence of $MgCl_2$–NaI (1 equiv.) and $N(C_2H_5)_3$ (2 equiv.) Yields are in the range 40–90%. Carboxylation of 2-butanone results in two isomeric β-keto acids in the ratio 45:55.[2]

[1] B. M. Trost and S. R. Angle, *Am. Soc.*, **107**, 6123 (1985).
[2] R. E. Tirpak, R. S. Olsen, and M. W. Rathke, *J. Org.*, **50**, 4877 (1985).

Carbon monoxide.
Carbonylation of organometallic reagents.[1] The review covers the carbonylation of organolithium, -magnesium, -boron, -mercury, and -palladium compounds (86 references).

[1] C. Narayana and M. Periasamy, *Synthesis*, 253 (1985).

1,1'-Carbonyldiimidazole.
Intramolecular C-acylation.[1] In a synthesis of (+)-actinobolin **4**, an antibiotic from *S. griseovirides*, the final carbon atom was introduced by O-acylation

2, PMS = p-CH$_3$C$_6$H$_5$CH$_2$SO$_2$

3

4 (R = L-alanine)

of **2** at C$_3$ with 1,1'-carbonyldiimidazole (**1**) followed by enolate formation with NaH and intramolecular C-acylation to provide **3**.

[1] R. S. Garigipati, D. M. Tschaen, and S. M. Weinreb, *Am. Soc.*, **107**, 7790 (1985).

(E)-(Carboxyvinyl)trimethylammonium betaine (1), m.p. 176–177° dec.
Preparation:

$$(CH_3)_3N + HC\equiv CCOOC_2H_5 \xrightarrow[76-82\%]{\overset{CH_2Cl_2, H_2O}{0 \to 25°}} (CH_3)_3\overset{+}{N}\diagup\diagdown COO^-$$

1

γ,δ-*Unsaturated aldehydes*. Alkoxides add to **1** to provide allyloxyacrylic acids, which undergo Claisen rearrangement at 160–180° with loss of CO_2 to provide γ,δ-unsaturated aldehydes.[1]
Example:

[1] G. Büchi and D. E. Vogel, *J. Org.*, **48**, 5406 (1983); *Org. Syn.*, submitted (1985).

Cerium(IV) ammonium nitrate (CAN).
   Nitration of arenes.   Reaction of 1-naphthol with CAN in acetic acid results in 2,4- and 4,6-dinitro derivatives. The reaction with CAN absorbed on silica gel results in 2-nitro- and 4-nitro-1-naphthol in 42 and 38% yield, respectively. Polynuclear arenes are not oxidized to quinones by CAN supported on silica but are converted mainly into mononitro derivatives. Thus phenanthrene is converted into 2-nitrophenanthrene (45% yield) and 3-nitrophenanthrene (28% yield).[1]
   Fragmentation of γ-hydroxyalkylsilanes.[2]   These silanes on treatment with 2–8 equiv. of CAN in aqueous CH_3CN at 25° undergo cleavage to unsaturated aldehydes or ketones.
   Examples:

2-Alkoxycephalosporins.[3]   Oxidation of cephalosporins with CAN (10 equiv.) in THF containing an alcohol results in alkoxylation at C_2 in ~40–55% yield. Example:

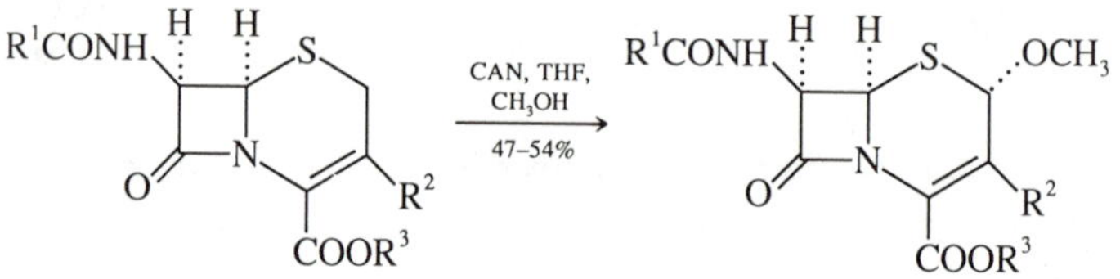

[1] H. M. Chawla and R. S. Mittal, *Synthesis*, 70 (1985).
[2] S. R. Wilson, P. A. Zucker, C. Kim, and C. A. Villa, *Tetrahedron Letters*, **26**, 1969 (1985).
[3] D. C. Humber and S. M. Roberts, *Syn. Comm.*, **15**, 681 (1985); R. A. Fletton, D. C. Humber, S. M. Roberts, and J. L. Wright, *J.C.S. Perkin I*, 1523 (1985).

Cesium fluoride.

Fluorodesilylation; sulfine; sulfene.   Sulfene, $CH_2=SO_2$, can be generated by reaction of $(CH_3)_3SiCH_2SO_2Cl$ with CsF.[1] Fluorodesilylation of trimethylsilyl-

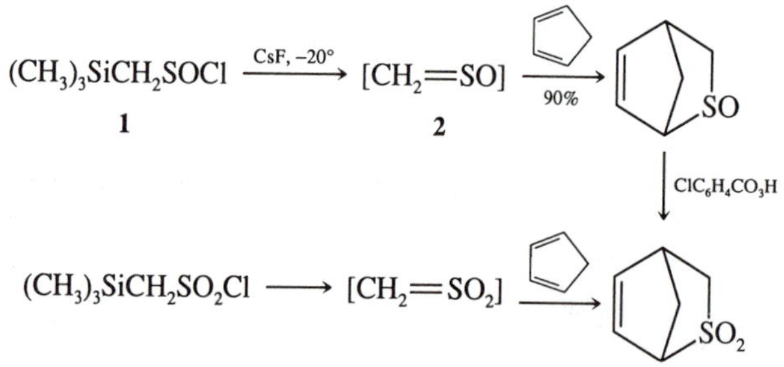

methanesulfinyl chloride (**1**) with CsF results in sulfine (**2**), as shown by the Diels-Alder adduct with cyclopentadiene (equation I).[2] The sulfonic anhydrides **3** are precursors to $RCH=SO_2$ (**4**), which form mainly *endo*-adducts **5** with cyclopentadiene.

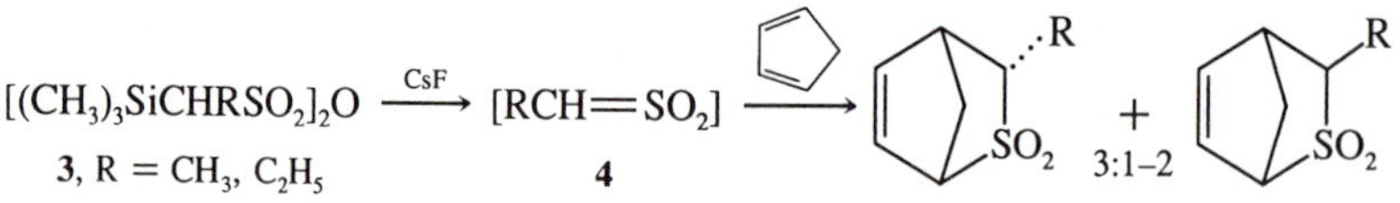

[1] E. Block and M. Aslam, *Tetrahedron Letters*, **23**, 4203 (1982).
[2] E. Block and A. Wall, *ibid.*, **26**, 1425 (1985).

Cesium fluoride–Tetraalkoxysilanes.

Michael addition to unsaturated amides.[1] This system (1 equiv. of each) effects Michael addition of ketones, nitro compounds, ethyl cyanoacetate, and diethyl malonate to α,β-unsaturated amides. Addition to methacrylamides is interesting because the final products are glutarimides or dihydropyridinones.

Examples:

$$N\equiv CCH_2COOC_2H_5 + CH_2=\underset{\underset{CH_3}{|}}{C}CONH_2 \xrightarrow[84\%]{}$$

$$C_6H_5COCH_3 + CH_2=\underset{\underset{CH_3}{|}}{C}CONH_2 \xrightarrow[55\%]{}$$

[1] C. Chuit, R. J. P. Corriu, R. Perz, and C. Reye, *Tetrahedron*, **42**, 2293 (1986).

Cetyltrimethylammonium permanganate, $CH_3(CH_2)_{15}N(CH_3)_3MnO_4$ **(1).**

Preparation.[1]

Selective oxidation of benzylic alcohols.[2] Benzylic alcohols are selectively oxidized in the presence of primary alcohols.

Example:

$$\xrightarrow[65\%]{1,\ CH_2Cl_2}$$

[1] V. Bhushan, R. Rathore, and S. Chandrasekaran, *Synthesis*, 431 (1984).
[2] R. Rathore, V. Bhushan, and S. Chandrasekaran, *Chem. Letters*, 2131 (1984).

Chloral.

Inversion of 7α-amino-1-oxacephems.[1] Synthetic cephalosporins are usually obtained as a mixture of 7α- and 7β-epimers, but only the latter are clinically useful antibiotics. A four-step method for epimerization of a 7α-amino-1-oxace-phem **(1)** to the epimer **(2)** is outlined in equation (I).

(I)

^{1} T. Aoki, N. Haga, Y. Sendo, T. Konoike, M. Yoshioka, and W. Nagata, *Tetrahedron Letters*, **26**, 339 (1985).

Chloramine-T–Sodium iodide.

Iodination of phenols.^{1} This combination (1.2 equiv. each) effects iodination of phenols in DMF, DMSO, or CH$_3$CN. Both electron-withdrawing or electron-donating groups are accommodated, but *para*-methoxy, -amino, or -acetamido groups inhibit the reaction.

Example:

^{1} T. Kometani, D. S. Watt, and T. Ji, *Tetrahedron Letters*, **26**, 2043 (1985).

α-Chloroacetamide, ClCH$_2$CONH$_2$ (**1**). Mol. wt. 93.5, m.p. 116–118°.

Carboxamidomethyl esters, RCOOCH$_2$CONH$_2$.^{1} These esters are prepared by successive reaction of the acid with Cs$_2$CO$_3$ and **1** (~80% yield). The esters are stable to TFA, diethylamine, and hydrogenolysis. They are readily hydrolyzed by Na$_2$CO$_3$ without racemization.

^{1} J. Martinez, J. Laur, and B. Castro, *Tetrahedron*, **41**, 739 (1985).

μ-Chlorobis(cyclopentadienyl)(dimethylaluminum)-μ-methylenetitanium (Tebbe reagent).

in situ Preparation. The reagent is available from Strem and Alfa, but is expensive. Fortunately it can be prepared *in situ* by reaction of dichlorobis(cyclopentadienyl)titanium and 2 equiv. of trimethylaluminum with stirring in toluene for 72 hours. Reagent prepared in this way can be used for methylenation of ketones or esters on a large scale in 75–85% yield.[1]

α,α'-Disubstituted oxocanes.[2] Naturally occurring oxocanes often are substituted by *cis*-α,α'-alkyl groups. A new route to this system involves methylenation of a substituted heptanolide (available by Baeyer-Villiger oxidation of a cycloheptanone) such as **2** with the Tebbe reagent. The unstable product **3** is converted to

4 by hydroboration and oxidation and then by standard reactions into lauthisan (**6**), the parent ring system found in several natural products.

[1] L. F. Cannizzo and R. H. Grubbs, *J. Org.*, **50**, 2386 (1985).
[2] R. W. Carling and A. B. Holmes, *J.C.S. Chem. Comm.*, 565 (1986).

2-Chloro-1,3-butadiene (Chloroprene), CH_2=CH(Cl)CH=CH_2 (**1**).

Diels-Alder reactions.[1] A key step in a synthesis of 4-demethoxydaunomycinone (**5**) involves a regioselective Diels-Alder reaction of anthracene-1,4,9,10-tetrone (**2**) with (**1**) to give the desired adduct (**3**) in 79% yield. A similar reaction with 2-acetoxy-1,3-butadiene affords only a 58% yield of the corresponding adduct. The product (rather unstable) is aromatized and hydrolyzed to give **4** in high yield.

2 + **1** $\xrightarrow[\text{79\%}]{\substack{C_6H_6, \\ \text{Xylene, }\Delta}}$ **3** $\xrightarrow[\text{2) H}_2\text{SO}_4\text{ (98\%)}]{\text{1) NaOAc, HOAc (91\%)}}$

4 $\xrightarrow[\text{steps}]{\text{Several}}$ **5**

This product was converted to **5** by known reactions.

[1] Y. Kimura, M. Suzuki, T. Matsumoto, R. Abe, and S. Terashima, *Bull. Chem. Soc. Japan*, **59**, 415 (1986).

Chlorodiisopinocampheylborane (Ipc$_2$BCl),

(**1**). The borane

(99% ee) can be prepared from (+)- or (−)-α-pinene (92% ee) by hydroboration with BH$_3$·S(CH$_3$)$_2$ followed by reaction with hydrogen chloride (75% yield).

Asymmetric reduction of ketones. Ipc$_2$BCl is somewhat superior to B-3-pin-anyl-9-borabicyclo[3.3.1]nonane (**12**, 397) for enantioselective reduction of alkyl aryl ketones at normal pressures to (S)-alcohols. In general, optical yields are 78–98%. It is also useful for asymmetric reduction of ketones in which one alkyl group is tertiary. Thus 3,3-dimethyl-2-butanone is reduced in 95% ee at 25°.[1]

[1] J. Chandrasekharan, P. V. Ramachandran, and H. C. Brown, *J. Org.*, **50**, 5446 (1985).

Chlorodiisopropylsilane, [(CH$_3$)$_2$CH]$_2$SiHCl (**1**).

anti-1,3-Diols. The diisopropylsilyl ethers (**2**) derived from β-hydroxy ke-tones on treatment with **1** are converted mainly to *anti*-siladioxanes **3** with SnCl$_4$. Desilylation of *anti*-**3** provides *anti*-1,3-diols (**4**).[1]

$$2 \qquad 3\text{-}(anti/syn = \\ 40\text{--}120{:}1) \qquad 4$$

Use of the related dimethylsilyl ethers, obtained from β-hydroxy ketones using commercially available chlorodimethylsilane, affords **4** with only slightly less diastereoselectivity. Other Lewis acids can be used in place of $SnCl_4$.[1]

[1] S. Anwar and A. P. Davis, *J.C.S. Chem. Comm.*, 831 (1986).

Chlorodimethoxyborane, $ClB(OCH_3)_2$ (**1**). The reagent is prepared from $B(OCH_3)_3$ with 2 equiv. of BCl_3.[1]

Carbonyl compounds from sulfones.[2] α-Sulfonyl carbanions react with **1** to form α-sulfonyl boronic esters, which are oxidized to aldehydes or ketones by *m*-chloroperbenzoic acid or sodium *m*-chloroperbenzoate ($ClC_6H_4CO_3H$ + NaH in CH_2Cl_2, $-60°$).

Example:

$$CH_3(CH_2)_5CH_2SO_2C_6H_5 \xrightarrow[\substack{2)\ \mathbf{1}}]{1)\ BuLi,\ hexane} \left[\begin{array}{c} SO_2C_6H_5 \\ | \\ CH_3(CH_2)_5CHB(OCH_3)_2 \end{array}\right] \xrightarrow[90\%]{ClC_6H_4CO_3H} CH_3(CH_2)_5CHO$$

Tandem aldol addition/allylmetal addition.[3]   Protected 1,3-diols can be prepared by the aldol condensation of the enol borate of an aldehyde with another aldehyde. These products can react with an allylic Grignard reagent to provide homoallylic 1,3-diols.

Example:

$$CH_3CHO \xrightarrow[\substack{2)\ ClB(OCH_3)_2}]{1)\ LDA} CH_2{=}CHOB(OCH_3)_2 \xrightarrow{CH_3CHO}$$

[1] E. Wibert and H. Smederud, *Z. Anorg. Chem.*, **225**, 204 (1935).
[2] J.-B. Baudin, M. Julia, and C. Rolando, *Tetrahedron Letters*, **26**, 2333 (1985).
[3] R. W. Hoffmann and S. Froech, *ibid.*, **26**, 1643 (1985).

Chlorodimethylthexylsilane (DTSCl), ClSi $-$ (1), 55–56°/10 mm. The

$$\begin{array}{ccc} CH_3 & CH_3 & CH_3 \\ | & | & | \\ ClSi & - & - \\ | & | & | \\ CH_3 & CH_3 & CH_3 \end{array}$$

chlorosilane is prepared in 93% yield by hydrosilylation of tetramethylethylene catalyzed by $AlCl_3$.

Protection of alcohols.[1] Dimethylthexylsilyl ethers are prepared from primary or secondary alcohols by reaction with **1** and either imidazole or $N(C_2H_5)_3$ in DMF. The ethers of tertiary alcohols are prepared by reaction with dimethylthexylsilyl trifluoromethanesulfonate in the presence of lutidine or $N(C_2H_5)_3$. Silylation of amines, amides, mercaptans, and acids is conducted under similar conditions.

Desilylation is effected with HCl, HF, or Bu_4NF. In all cases the cleavage is 2–3 times slower than that of t-butyldimethylsilyl derivatives. The dimethylthexylsilanol formed can be recycled to **1** with thionyl chloride or PCl_5 (60–85% yield).

[1] K. Oertle and H. Wetter, *Tetrahedron Letters*, **26**, 5511 (1985); H. Wetter and K. Oertle, *ibid.*, **26**, 5515 (1985).

Chlorodiphenylsilane, polymeric.

Protection of hydroxyl groups.[1] The reagent is obtained from styrene-divinylbenzene copolymer by lithiation (n-BuLi, TMEDA) and reaction with dichlorodiphenylsilane. Silylation of an alcohol is effected by reaction with **1** and diisopropylethylamine. The silane is useful for preferential silylation of primary alcohols in the presence of secondary alcohols.

[1] T.-H. Chan and W.-Q. Huang, *J.C.S. Chem. Comm.*, 909 (1985).

Chloromethyldiphenylsilane (1).

α-Alkylidene-γ-lactones.[1] The products of C-silylation of γ-lactones with **1** on deprotonation and reaction with aldehydes or ketones furnish α-alkylidene-γ-lactones in about 30–90% yield, with moderate to high selectivity for the (E)-isomer. This reaction was used to prepare a natural bisbutenolide ancepsenolide (**2**, second example).

Examples:

(100% E)

(E,E/E,Z = 80:20)

80% | Raney Ni

2

1,1-Dimethylalkenes.[2]   The readily available α-silyl esters, obtained by C-silylation of lithium ester enolates with **1**, are useful precursors to trisubstituted alkenes, including 1,1-dimethylalkenes.

Example:

$$n\text{-}C_8H_{17}\underset{\underset{CH_3Si(C_6H_5)_2}{|}}{C}HCOOC_2H_5 \xrightarrow[51.8\%]{\substack{1)\ CH_3MgBr \\ 2)\ CH_3Li \\ 3)\ KOC(CH_3)_3}} n\text{-}C_8H_{17}CH{=}C(CH_3)_2$$

[1] G. L. Larson and R. M. Betancourt de Perez, *J. Org.*, **50**, 5257 (1985).
[2] G. L. Larson, I. Montes de Lopez-Capero, and L. R. Mieles, *Org. Syn*, submitted (1986).

Chloromethyl ethyl ether, $C_2H_5OCH_2Cl$ (**1**). Mol. wt. 94.54, b.p. 82°. Supplier: Aldrich.

Isoflavanones.[1]   A synthesis of these products from a polyhydroxy desoxybenzoin such as **2** involves conversion of the nonhydrogen-bonded hydroxyl group to the protected ether by reaction with **1** and K_2CO_3 in acetone at 25°. Further reaction with **1** at 60–70° gives an α-hydroxymethyl ketone **3** in 85–95% yield. This product cyclizes to the isoflavanone **4** in the presence of Na_2CO_3. The final step involves deprotection of the ethoxymethyl group to give **5** (92–94% yield).

2 (R = H, OCH$_3$)

3

4, R^2 = C$_2$H$_5$OCH$_2$
5, R^2 = H

[1] A. C. Jain and A. Sharma, *J.C.S. Chem. Comm.*, 338 (1985).

Chloromethyl methyl ether. A new synthesis[1] from methoxyacetic acid involves chlorination with thionyl chloride followed by decarbonylation of the methoxyacetyl chloride catalyzed by aluminum chloride:

$$CH_3OCH_2COOH \xrightarrow[\substack{81\%}]{\substack{SOCl_2, \\ 110-120°}} CH_3OCH_2COCl \xrightarrow[\substack{78\%}]{\substack{AlCl_3 \\ 20-80°}} CH_3OCH_2Cl + CO$$

[1] J. Stadlwieser, *Synthesis*, 490 (1985).

Chloromethyltrimethylsilane.

   Review. Anderson[1] has reviewed synthetic applications of this and related silanes (122 references).

[1] R. Anderson, *Synthesis*, 717 (1985).

***m*-Chloroperbenzoic acid.**

   α-Hydroxy ketones.[1]   Oxidation of the (vinyl)alkoxysilane 1 with ClC$_6$H$_4$-CO$_3$H and KHF$_2$ in DMF gives the ketone 2 in high yield. In contrast, oxidation of 1 with ClC$_6$H$_4$CO$_3$H in CH$_2$Cl$_2$ gives an epoxide (3) in quantitative yield. This product undergoes C—Si oxidative cleavage on treatment with H$_2$O$_2$ (30%), KHF$_2$, and KHCO$_3$ to give the α-hydroxy ketone 4. Since 1 can be obtained by hydro-silylation [HSi(OC$_2$H$_5$)$_2$CH$_3$, H$_2$PtCl$_6$] of a dialkylalkyne, this oxidation provides a useful method for conversion of alkynes to α-hydroxy ketones in about 80% overall yield.

Oxidation of a β,γ-enal.[2] The gradual addition of $ClC_6H_4CO_3H$ (2 equiv.) to the β,γ-enal **1** in refluxing $CHCl_3$ (17 hours) results in formation of the phytotoxin dihydroactinidiolide (**2**) in 83% yield. However, if the reaction is conducted in

CCl_4 at 0° (3 hours), **2** and **3** are formed in a 14:86 ratio, but **3** is probably not a precursor to **2** since dehydration of **3** to **2** requires more forcing conditions.

Enantioselective oxidation of sulfides. *m*-Chloroperbenzoic acid attacks the sulfide **1** preferentially from the less hindered direction to introduce an oxygen atom with high diastereoselectivity.[3]

This chiral peracid oxidation was used to convert the vinyl sulfide **3**, derived from (S)-(+)-camphor-10-sulfonyl chloride, to a chiral sulfoxide (96:4 selectivity),

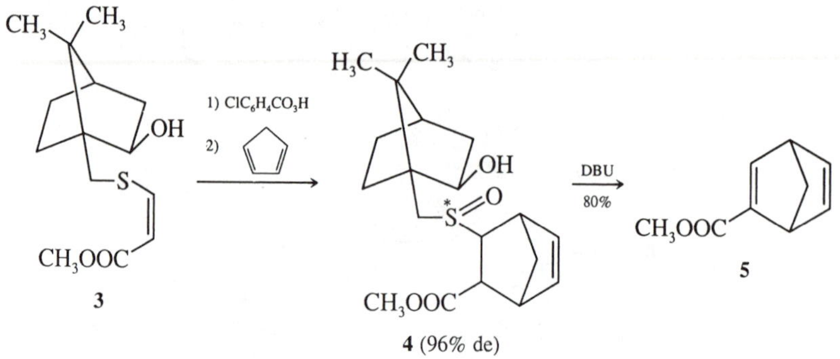

3

1) ClC₆H₄CO₃H

2) (cyclopentadiene)

4 (96% de)

DBU
80%

CH₃OOC

5

which reacts with cyclopentadiene to form the *endo*-adduct exclusively in the dia-stereoisomeric ratio >98:2. Elimination of the chiral adjuvant to give the chiral 2-methoxycarbonylnorbornadiene **5** is effected with DBU.[4]

Telluroxide elimination. *sec*-Alkyl phenyl tellurides on oxidation with ClC₆H₄CO₃H in ether afford alkenes as the major product. In some cases a stable Te(IV) compound is formed, which undergoes elimination on pyrolysis.[5]

Examples:

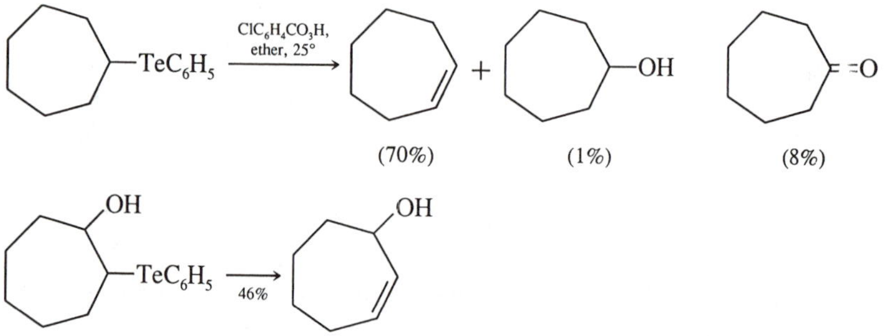

ClC₆H₄CO₃H,
ether, 25°

(70%)   (1%)   (8%)

TeC₆H₅   46%

α-Hydroxylation of β-dicarbonyl compounds (cf., **11**,122)[6] The silyl enol ether of β-keto esters and β-diketones is hydroxylated at the α-position by *m*-chloroperbenzoic acid in CH₂Cl₂ generally in 70–90% yield.

Example:

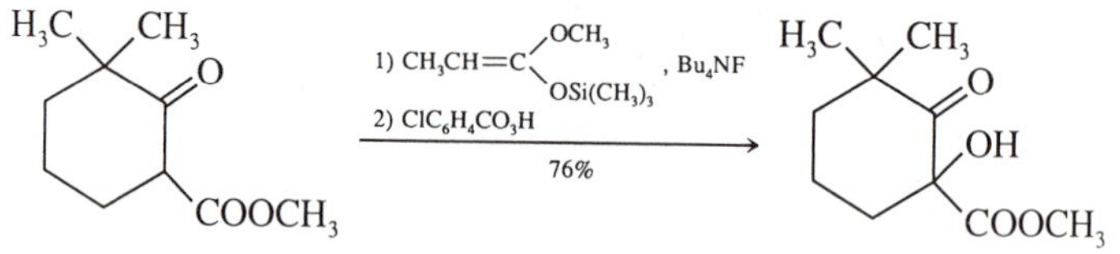

1) CH₃CH=C(OCH₃)(OSi(CH₃)₃), Bu₄NF

2) ClC₆H₄CO₃H

76%

[1] K. Tamao and K. Maeda, *Tetrahedron Letters*, **27**, 65 (1986).
[2] T. E. Nickson, *ibid.*, **27**, 1433 (1986).

[3] R. S. Glass, W. N. Setzer, U. S. G. Prabhu, and G. S. Wilson, *ibid.*, **23**, 2335 (1982).
[4] O. DeLucchi, C. Marchioro, G. Valle, and G. Modena, *J.C.S. Chem. Comm.*, 878 (1985).
[5] S. Uemura, K. Ohe, and S. Fukuzawa, *Tetrahedron Letters*, **26**, 895 (1985); S. Uemura and S. Fukuzawa, *J.C.S. Perkin I*, 471 (1985).
[6] R. Z. Andriamialisoa, N. Langlois, and Y. Langlois, *Tetrahedron Letters*, **26**, 3563 (1985).

N-Chlorosuccinimide.

Allylic amines (**12**,121).   Coupling of allylic phenyl selenides with aliphatic or aromatic primary amines is possible if the selenide is treated with NCS and $N(C_2H_5)_3$ in CH_3OH at $-25°$ prior to the addition of the amine. Yields are only modest in coupling of highly substituted selenides or amines.[1]

Example:

β,γ-Unsaturated α-amino esters.   γ-Phenylseleno-α,β-unsaturated esters on treatment with NCS (3 equiv.), N,N-diisopropylethylamine (6 equiv.), and an alkyl carbamate (3 equiv.) in methanol at 25° rearrange to derivatives of β,γ-unsaturated α-amino acids, generally in 60–80% yield.[2]

Example:

Optically active amino acids.[3]   The NCS-promoted rearrangement of allylic phenyl selenides (**12**,121) when applied to an optically active substrate (**1**), available from ethyl (S)-lactate, results in an allylic amine (**2**), which can be converted into an optically active N-protected D-amino acid (**3**) in 78–84% ee.

[1] A. Spaltenstein, P. A. Carpino, and P. B. Hopkins, *Tetrahedron Letters*, **27**, 147 (1986).
[2] J. N. Fitzner, D. V. Pratt, and P. B. Hopkins, *ibid.*, **26**, 1959 (1985).
[3] J. N. Fitzner, R. G. Shea, J. E. Fankhauser, and P. B. Hopkins, *J. Org.*, **50**, 417 (1985).

N-Chlorosuccinimide–Silver nitrate.

Oxocenes.[1]   A general route to these unsaturated eight-membered cyclic ethers involves an intramolecular cyclization of a hydroxyl group with a sulfonium ion, generated from a dithioketal by NCS-AgNO$_3$ (**4**,216) and a base (2,6-lutidine).
Example:

[1] K. C. Nicolaou, M. E. Duggan, and C.-K. Hwang, *Am. Soc.*, **108**, 2468 (1986).

Chlorosulfonyl isocyanate (CSI).

α-Alkylidene-β-lactams (**3**, 52).[1]   Allenyl aryl sulfides, which can be prepared by reaction of propargylic alcohols with arenesulfonyl chlorides, react with CSI regioselectively at the vinyl sulfide group to form α-alkylidene-β-lactams after reduction.
Example:

Review.   Dhar and Murthy[2] have reviewed reactions of CSI reported during 1977–1984 (121 references).

[1] J. D. Buynak, M. N. Rao, R. Y. Chandrasekaran, E. Haley, P. de Meester, and S. C. Chu, *Tetrahedron Letters*, **26**, 5001 (1985).
[2] D. N. Dhar and K. S. K. Murthy, *Synthesis*, 437 (1986).

Chlorotrimethylsilane–Lithium.

Allylsilanes.[1]   Allyl trimethylsilyl ethers are converted into allylsilanes by reaction with chlorotrimethylsilane (5 equiv.) and lithium sand in THF.

Example:

$$(CH_3)_2C{=}CHCH_2OSi(CH_3)_3 \xrightarrow[\substack{85\%}]{\substack{ClSi(CH_3)_3,\ Li, \\ THF,\ 0 \to 25°}} (CH_3)_2C{=}CHCH_2Si(CH_3)_3$$

or

$$CH_2{=}CHC(CH_3)_2OSi(CH_3)_3 \qquad + [(CH_3)_3Si]_2O + LiCl$$

[1] C. Biran, J. Gerval, and J. Dunogues, *Org. Syn.*, submitted (1985).

Chlorotrimethylsilane–Sodium iodide.

Cleavage of 4-tetrahydropyrones.[1]   The terpene *ar*-atlantone (**2**) and the dihydro derivative *ar*-turmerone (**3**) have been obtained by cleavage of the 4-tetrahydropyrone **1** with $ClSi(CH_3)_3$/NaI. Reaction of **1** with $ClSi(CH_3)_3$/NaI in DMF

gives **2** exclusively. The reaction in CH_3CN furnishes **3** in low yield, but reaction in refluxing PrCN gives **3** in 68% yield. Presumably **2** is an intermediate, since **2** can be reduced to **3** by the reagent. Such a 1,4-reduction with $ClSi(CH_3)_3$/NaI has not been observed previously.

Cleavage of tetrahydrofurfuryl alcohols.[2]   The reaction of the alcohol **1** with this reagent in purified acetone gives the iodoacetonide **2** in 75% yield.

[1] T. Sakai, K. Miyata, M. Ishikawa, and A. Takeda, *Tetrahedron Letters*, **26**, 4727 (1985).
[2] M. Jatczak, R. Amouroux, and M. Chastrette, *ibid.*, **26**, 2315 (1985).

Chlorotrimethylsilane–Zinc.

Reductive silylation of α-diketones or quinones.[1]   α-Diketones are conveniently converted into 1,2-bis(silyloxy)alkenes by reaction with $ClSi(CH_3)_3$ and zinc. Yields are increased by use of THF rather than ether as solvent and by ultrasonic irradiation.

[1] P. Boudjouk and J. H. So, *Syn. Comm.*, **16**, 775 (1986).

Chromium carbene complexes.

Aldol reactions.[1]   The anion generated (BuLi) from chromium carbene complexes undergoes aldol reactions with aldehydes or ketones activated by a Lewis acid. Best results are obtained with ketones in the presence of BF_3 etherate, whereas $TiCl_4$ is the preferred catalyst for aldehydes and acetals.
Example:

Reaction with diynes; indanols.[2]   The reaction of the complex **1** with the diyne **2** results in the indanol **3** as the major product.

Cyclobutanones.[3]   The reaction of pentacarbonyl(methoxymethylmethylene)-chromium (**1**) with the enyne **2** in acetonitrile results in two isomeric bicycloheptanones [(E)-**3** and (Z)-**3**] as the major products. These probably arise by initial reaction with the triple bond to give a vinylcarbene complex (**a**), which undergoes insertion of CO to give a vinylketene complex (**b**). An intramolecular [2 + 2] cycloaddition results in the cyclobutanones (**3**).

[1] W. D. Wulff and S. R. Gilbertson, *Am. Soc.*, **107**, 503 (1985).
[2] W. D. Wulff, R. W. Kaesler, G. A. Peterson, and P.-C. Tang, *ibid.*, **107**, 1060 (1985).
[3] W. D. Wulff and R. W. Kaesler, *Organometallics*, **4**, 1461 (1985).

Chromium(II) chloride.

Alkynylchromium compounds.[1]   Alkynyl halides react with $CrCl_2$ (or $CrCl_3/$
$LiAlH_4$) to form unisolable alkynylchromium compounds, which undergo selective
1,2-addition to aldehydes.

Example:

$$OHC(CH_2)_4\overset{O}{\underset{\|}{C}}(CH_2)_3CH_3 \xrightarrow[76\%]{\substack{BuC{\equiv}Cl \\ CrCl_2, DMF}} BuC{\equiv}C\underset{OH}{\underset{|}{C}}H(CH_2)_4\overset{O}{\underset{\|}{C}}(CH_2)_3CH_3$$

α-Allenic aldehydes.[2]   Reduction of γ-bromo-α-acetylenic acetals (**1**) with
$CrCl_2$ in THF and HMPT results in α-allenic acetals (**2**), which can be hydrolyzed
under very mild conditions to α-allenic aldehydes (**3**).

$$\underset{\mathbf{1}}{RCHBrC{\equiv}CCH(OC_2H_5)_2} \xrightarrow[50-90\%]{\substack{CrCl_2, \\ THF, HMPT}}$$

$$\underset{\mathbf{2}}{RCH{=}C{=}CHCH(OC_2H_5)_2} \xrightarrow[\substack{80-95\% \\ crude}]{\substack{HOOCCOOH, \\ ROH, 20°}} \underset{\mathbf{3}}{RCH{=}C{=}CHCHO}$$

[1] K. Takai, T. Kuroda, S. Nakatsukasa, K. Oshima, and H. Nozaki, *Tetrahedron Letters*, **26**, 5585 (1985).
[2] B. Ledoussal, A. Gorgues, and A. Le Coq, *ibid.*, **26**, 51 (1985).

Chromium(II) sulfate.

trans-α,β-Unsaturated ketones.[1]   α-Alkynyl ketones are reduced selectively
by $CrSO_4$ to *trans*-enones in 40–85% yield. The reaction is compatible with epoxy,
ester, and unsaturated ester groups.

Example:

$$C_2H_5OOCCH_2\underset{CH_3}{\overset{CH_3}{\underset{|}{\overset{|}{C}}}}{-}C{\equiv}C\overset{O}{\underset{\|}{C}}CH_3 \xrightarrow[78\%]{\substack{CrSO_4, DMF, \\ H_2O}} C_2H_5OOCCH_2\underset{(CH_3)_2}{\underset{|}{C}}{\diagup}\overset{H}{\underset{}{C}}{=}C{\diagdown}\overset{\overset{O}{\underset{\|}{C}}CH_3}{H}$$

(<2% *cis*)

[1] A. B. Smith, III, P. A. Levenberg, and J. Z. Suits, *Synthesis*, 184 (1986).

Copper(II) acetate–Iron(II) sulfate.

Fragmentation of α-alkoxy hydroperoxides (**10**, 103–104). The α-alkoxy hydroperoxide **2**, obtained by ozonization of the $\Delta^{9,10}$ -octalin **1**, on fragmentation with Cu(OAc)$_2$ and FeSO$_4$ in CH$_3$OH at 25° undergoes ring enlargement to the 14-membered macrolide **3** with ≥20:1 regio- and stereocontrol.[1]

[1] S. L. Schreiber and W.-F. Liew, *Am. Soc.*, **107**, 2980 (1985).

Copper(I) chloride.

Bicyclic γ-lactams.[1]   N-Allyltrichloroacetamides in which the double bond is associated with a cyclopentenyl or -hexenyl group undergo cyclization in the presence of CuCl in CH$_3$CN or of Cl$_2$Ru[P(C$_6$H$_5$)$_3$]$_3$ in C$_6$H$_6$ or C$_6$H$_4$(CH$_3$)$_2$.

Example:

[1] H. Nagashima, K. Ara, H. Wakamatsu, and K. Itoh, *J.C.S. Chem. Comm.*, 518 (1985).

Copper(II) chloride–Copper(II) oxide.

1,2-Cycloalkanediones.[1]   The shortest synthesis of these diones is the SeO$_2$ oxidation of a cycloalkanone (**1**, 993), but this method has disadvantages for large-scale reactions. A new two-step method involves α-methylthiolation of the ketone followed by oxidation of the methylthio ketones with $CuCl_2/CuO$ in aqueous acetone at room temperature.

Example:

[1] B. Gregoire, M.-C. Carre, and P. Caubere, *J. Org.*, **51**, 1419 (1986).

Crotyldiisopinocampheylborane. $Ipc_2BCH_2CH=CHCH_3$ (**1**). The (Z)-isomer is prepared by reaction of (Z)-crotylpotassium with methoxydiisopinocampheylborane [obtained from (−)-α-pinene] to form an ate complex, which is converted into (Z)-**1** on reaction with BF_3 etherate at −78°. The (E)-isomer is prepared similarly from (E)-crotylpotassium. Reaction of acetaldehyde with (Z)-**1** at −78° results in (2S,3S)-3-methyl-4-pentene-2-ol (**2**) in 99% diastereoselectivity and 95% enantioselectivity (equation I). A similar reaction with (E)-**1** provides (2S,3R)-**2**.

The two remaining stereoisomers of **2** are prepared by a similar reaction with **1** derived from (+)-α-pinene.[1]

[1] H. C. Brown and K. S. Bhat, *Am. Soc.*, **108**, 293 (1986).

Crotyltrimethylsilanes, (E)- and (Z)-**1**.

Diastereoselective reaction with acetals.[1] Both (E)- and (Z)-**1** react with acetals of aliphatic aldehydes with high *syn*-selectivity irrespective of the geometry of **1** or of the nature of the Lewis acid catalyst. However, the selectivity of reactions with acetals of aromatic aldehydes does depend on the geometry of **1**, being more *anti*-selective with (Z)-**1** and more *syn*-selective with (E)-**1**. Moreover, electron-withdrawing groups on (Z)-**1** increase the *anti*-selectivity, whereas these groups attached to (E)-**1** increase the *syn*-selectivity. The reason for these electronic effects is not clear at present.

[1] A. Hosomi, M. Ando, and H. Sakurai, *Chem. Letters*, 365 (1986).

Cyano-*t*-butyldimethylsilane. This reagent can be prepared[1] in 75% yield by re-action of *t*-butyldimethylchlorosilane with KCN in refluxing CH_2Cl_2 catalyzed by 18-crown-6.

[1] J. R. Hwu, J. G. Lazar, and P. F. Corless, *Synthesis*, 1020 (1984).

2-Cyanopyridinium chlorochromate, $\cdot \, CrO_3Cl^-$ **(1).** This chlorochromate

is prepared by addition of 2-cyanopyridine in 12 *N* HCl to a solution of CrO_3 in 6 *N* HCl at 0°. The product separates as yellow-orange needles that can be stored indefinitely at 4° (80% yield).

1,4-Oxygenation of 1-alkylcyclopentadienes.[1]   This reagent is superior to PCC for oxidation of **2** to the cyclopentenone **3**. This product is converted into the ketol **5** by epoxidation followed by rearrangement with DBU. This conversion of **2** to **5**

2, R = *n*-C_5H_{11} · · · 3 · · · 4 · · · 5

is more efficient than an earlier process involving oxygenation, reduction of the resulting endoperoxide to a diol, and finally oxidation with PDC (58% overall yield).

[1] E. J. Corey and M. M. Mehrotra, *Tetrahedron Letters*, **26**, 2411 (1985).

Cyanotrimethylsilane.
    Phenylacetic acids (**11**, 148).[1]   Complete details are available for conversion

of aryl ketones to phenylacetic acids with $CNSi(CH_3)_3$ followed by reductive hydrolysis of the cyanohydrin with $SnCl_2 \cdot 2H_2O$.

Example:

$$p\text{-}HOC_6H_4\overset{O}{\overset{\|}{C}}C_2H_5 \xrightarrow[ZnI_2]{(CH_3)_3SiCN} p\text{-}(CH_3)_3SiOC_6H_4\overset{C_2H_5}{\underset{CN}{\overset{|}{C}}}\text{—}OSi(CH_3)_3 \xrightarrow[60–70\%]{\underset{HOAc}{SnCl_2,\ HCl}} p\text{-}HOC_6H_4\overset{C_2H_5}{\overset{|}{C}}HCOOH$$

Cleavage of oxetanes.[2]   In the presence of ZnI_2, oxetanes are opened by cyanotrimethylsilane regioselectively to γ-hydroxy isocyanides, which undergo acid-catalyzed hydrolysis to γ-amino alcohols.

Example:

$$\xrightarrow[76\%]{\underset{ZnI_2,\ CH_2Cl_2}{(CH_3)_3SiCN}} (CH_3)_3SiO(CH_2)_2\underset{CH_3}{\overset{|}{C}}HN\equiv C \xrightarrow[71\%]{\underset{CH_3OH}{HCl,}} HO(CH_2)_2\underset{CH_3}{\overset{|}{C}}HNH_2$$

[1] J. L. Belletire, *Org. Syn.*, submitted (1985).
[2] P. G. Gassman and L. M. Haberman, *Tetrahedron Letters*, **26**, 4971 (1985).

(1,5-Cyclooctadiene)(pyridine)(tricyclohexylphosphine)iridium(I) hexafluorophosphate (1). Preparation.[1]

Stereoselective hydrogenations.   The stereochemistry of the hydrogenation of a double bond catalyzed by this Ir(I) complex is markedly controlled by the presence of a carboxamide group. The effect is attributed to coordination between the CONH group and iridium. Reductions of the same substrates with Pd/C show no stereoselection.[2]

Examples:

$$\xrightarrow[89\%]{\underset{CH_2Cl_2}{H_2,\ 1,}}$$

(130:1)

(530:1)

Brown *et al.*[3] have examined the directed hydrogenation of 6-methylenebicyclo-[2.2.2]octane-2-ols (**3**) catalyzed by **1** or a related rhodium catalyst (**2**). Catalytic hydrogenation of the *exo*-alcohol **3** is not stereoselective, but the hydrogenation of the *endo*-alcohol is highly stereoselective, particularly when catalyzed by **1**.

$[(C_6H_5)_2P(CH_2)_4P(C_6H_5)_2](C_7H_8)Rh^+BF_4^-$ (**2**)   51:49

1   55:45

2

1

15:85

0.3:99.7

Crabtree and Davis[4] have discussed factors involved in directed homogeneous hydrogenation of double bonds by hydroxyl, methoxyl, ester, and keto groups. Stereoselectivity requires binding of the directing group, the double bond, and hydrogen to the metal. Useful catalysts for directed hydrogenation have the 12-electron configuration of the metal, which permits complexing to an 18-electron configuration. In addition the directing group appears to protect the catalyst from deactivation. They also note that the minor product of directed hydrogenation is usually the major product of hydrogenations catalyzed by Pd/C in ethanol, but in the latter case the directing effect is slight.

[1] D. A. Evans and M. M. Morrissey, *Am. Soc.*, **106**, 3866 (1984).
[2] A. G. Schultz and P. J. McCloskey, *J. Org.*, **50**, 5905 (1985).
[3] J. M. Brown and S. A. Hall, *Tetrahedron*, **41**, 4639 (1985); J. M. Brown, A. E. Derome, and S. A. Hall, *ibid.*, **41**, 4647 (1985).
[4] R. H. Crabtree and M. W. Davis, *J. Org.*, **51**, 2655 (1986).

Cyclopropenone 1,3-propanediyl ketal (1).

Diels-Alder reactions.[1] This cyclopropenone ketal undergoes [4 + 2] cycloaddition with electron-deficient or electron-rich dienes at 25° when the reaction is conducted without solvent. The adducts are usually obtained as a single stereoisomer, probably the *trans*-isomer formed from an *exo*-transition state.

This cycloaddition can be used for two approaches to tropones. One involves cycloaddition of **1** to methyl 5-methoxy-2,4-pentadienoate, as outlined in equation (I). The other is a pressure-promoted cycloaddition of **1** with α-pyrone, outlined

in equation (II). The *exo*-adduct and the product of decarboxylation are formed in about equal amount. Both can be converted into tropone.

[1] D. L. Boger and C. E. Brotherton, *Tetrahedron*, **42**, 2777 (1986).

D

Diacetatobis(triphenylphosphine)palladium(II), $(CH_3CO_2)_2Pd[P(C_6H_5)_3]_2$ **(1).**
Preparation.[1] Supplier: Strem.

Cyclization of 1,6-enynes to 1,3- and 1,4-dienes.[2]   The 1,6-enynes are prepared
by Pd(0)-catalyzed alkylation of allylic carboxylates with the anion of dimethyl
propargylmalonate. The products cyclize in the presence of this Pd(II) catalyst or
$[(o\text{-}CH_3C_6H_4)_3P]_2Pd(OAc)_2$ to 1,3- or 1,4-dienes.

Examples:

E = COOCH₃

[1] T. A. Stephenson, S. M. Morehouse, A. R. Powell, J. P. Heffer, and G. Wilkinson, *J. Chem. Soc.*, 3632 (1965).
[2] B. M. Trost and M. Lautens, *Am. Soc.*, **107**, 1781 (1985).

1,5-Diazabicyclo[4.3.0]nonene-5 (DBN).
Marschalk reaction.[1]   Leucoquinizarin (**1**) reacts with propionaldehyde and
DBN (or DBU) in DMF to form 2-(1-hydroxypropyl)quinizarin (**2**) in high yield.
This nonbasic and nonaqueous version of the Marschalk reaction was developed
in order to permit use of aldehydocarbohydrates in a diastereoselective route to
anthracyclinones.

[1] S. Qureshi, G. Shaw, and G. E. Burgess, *J.C.S. Perkin I*, 1557 (1985).

1,4-Diazabicyclo[2.2.2]octane (DABCO).

α-Methylene-β-hydroxyalkanones.[1]   In the presence of DABCO, aldehydes react with methyl vinyl ketone to form α-methylene-β-hydroxyalkanones in 50–75% yield.

Example

$$n\text{-}C_7H_{15}CHO + CH_2{=}CHCOCH_3 \xrightarrow[63\%]{\text{DABCO,}\ \text{THF, }25°} n\text{-}C_7H_{15}CH(OH)\underset{\underset{CH_2}{\|}}{C}COCH_3$$

[1] D. Basavaiah and V. V. L. Gowriswari, *Tetrahedron Letters*, **27**, 2031 (1986).

1,8-Diazabicyclo[5.4.0]undecene-7 (DBU).

t-Butyldimethylsilylation.[1]   This base promotes silylation of alcohols, thiols, amines, carboxylic acids, and phenols with *t*-butyldimethylchlorosilane in 80–98% yield. Benzene, CH_2Cl_2, or CH_3CN can be used as solvents.

[1] J. M. Aizpurua and C. Palomo, *Tetrahedron Letters*, **26**, 475 (1985).

$$\overset{\text{NHBz}}{|}$$

N,N′-Dibenzoylcystine HOOCCHCH$_2$S)$_2$ (1).

Asymmetric reduction of β-arylcarbonyl esters.[1]   Reduction of these esters with lithium borohydride and (R,R′)-**1** and *t*-butyl alcohol affords the corresponding 3-hydroxy esters in 80–92% ee (equation I).

$$(I)\quad Ar\overset{\overset{O}{\|}}{C}CH_2COOR \xrightarrow[65-95\%]{\text{LiBH}_4,\ \mathbf{1},\ (CH_3)_3COH} Ar\overset{\overset{OH}{|}}{C}HCH_2COOR$$

(80–92% ee)

[1] K. Soai, T. Yamanoi, H. Hikima, and H. Oyamada, *J.C.S. Chem. Comm.*, 138 (1985).

Dibromoborane–Dimethyl sulfide.

Hydroboration.[1]   Hydroborations with this reagent in CH_2Cl_2 or THF (heterogeneous) are relatively slow (5–12 hours at 25°), but application of ultrasound permits complete reaction in 1–2 hours. Ultrasound also increases the rate of hydroboration with catecholborane or with 9-BBN (THF or neat).

[1] H. C. Brown and U. S. Racherla, *Tetrahedron Letters*, **26**, 2187 (1985).

Dibromomethane–Zinc/Copper(I) chloride.

Cyclopropanation.[1] CH_2I_2 is used in cyclopropanation of the Simmons-Smith type, but it is about 20 times more expensive than CH_2Br_2. CH_2Br_2 in combination with Zn/CuCl can be used for cyclopropanation, particularly when sonication is used to promote the heterogeneous reaction in ether. Reaction times are 2–4 hours, and yields range from 30–70%.

[1] E. C. Friedrich, J. M. Domek, and R. Y. Pong, *J. Org.*, **50**, 4640 (1985).

Dibromomethyllithium, LiCHBr₂ (1).

Bromomethyl ketones.[1] Reaction of (dibromomethyl)lithium (**1**) with esters and then with BuLi (excess) generates the enolate (**a**) of a bromomethyl ketone. Quenching with acidic ethanol gives the bromomethyl ketone (**2**) in 70–85% yield. Quenching with Ac₂O generates bromo enol acetates.

Example:

Ester homologation (**11**, 104).[2] In an improved method for ester homologation via an α-bromo-α-keto dianion, an ester (RCO₂C₂H₅) is treated sequentially with LiCHBr₂, LiTMP (1 equiv.), and then an excess of BuLi to provide RC≡C—OLi. Rearrangement occurs with complete retention of stereochemistry. This method provides a stereospecific synthesis of the antibiotic oudemansin (**2**).

[1] C. J. Kowalski and M. S. Haque, *J. Org.*, **50**, 5140 (1985).
[2] C. J. Kowalski, M. S. Haque, and K. W. Fields, *Am. Soc.*, **107**, 1429 (1985).

Di-*t*-butyl dicarbonate, (Boc)$_2$O (1).

Protection of pyrroles and indoles. These heterocycles undergo *t*-butoxycarbonylation in 80–95% yield on reaction with 1 and a catalytic amount of DMAP in acetonitrile. The Boc group is cleaved almost instantaneously by TFA at 25°.[1] The same conditions are useful for protection of the indole group of tryptophan.[2] Primary amino acid esters can be converted into bis(Boc) derivatives, and amino groups of peptides also undergo *t*-butoxycarbonylation.[3]

t-Butoxycarbonylation under phase-transfer conditions.[4] *t*-Butoxycarbonylation of phenols, alcohols, enols, and thiols with 1 proceeds in satisfactory yield under phase-transfer conditions. K$_2$CO$_3$ in combination with 18-crown-6 in THF is particularly useful for phenols, enols, and thiols. Alcohols are generally less reactive; in this case Bu$_4$NHSO$_4$ in CH$_2$Cl$_2$/aqueous NaOH is generally satisfactory.

[1] L. Grehn and U. Ragnarsson, *Angew. Chem. Int. Ed.*, **23**, 296 (1984).
[2] H. Franzén, L. Grehn, and U. Ragnarsson, *J.C.S. Chem. Comm.*, 1699 (1984).
[3] L. Grehn and U. Ragnarsson, *Angew. Chem. Int. Ed.*, **24**, 510 (1985).
[4] F. Houlihan, F. Bouchard, J. M. J. Fréchet, and C. G. Willson, *Can. J. Chem.*, **63**, 153 (1985).

Di-*t*-butyliminoxyl, [(CH$_3$)$_3$C]$_2$C=NO· (1). The blue radical is obtained on oxidation of di-*t*-butyl ketoxime with CAN in CH$_3$OH. It is extracted with pentane or hexane and stored at −15°.

Oxidation of amines.[1] The radical effects oxidation of primary or secondary amines to imines at 25° (4–8 hours). The products are generally isolated as the dinitrophenylhydrazones of the corresponding carbonyl compounds.

[1] J. J. Cornejo, K. D. Larson, and G. D. Mendenhall, *J. Org.*, **50**, 5382 (1985).

2,6-Di-*t*-butyl-4-methoxyphenol,   (1),   m.p.   105–106°.

Supplier: Aldrich.

Conjugate addition to α,β-unsaturated esters.[1] The unsaturated esters of this phenol undergo ready conjugate addition with a variety of organolithium reagents, with *t*-butyllithium providing the only exception. The adducts are oxidatively hydrolyzed to carboxylic acids by CAN (**10**, 162).

Example:

[1] M. P. Cooke, Jr., *J. Org.*, **51**, 1637 (1986).

Di-*t*-butylmethylsilyl trifluoromethanesulfonate, *t*-Bu$_2$MeSiOTf. The triflate is obtained as an oil, b.p. 63–65°/15 mm, by reaction of *t*-Bu$_2$MeSiH with triflic acid (95% yield). The reagent is used to prepare DTBMS esters and ethers, which are more stable than the *t*-butyldimethylsilyl counterparts. Thus the esters are not reduced by lithium *t*-butyldiisobutylaluminum hydride (**10**, 239–240) or hydrolyzed by acid, but are cleaved by Bu$_4$NF.[1]

[1] R. S. Bhide, B. S. Levison, R. B. Sharma, S. Ghosh, and R. G. Salomon, *Tetrahedron Letters*, **27**, 671 (1986).

Dibutyltin oxide.

Regioselective oxidation of **1,2-diols.**[1]   The oxidation of di-secondary glycols to acyloins (**5**, 188) can be extended to oxidation of other glycols. Thus the stannylene of **1** is oxidized by bromine to **2** in high yield. The reaction is regioselective with unsymmetrical diols (**3 → 4**).

Dibutylstannylenes of glycols can also undergo glycol cleavage (equation I).

Bu$_4$NIO$_4$	82%	99%
Pb(OAc)$_4$	85%	91%
(C$_6$H$_5$)$_3$Bi(OAc)$_2$	66%	90%

Regioselective acylation of glycols.[2]   Direct reaction of propane-1,2-diol with benzoyl chloride and pyridine followed by silylation results in **3** and **4** in the ratio 9:91. The regioselectivity is reversed by conversion of **1** into the stannylene **2** prior

to acylation. In this case benzoylation occurs more readily at the more substituted hydroxyl group.

[1] S. David and S. Hanessian, *Tetrahedron*, **41**, 643 (1985).
[2] A. Ricci, S. Roelens, and A. Vannucchi, *J.C.S. Chem. Comm.*, 1457 (1985).

Dicarbonylcyclopentadienylcobalt.

[2+2+2]*Cycloaddition of enediynes.*   Photolysis of the enediyne **1** in the presence of catalytic quantities of CpCo(CO)₂ in refluxing toluene results in the tricyclic diene **2**. The reaction involves a diene rearrangement in the expected product (**a**).[1]

Benzhydrindanes.[2]   The reaction of **1** and **2** in the presence of CpCo(CO)₂ provides the benzocyclobutene **3** in high yield. As expected, **3** is converted at 175° into the isomeric benzhydrindanes **4** and **5** via an intermediate *o*-xylylene.

Surprisingly the benzocyclobutene **6** is stable at 175°, and does not decompose to a benzhydrindane even at higher temperatures, possibly owing to steric effects of the silyl group adjacent to the four-membered ring.

Reaction with cyclobutenediones. This cobalt complex inserts into di-methylcyclobutenedione (**1**) to give the cobalt complex **2** in high yield. The complex **2** reacts with alkynes slowly at 120° to give complexes of benzoquinones such as **4**. The reaction is facilitated by replacement of the CO ligand of **2** by $S(C_2H_5)_2$,

Py, or $P(C_6H_5)_3$. In this way the complex **4** can be obtained in 76% yield and converted to the benzoquinone **5** in good yield. The complex **2** on irradiation loses CO to give the complex **3** of dimethylbisketene. This complex reacts with a variety of alkynes to give complexes (**6**) of benzoquinones.[3]

The reaction of $CpCo(CO)_2$ with benzocyclobutenedione gives a phthaloylcobalt complex in high yield (98%); reaction of this complex with alkynes results in complexes of naphthoquinones.[4]

[1] E. Duñach, R. L. Halterman, and K. P. C. Vollhardt, *Am. Soc.*, **107**, 1664 (1985).
[2] R. L. Halterman, N. H. Nguyen, and K. P. C. Vollhardt, *Am. Soc.*, **107**, 1379 (1985).
[3] L. S. Liebeskind and C. F. Jewell, Jr., *J. Organomet. Chem.*, **285**, 305 (1985); C. F. Jewell, Jr., L. S. Liebeskind, and M. Williamson, *Am. Soc.*, **107**, 6715 (1985).
[4] L. S. Liebeskind, S. L. Baysdon, V. Goedken, and R. Chidambaram, *Organometallics*, **5**, 1086 (1986).

Dicarbonyl(cyclopentadienyl)[(dimethylsulfonium)methyl]iron(II) tetrafluoroborate, $Cp(CO)_2FeCH_2\overset{+}{S}(CH_3)_2BF_4^-$ (**1**).

Cyclopropanation.[1]   Full details are available for preparation and use of this methylene transfer reagent. It is useful for selective cyclopropanation of 1,1-disubstituted double bonds.

Example:

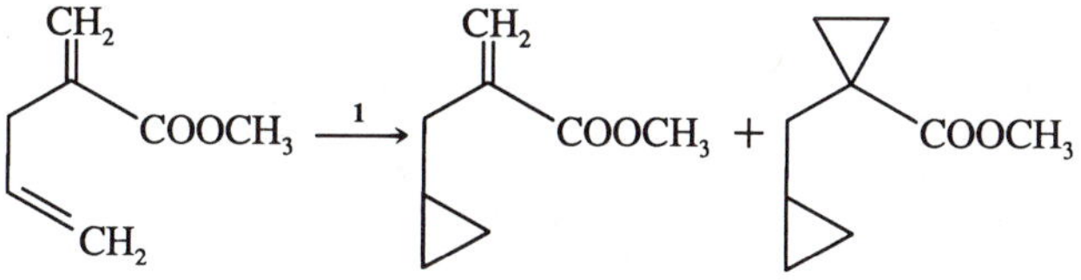

(62%)   (5%)

[1] E. J. O'Connor and P. Helquist, *Org. Syn.*, submitted (1985).

Di-μ-carbonylhexacarbonyldicobalt.

Carbonylation of mercaptans.[1]   Benzylic and aryl mercaptans in the presence of $Co_2(CO)_8$ undergo desulfuration and carbonylation to give esters.
Examples:

$$p\text{-}CH_3C_6H_4CH_2SH + CO \xrightarrow[83\%]{\substack{Co_2(CO)_8, \\ CH_3OH}} p\text{-}CH_3C_6H_4CH_2COOCH_3$$

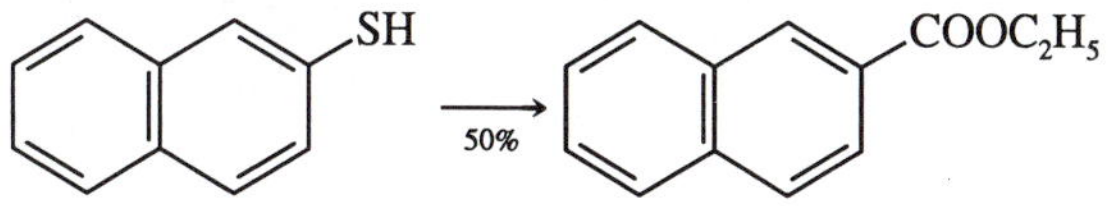

Propargylation (**8**, 148–149; **10**, 129).   In the presence of a Lewis acid, cobalt-complexed propargylic ethers alkylate silyl enol ethers with high *syn*-selectivity.
Example:

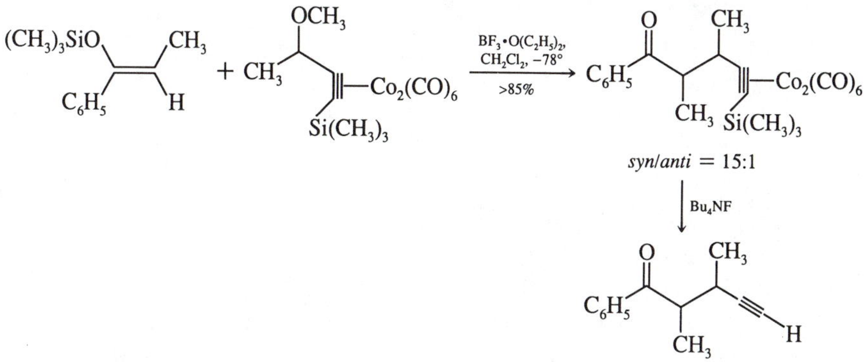

syn/anti = 15:1

This reaction can be used for intramolecular alkylation of allylic silanes to obtain cycloalkynes.[2]

Example:

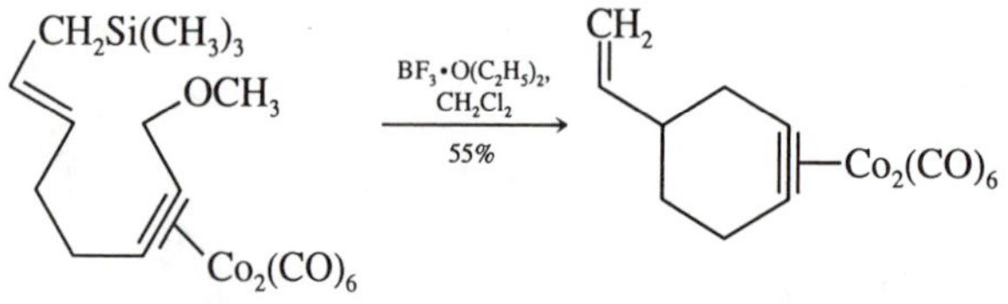

Hydrocarbonylation of 1,4-dienes.[3] Hydrocarbonylation of 1,4-pentadiene catalyzed by $Co_2(CO)_8$ results in 2-methylcyclopentanone as the only cyclic product. As expected, a *gem*-dimethyl group at C_3 enhances the tendency to form cyclopentanones. Thus hydrocarbonylation of the 1,4-diene **1** leads to the tetramethylcyclopentanone **2** with a marked preference for the *cis*-isomer. The reaction has

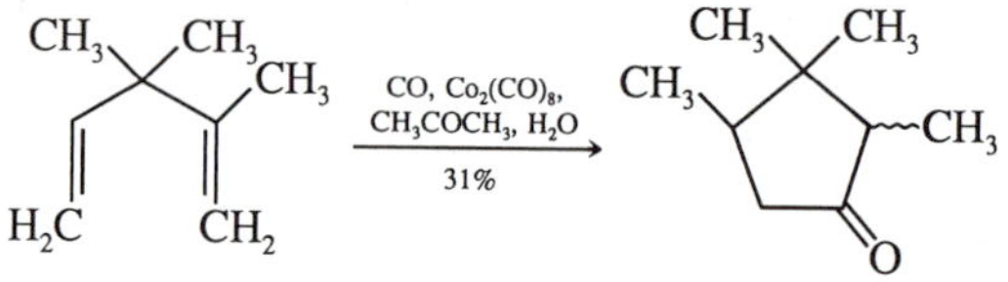

been used for a regioselective synthesis of α-cuparenone (**5**) by hydrocarbonylation of **3** to form **4**.

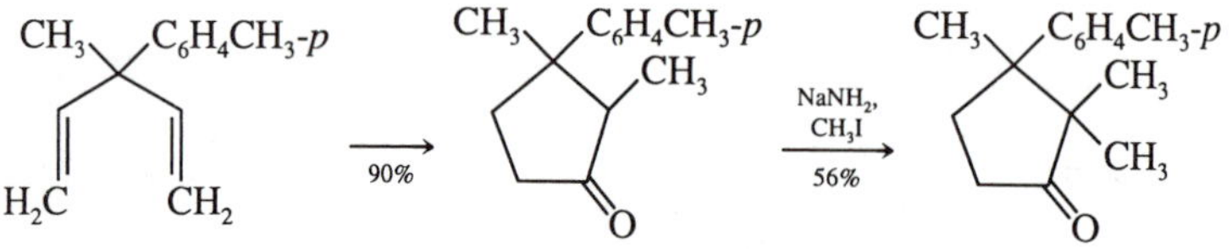

Carbonylation of epoxides.[4] The epoxide of a 1-alkene undergoes carbonylation under catalysis with $Co_2(CO)_8$ in the presence of K_2CO_3 (1 equiv.) in ethanol at moderate temperatures to afford β-hydroxy esters. A by-product is the ketone formed by rearrangement of the epoxide. This reaction provides an essential step in the synthesis of the cyclopentenone **3** from the epoxide (**1**) of ethyl 10-undecenoate.

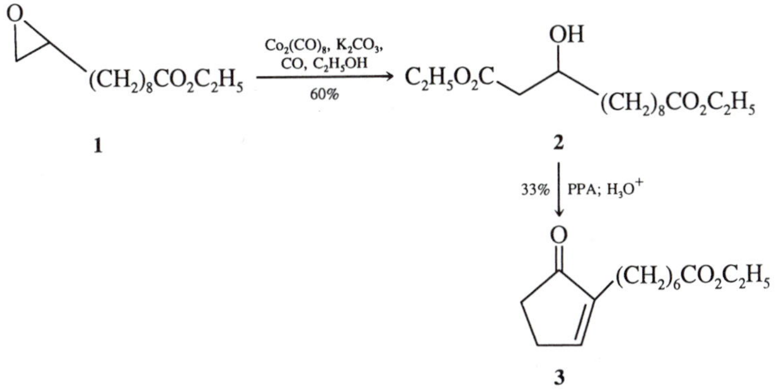

Hydrosilylation of aryl nitriles.[5] This metal carbonyl catalyzes hydrosilylation of aryl nitriles to provide N,N-disilylamines. The rate is slower in reactions of substrates substituted by electron-withdrawing groups. The actual reagent may be $(CH_3)_3SiCo(CO)_4$.

Example:

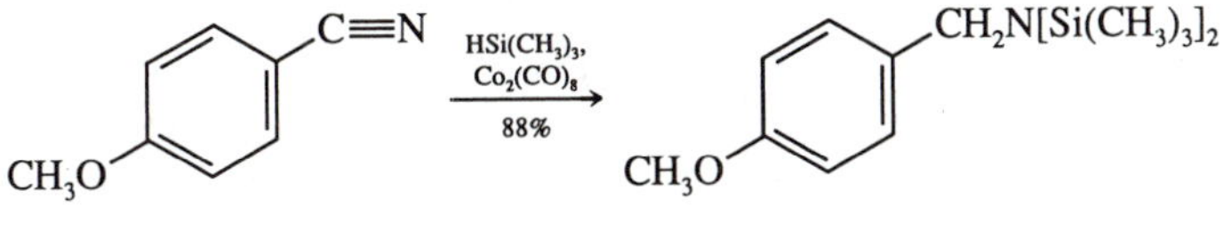

[1] S. C. Shim, S. Antebi, and H. Alper, *J. Org.*, **50**, 147 (1985).
[2] S. L. Schreiber, T. Sammakia, and W. E. Crowe, *Am. Soc.*, **108**, 3128 (1986).
[3] P. Eilbracht, E. Balss, and M. Acker, *Ber.*, **118**, 825 (1985).
[4] E. Dalcanale and M. Foà, *Synthesis*, 492 (1986).
[5] T. Murai, T. Sakane, and S. Kato, *Tetrahedron Letters*, **26**, 5145 (1985).

Dicarbonyl(nitrosyl)triphenylphosphinecobalt, $(CO)_2(NO)CoP(C_6H_5)_3$ (**1**), m.p. 135–136°. This air-stable complex is prepared from $Co_2(CO)_8$ and $NaNO_2$ to give $Co(NO)(CO)_3$, which reacts with $P(C_6H_5)_3$ to form the complex **1**.

Acylcobaltate complexes, $[\overset{O}{\overset{\|}{R}C}Co(NO)(CO)P(C_6H_5)_3]^-Li^+$ (**2**). These are prepared by reaction of RLi with **1** in THF. They are stable at $-40°$ but decompose at 25°. The acyl group is transferred to α,β-enones, quinones, and allylic halides.[1]

Examples:

[1] L. S. Hegedus and R. J. Perry, *J. Org.*, **50**, 4955 (1985).

Dichlorobis(cyclopentadienyl)titanium.

Reduction of **1,2-dibromides.** Chlorobis(cyclopentadienyl)titanium, generated *in situ* by zinc reduction of $(C_5H_5)_2TiCl_2$ in THF, effects rapid reductive debromination of *vic*-dibromides to alkenes by an *anti*-elimination.[1]
Examples:

[1] S. G. Davies and S. E. Thomas, *Synthesis*, 1027 (1984).

Dichlorobis(trifluoromethanesulfonato)titanium(IV).

1,3-Diketones.[1] 1-Alkynes react with acetic anhydride in the presence of $Cl_2Ti(OTf)_2$ to provide 1,3-diketones.
Example:

$$HC{\equiv}C(CH_2)_3OAc + (CH_3CO)_2O \xrightarrow[80\%]{\substack{Cl_2Ti(OTf)_2 \\ 0 \to 25°}} CH_3COCH_2CO(CH_2)_3OAc$$

[1] Y. Tanabe and T. Mukaiyama, *Chem. Letters*, 673 (1985).

Dichlorobis(triphenylphosphine)palladium(II).

Coupling of vinyl triflates with alkenes.   Vinyl triflates undergo Heck coupling with activated alkenes under catalysis with $Cl_2Pd[P(C_6H_5)_3]_2$ in DMF or DMSO. Olefination with styrenes proceeds in high yields with $Pd[P(C_6H_5)_3]_4$ and 3 equiv. of $N(C_2H_5)_3$. Addition of LiCl is helpful in coupling of vinyl triflates with 1-alkynes to afford 1,3-enynes.[1]

Examples:

α-Diketones.[2]   Unsymmetrical α-diketones can be prepared in 40–65% yield by coupling of acyltributyltin reagents with acyl chlorides in toluene (100°) catalyzed by $PdCl_2[P(C_6H_5)_3]_2$.

Example:

$$(CH_3)_2CHC\overset{\underset{\|}{O}}{}SnBu_3 + p\text{-}CH_3C_6H_4COCl \xrightarrow[63\%]{Pd(II)} (CH_3)_2CHC\overset{\underset{\|}{O}}{}-C\overset{\underset{\|}{O}}{}C_6H_4CH_3\text{-}p$$

Symmetrical α-diketones can be prepared in comparable yields by reaction of the acyl halide with hexabutylditin in the presence of the Pd(II) catalyst.

Example:

$$2C_6H_5COCl \xrightarrow[Pd(II)]{Bu_6Sn_2} [C_6H_5COSnBu_3]_2 \xrightarrow{52\%} C_6H_5C\overset{\underset{\|}{O}}{}-C\overset{\underset{\|}{O}}{}C_6H_5$$

Carbonylation of 1-iodo-1,4- and -1,5-dienes.[3]   Carbonylation of these iododienes catalyzed by $Cl_2Pd[P(C_6H_5)_3]_2$, $Pd(OAc)_2$, or bis(dibenzylideneacetone)-

palladium [Pd(dba)$_2$] and in the presence of N(C$_2$H$_5$)$_3$ (1.5 equiv.) and CH$_3$OH (4 equiv.) results in cyclopentenones and cyclohexenones.

Examples:

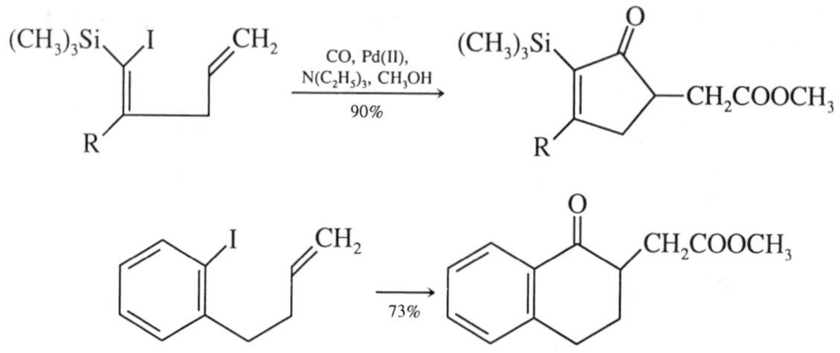

[1] W. J. Scott, M. R. Peña, K. Swärd, S. J. Stoessel, and J. K. Stille, J. Org., 50, 2302 (1985).
[2] J.-B. Verlhac, E. Chanson, B. Jousseaume, and J.-P. Quintard, Tetrahedron Letters, 26, 6075 (1985).
[3] J. M. Tour and E. Negishi, Am. Soc., 107, 8289 (1985).

2,3-Dichloro-5,6-dicyano-1,4-benzoquinone (DDQ).

Oxidation of enediols.[1]   The cis-enediol 1 undergoes facile oxidative cleavage, but is oxidized by DDQ in quantitative yield to the hydroxy enone 2.

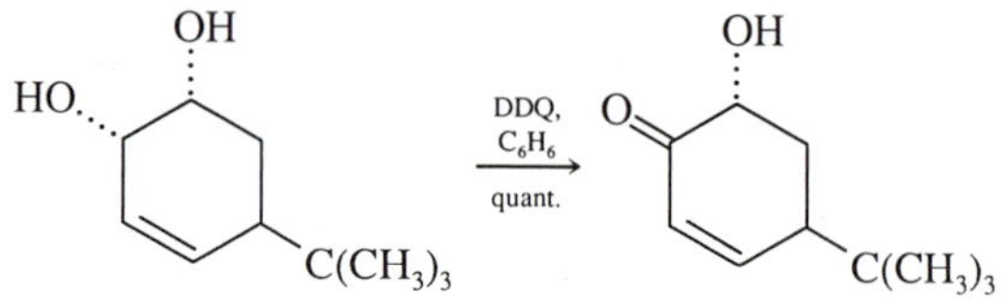

Protection of carboxylic acids.[2]   2,6-Dimethoxybenzyl esters on treatment with DDQ (1 equiv.) at 25° in CH$_2$Cl$_2$ containing some water are slowly converted into the corresponding acids in 90–95% yield. 4-Methoxybenzyl esters are stable to this oxidation, whereas 2,4,6-trimethoxybenzyl esters are hydrolyzed in the presence of silica gel.

Cerium(IV) ammonium nitrate (CAN) is effective for oxidative hydrolysis of phenyl esters substituted by hydroxyl, methoxy, and dimethylamino groups in CH$_3$CN–H$_2$O at 0°.

Regeneration from DDHQ[3] (1, 215; 7, 96–97).   Oxidation of a stirred suspension of the hydroquinone in water with nitric acid at 20–25° provides the quinone in 88% yield.

[1] B. A. McKittrick and B. Ganem, *J. Org.*, **50**, 5897 (1985).
[2] C. U. Kim and P. F. Misco, *Tetrahedron Letters*, **26**, 2027 (1985).
[3] M. S. Newman and V. K. Khanna, *Org. Prep. Proc. Int.*, **17**, 422 (1985).

1,1-Dichloro-2,2-difluoroethylene, $CF_2{=}CCl_2$ **(1).** Supplier: PCR Research Chem. Inc., Gainesville, Florida.

(2-Hydroxyaryl)alkynes.[1]   Phenyl 1,1-difluoro-2,2-dichloroethyl ethers **(1)** are prepared from phenols with **1** in aqueous KOH/CH_2Cl_2 catalyzed by Bu_4NOH (~90% yield). These ethers on treatment with BuLi (4 equiv.) rearrange to *o*-acetylenic phenols **(a)**, which are usually isolated as acetylenic silanes **(2)** in 70–80% yield.

[1] R. Subramanian and F. Johnson, *J. Org.*, **50**, 5430 (1985).

Dichloroketene.

vic-Dicarboxylation of alkenes.[1]   This dicarboxylation can be effected in two steps. The α,α-dichlorocyclobutanone, obtained by cycloaddition of dichloroketene, is converted to the β-chloro enol acetate by reaction with BuLi and then with acetic anhydride. This intermediate is oxidized to the *vic*-dicarboxylic acid by $RuCl_3$ and $NaIO_4$ **(11,** 462–463).

Example:

[2 + 2]Cycloaddition to a cyclic allyl ether.[2]   The key step in a synthesis of lineatin **(3)**, the aggregation pheromone of the bark beetle, is the addition of dichloroketene to the alkene **1**. Under usual conditions (POCl₃, **8,** 156) the desired adduct is obtained in 7% yield. Fortunately, substitution of 1,2-dimethoxyethane for POCl₃ increases the yield of **2** to 50–60%.

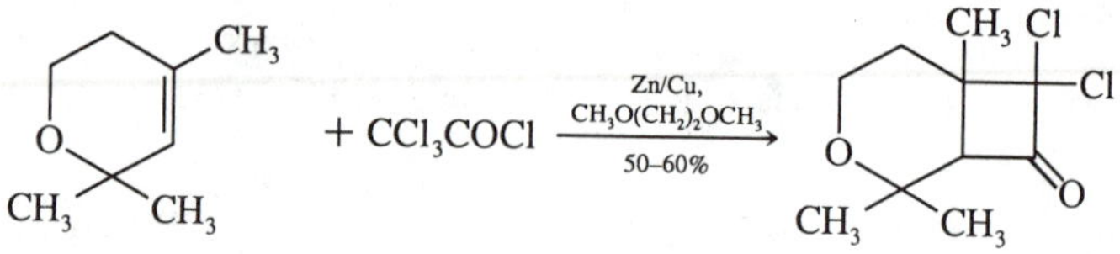

1   **2**

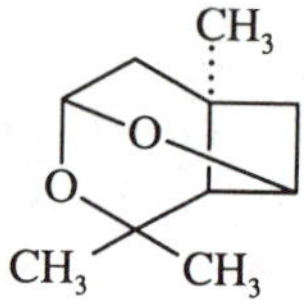

3

[1] J.-P. Depres and A. E. Greene, *Org. Syn.*, submitted (1986).
[2] B. D. Johnston, K. N. Slessor, and A. C. Oehlschlager, *J. Org.*, **50**, 114 (1985).

Dichloromethylenedimethylammonium chloride.

Cyanation of aromatics.[1]   Viehe's salt (**1**) effects electrophilic cyanation of aromatic or heteroaromatic systems.

Example:

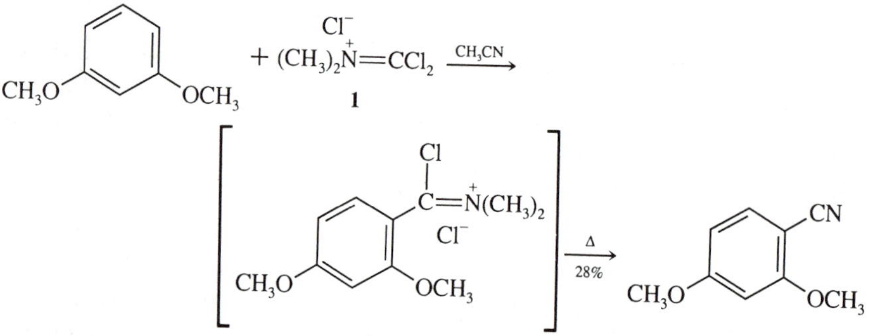

[1] J. Bergman and B. Pelcman, *Tetrahedron Letters*, **27**, 1939 (1986).

Dichloromethyllithium (1).

Homologation of boronic esters.   The reagent **1** can be formed *in situ* by reaction of CH_2Cl_2 with lithium 2,2',6,6'-tetramethylpiperidide (LiTMP) or even LDA. The product obtained on reaction of **1** with alkylboronic esters, an α-chloroboronic ester, is reduced by potassium triisopropoxyborohydride (KIPBH) to the homologated boronic ester.[1]

Example:

$$n\text{-}C_6H_{13}B(OCH_3)_2 + CH_2Cl_2 \xrightarrow{\text{base}} \left[n\text{-}C_6H_{13}\underset{\underset{Cl}{|}}{C}HB(OCH_3)_2 \right] \xrightarrow{\text{KIPBH}} n\text{-}C_7H_{15}B(OCH_3)_2$$

LDA   45%
LiTMP   83%

[1] H. C. Brown and S. M. Singh, *Organometallics*, **5**, 994, 998 (1986).

Dichlorotris(triphenylphosphine)ruthenium(II).

N-*Heterocyclization*.[1] 1,5-Pentanediol reacts with primary amines at 150–180° in the presence of a ruthenium catalyst to form N-substituted piperidines. For the reaction with aromatic amines $RuCl_2[P(C_6H_5)_3]_3$ is the catalyst of choice. On the other hand, the most effective catalyst for the reaction with aliphatic amines is $RuCl_3$ combined with either tributylphosphine or triethylphosphine.

Examples:

$$C_6H_5NH_2 + HO(CH_2)_5OH \xrightarrow{\text{Ru(II)}} [C_6H_5NH(CH_2)_5OH] \xrightarrow[87\%]{} C_6H_5N\langle\text{ }\rangle$$

$$(C_6H_5)_2CHNH_2 + HO(CH_2)_2\underset{\underset{CH_3}{|}}{N}(CH_2)_2OH \xrightarrow[19\%]{\overset{RuCl_3}{PBu_3}} (C_6H_5)_2CHN\langle\text{ }\rangle NCH_3$$

The ruthenium-catalyzed reaction of amines with bis(2-hydroxyethyl) ether or bis(2-hydroxyethyl)amines gives N-substituted morpholines or piperazines, respectively.

Selective reduction of aldehydes.[2] This reduction can be effected with formic acid (1.25 equiv.) and triethylamine (1 equiv.) and a catalytic amount of $RuCl_2[P(C_6H_5)_3]_3$ at 25° in THF. Under these conditions nitro, keto, ester, and *t*-amide groups are not reduced. The reduction of aldehydes, however, is sensitive to steric hindrance. Thus mesitaldehyde is reduced very slowly.

[1] Y. Tsuji, K.-T. Huh, Y. Ohsugi, and Y. Watanabe, *J. Org.*, **50**, 1365 (1985).
[2] B. T. Khai and A. Arcelli, *Tetrahedron Letters*, **26**, 3365 (1985).

Dicyclohexylcarbodiimide–4-Dimethylaminopyridine.

Macrolactonization.[1] The original conditions of Steglich esterification are ineffective for lactonization of ω-hydroxy carboxylic acids. However, addition of a soluble amine hydrochloride as a proton-transfer agent results in a useful method for macrolactonization. The most effective additive is DMAP·HCl, prepared from DMAP and anhydrous HCl in THF. Use of this variation and high dilution (slow addition of the hydroxy acid) results in yields of 32–95% for lactones of ring size ranging from 13–17.

[1] E. P. Boden and G. E. Keck, *J. Org.*, **50**, 2394 (1985).

Dicyclopentadienyldihydridozirconium, Cp_2ZrH_2. This zirconium reagent is obtained as a white solid by reduction of $(Cp_2ZrCl)_2O$ with $LiAlH_4$ in THF in 66% yield.[1]

Oppenauer oxidation and Meerwein-Ponndorf-Verley reduction.[2] These reactions use aluminum alkoxides for transfer of hydrogen to a ketone (generally acetone) or transfer of hydrogen from an alcohol (generally isopropanol). Cp_2ZrH_2 can serve as the transfer catalyst for Oppenauer oxidation using either benzaldehyde or benzophenone. This variation is particularly useful for oxidation of long-chain primary alcohols and allylic alcohols (90–95% yield). Reduction of ketones can be effected with Cp_2ZrH_2 as catalyst and isopropanol as hydride donor. Decreasing ease of reduction of ketones is in the order: aromatic, alicyclic, and aliphatic. α,β-Enones and -enals are reduced slowly to the unsaturated alcohols. Recovered catalyst can be reused repeatedly.

[1] P. C. Wailes and H. Weigold, *J. Organomet, Chem.*, **24**, 405 (1970).
[2] Y. Ishii, T. Nakano, A. Inada, Y. Kishigami, K. Sakurai, and M. Ogawa, *J. Org.*, **51**, 240 (1986).

1,2-Diethoxy-1,2-bis(trimethylsilyloxy)ethylene,

$$\underset{(CH_3)_3SiO}{\overset{C_2H_5O}{\diagdown}} C = C \underset{OC_2H_5}{\overset{OSi(CH_3)_3}{\diagup}} \qquad (1).$$

α,β-*Diketo esters.*[1] Acid chlorides react with **1** in the presence of $ZnCl_2$ to give protected derivatives of α,β-diketo esters (equation I).

$$(I) \; RCOCl + 1 \xrightarrow[\substack{72-93\%}]{\substack{ZnCl_2, \\ CH_2Cl_2}} RC\overset{\overset{O}{\|}}{-}\overset{\overset{OC_2H_5}{|}}{C}\underset{\underset{OSi(CH_3)_3}{|}}{-}COOC_2H_5 \xrightarrow{H_3O^+} RC\overset{\overset{O}{\|}}{-}\overset{\overset{O}{\|}}{C}COOC_2H_5$$

[1] M. T. Reetz and S.-H. Kyung, *Tetrahedron Letters*, **26**, 6333 (1985).

Diethoxycarbenium tetrafluoroborate.

α,β-*Enals.*[1] β-Keto diethylacetals, obtained by reaction of ketones with this reagent, can be converted into one-carbon-homologated α,β-enals by reduction ($NaBH_4$) and acid-catalyzed rearrangement.

Example:

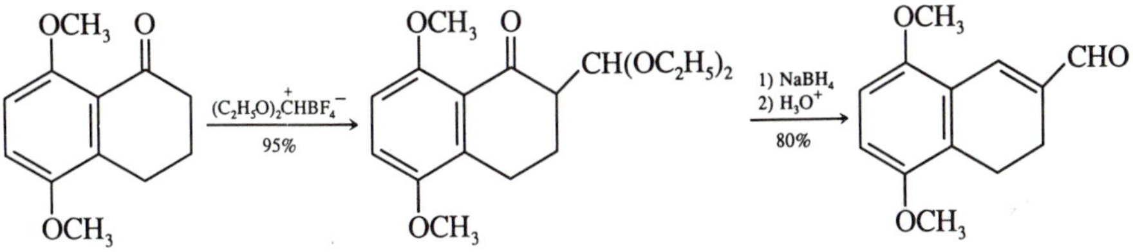

[1] R. Dasgupta and U. R. Ghatak, *Tetrahedron Letters*, **26**, 1581 (1985).

Diethoxytriphenylphosphorane, $(C_6H_5)_3P(OC_2H_5)_2$ (**1**). The reagent is obtained as a viscous oil by reaction of $(C_6H_5)_3P$ with diethyl peroxide in CH_2Cl_2. It can be prepared conveniently and stored indefinitely in toluene at 25° (N_2).

Dehydration of diols to cyclic ethers.   The reagent dehydrates a variety of diols to cyclic ethers with formation of triphenylphosphine oxide as the co-product. Yields of cyclic ethers are high from 1,2-,1,4- and 1,5-diols. Although (Z)-2-butene-1,4-diol is converted into 2,5-dihydrofuran in 95% yield, the (E)-isomer is converted in 35–40% yield into 3,4-epoxy-1-butene.[1]

Other examples:

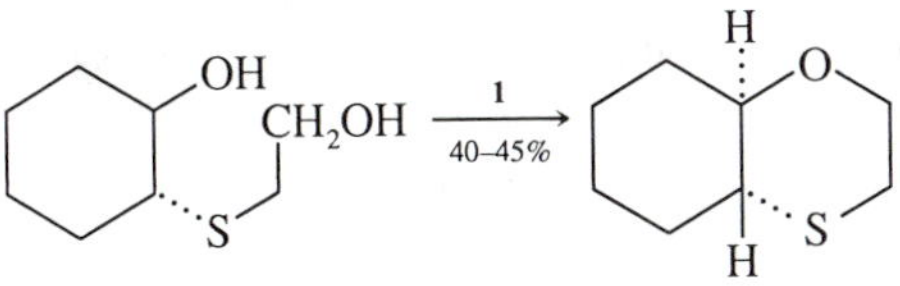

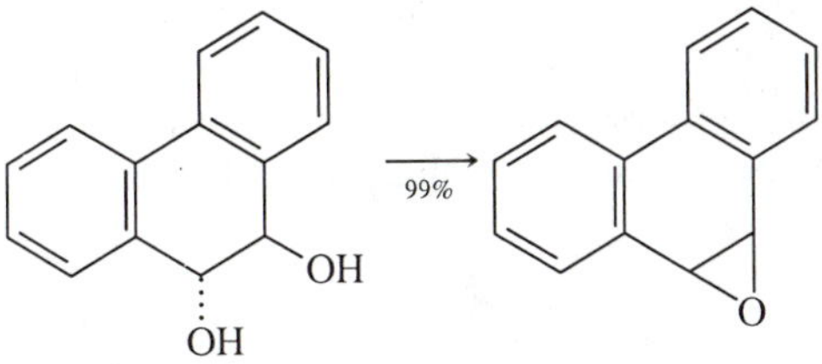

The reaction of cyclic 1,2-diols with **1** can result in epoxides or ketones, as illustrated for the case of 3-methylcyclohexane-1,2-diols (equations I and II).[2]

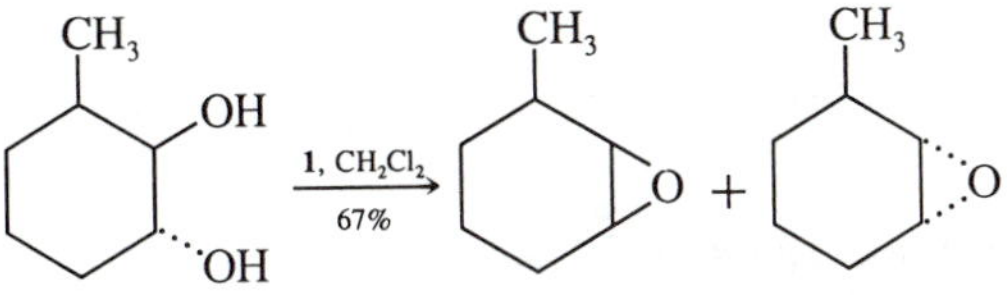

78.8:21.2

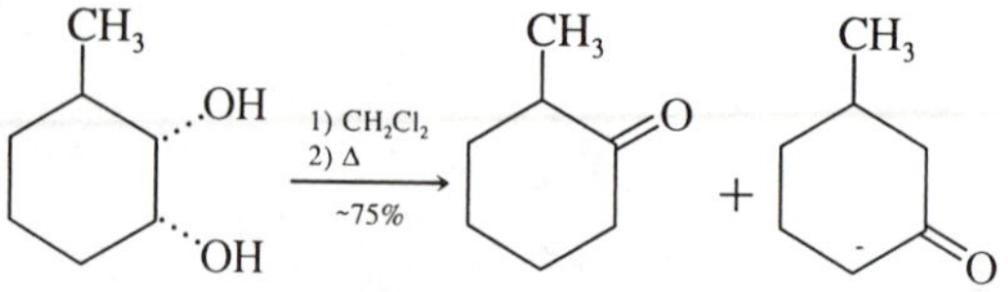

70.4:29.6

The reaction of a polymeric form of **1**, diethoxydiphenylpolystyrylphosphorane, (P)—P(OC$_2$H$_5$)$_2$(C$_6$H$_5$)$_2$, with simple diols provides high yields of easily isolated cyclic ethers.[3]

Aziridines.[4]   This reagent (**1**) effects cyclodehydration of β-amino alcohols to aziridines in 85–95% yield.

Example:

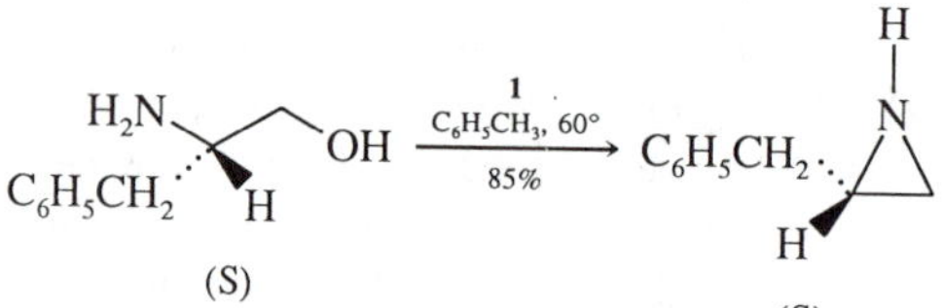

[1] P. L. Robinson, C. N. Barry, J. W. Kelly, and S. A. Evans, Jr., *Am. Soc.*, **107**, 5210 (1985).
[2] P. L. Robinson and S. A. Evans, Jr., *J. Org.*, **50**, 3860 (1985).
[3] J. W. Kelly, P. L. Robinson, and S. A. Evans, Jr., *ibid.*, **50**, 5007 (1985).
[4] J. W. Kelly, N. L. Eskew, and S. A. Evans, Jr., *ibid.*, **51**, 95 (1986).

Diethyl dicarbonate, O(COOC$_2$H$_5$)$_2$ (**1**).

β-Keto esters.[1]   Ethoxycarbonylation of saturated cycloalkanones can be effected in good yield by reaction of the potassium enolate with **1** in benzene. Ethoxycarbonylation of unsaturated cycloalkanones is best effected by reaction of **1** with the enolate generated with lithium dicyclohexylamide in ether.

Example:

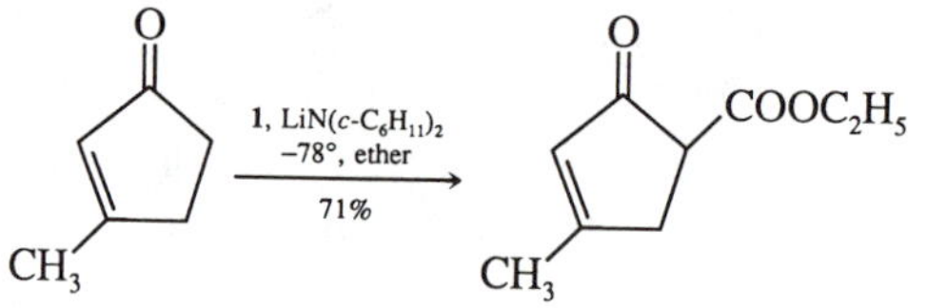

[1] J. Hellou, J. F. Kingston, and A. G. Fallis, *Synthesis*, 1014 (1984).

(Diethylamino)sulfur trifluoride (DAST).

Glycosyl fluorides.   Two laboratories[1,2] have reported that the free anomeric hydroxyl group of carbohydrates is replaced by fluorine on treatment with DAST.

Reaction with neat reagent proceeds mainly with inversion. A marked solvent effect is observed in the reaction of **1** with DAST.

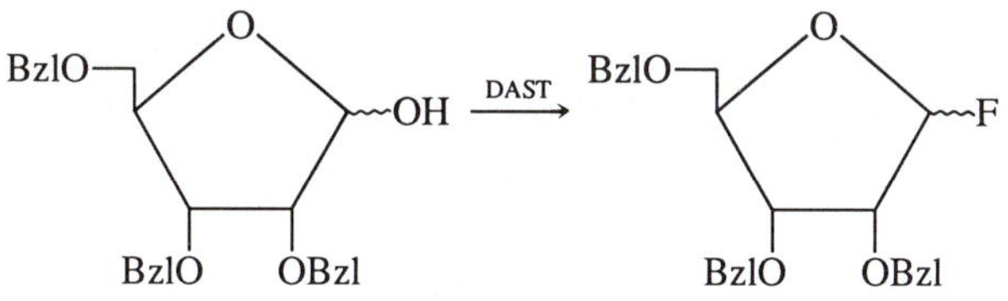

	1	CH$_2$Cl$_2$	2, β/α = 2.0
		(C$_2$H$_5$)$_2$O	4.2
		THF	9.9

2-Deoxyglycosides.[3] The reaction of DAST with a pyranoside having a free hydroxyl group at C$_2$ and substituted at C$_1$ with a methoxy, acetoxy, phenylthio, or azido group results in a glycosyl fluoride with the original C$_1$-substituent now at C$_2$. The rearrangement involves inversion of stereochemistry at both centers.
    Example:

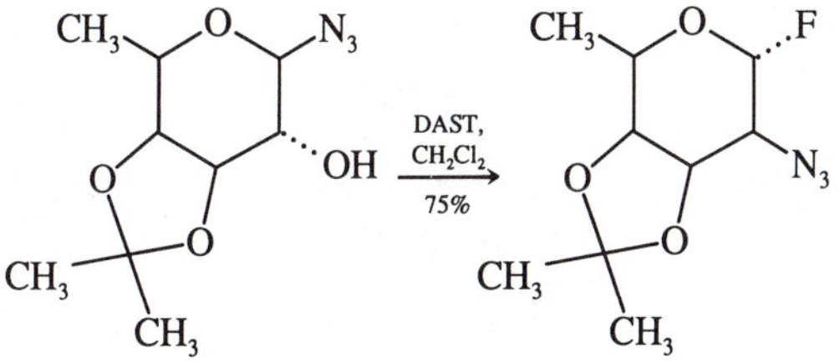

The reaction produces a stereoselective synthesis of α- or β-2-deoxyglycosides. α-Linked disaccharides such as **1** can be prepared by a three-step sequence with high stereocontrol.
    Example:

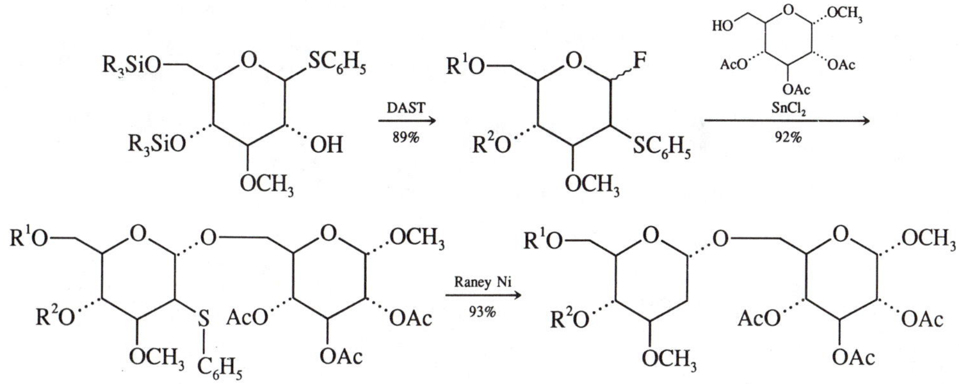

α-*Fluorothio ethers*.[4]   The reaction of DAST with methyl phenyl sulfoxide (**1**) in $CHCl_3$ results in fluoromethyl phenyl sulfide (**2**) in 85% yield, which can be oxidized to the sulfoxide (**3**) by $ClC_6H_4CO_3H$. The reaction is general for primary-

$$C_6H_5\overset{\overset{\displaystyle O}{\|}}{S}CH_3 \xrightarrow[85\%]{DAST,\ 25°} C_6H_5SCH_2F \xrightarrow[80\%]{ClC_6H_4CO_3H} C_6H_5\overset{\|}{\underset{\displaystyle O}{S}}CH_2F$$

$$\quad\quad\quad\mathbf{1}\quad\quad\quad\quad\quad\quad\quad\quad\quad\mathbf{2}\quad\quad\quad\quad\quad\quad\quad\quad\quad\mathbf{3}$$

alkyl aryl sulfoxides and methyl alkyl sulfoxides. The reaction can be catalyzed by ZnI_2.

[1] W. Rosenbrook, Jr., D. A. Riley, and P. A. Lartey, *Tetrahedron Letters*, **26**, 3 (1985).
[2] G. H. Posner and S. R. Haines, *ibid.*, **26**, 5 (1985).
[3] K. C. Nicolaou, T. Ladduwahetty, J. L. Randall, and A. Chucholowski, *Am. Soc.*, **108**, 2466 (1986).
[4] J. R. McCarthy, N. P. Peet, M. E. LeTourneau, and M. Inbasekaran, *Am. Soc.*, **107**, 735 (1985).

2,3-Dihydro-1,4-dioxin, (**1**). This reagent is prepared by treatment of $(HOCH_2CH_2)_2O$ with copper chromite and $KHSO_4$.[1]

***Protection of alcohols*.**[2]   Alcohols react with **1** in the presence of a Lewis acid catalyst to form 1,4-dioxan-2-yl derivatives, which are stable to $LiAlH_4$ and organolithiums.

α-*Hydroxymethyl ketones*.[3]   2,3-Dihydro-1,4-dioxin-5-yllithium (**2**), prepared with *t*-BuLi at $-30°$, adds to aldehydes or ketones to form alcohols (**3**) that undergo allylic rearrangement to **a** in contact with SiO_2 containing a trace of oxalic acid. Reduction of **a** followed by hydrolysis provides α-hydroxymethyl ketones (**4**).

α,α′-Dihydroxy ketones.[4] The allylic alcohols **3** on epoxidation in CH_3OH are converted into **5**, obtained as two diastereomers. These are converted into α,α′-dihydroxy ketones (**6**) by reduction ($NaBH_4$) and hydrolysis (equation I).
Example:

(I)

$$ 3 \xrightarrow[\text{50–60\%}]{\substack{ClC_6H_4CO_3H, \\ CH_3OH}} 5 \xrightarrow[\text{2) } H_3O^+ \text{ (45–90\%)}]{\text{1) } NaBH_4 \text{ (55–80\%)}} 6 $$

[1] R. D. Moss and J. Page, *J. Chem. Eng. Data*, **12**, 452 (1967).
[2] M. Fetizon and I. Hanna, *Synthesis*, 806 (1985).
[3] M. Fetizon, I. Hanna, and J. Rens, *Tetrahedron Letters*, **26**, 3453 (1985).
[4] M. Fetizon, P. Goulaouic, and I. Hanna, *Tetrahedron Letters*, **26**, 4925 (1985).

2,2′-Dihydroxy-1,1′-binaphthyl (1).

Asymmetric cyclization of nerol and farnesol.[1] A biomimetic cyclization of nerol to limonene can be effected with an aluminum reagent **2** and a chiral leaving group. Thus the neryl ether **3** of R-(+)-2,2′-dihydroxy-1,1′-binaphthyl is cyclized

$$ i\text{-BuAl} \begin{cases} OC_6H_2\text{-2,4,6-}t\text{-Bu} \\ OTf \end{cases} \quad (2) $$

in the presence of **2** to D-limonene **4** (77% ee) and **5**. The farnesyl ether of **1** is cyclized to β-bisabolene in 76% ee.

[1] S. Sakane, J. Fujiwara, K. Maruoka, and H. Yamamoto, *Tetrahedron*, **42**, 2193 (1986).

Diiodomethane–Samarium.

Iodohydrins.[1] Ketones or aldehydes undergo iodomethylation when treated with CH_2I_2 and SmI_2 (2 equiv.) to form iodohydrins (60–95% yield). A more convenient procedure uses CH_2I_2 and samarium(0) in THF at 0°.

Examples:

$$n\text{-}C_6H_{13}COCH_3 + CH_2I_2 \xrightarrow[93\%]{\substack{Sm, THF, \\ 0°}} n\text{-}C_6H_{13}\underset{\underset{OH}{|}}{\overset{\overset{CH_2I}{|}}{C}}CH_3$$

$$n\text{-}C_{11}H_{23}CHO + CH_2I_2 \xrightarrow[72\%]{} n\text{-}C_{11}H_{23}\underset{\underset{OH}{|}}{C}HCH_2I$$

A similar reaction with α-halo ketones provides cyclopropanols.
Example:

$$C_6H_5COCH_2Br + CH_2I_2 \xrightarrow[81\%]{Sm} \underset{HO}{\overset{C_6H_5}{}}\!\!\!\bigtriangledown$$

[1] T. Imamoto, T. Takeyama, and H. Koto, *Tetrahedron Letters*, **27**, 3243 (1986).

Diiodomethane–Triisobutylaluminum.

Cyclopropanation of unsaturated alcohols.[1]   The reaction of perillyl alcohol (**1**) with this reagent results in regioselective cyclopropanation of the isolated double bond and is therefore complementary to the regioselectivity of the Simmons-Smith reaction. The actual reagent is probably diisobutyl(iodomethyl)aluminum.

[1] K. Maruoka, Y. Fukutani, and H. Yamamoto, *J. Org.*, **50**, 4412 (1985); K. Maruoka, S. Sakane, and H. Yamamoto, *Org. Syn.*, submitted (1986).

Diiodomethane–Zinc–Titanium(IV) chloride.

Methylenation of aldehydes and ketones.[1]   A reagent prepared in CH_2Cl_2 from CH_2I_2, Zn, and $TiCl_4$ is considerably more reactive than one prepared from CH_2Br_2, Zn, and $TiCl_4$ (**8**, 339, **11**, 337, **12**, 322). Moreover, it affords less of the coupled products. This new reagent does not react with esters. It is particularly useful for methylenation of easily enolizable ketones, for which the Wittig reagent is not useful. The isolated yields of some methylenated products are shown.

(64%)

$CH_3(CH_2)_{10}CH{=}CH_2$

(72%)

$$\underset{(73\%)}{C_6H_5\overset{\overset{\displaystyle CH_2}{\|}}{C}CH_2\overset{\overset{\displaystyle OAc}{|}}{C}HR}$$

[1] J. Hibino, T. Okazoe, K. Takai, and H. Nozaki, *Tetrahedron Letters*, **26**, 5579 (1985).

Diiodomethane–Zinc–Titanium(IV) isopropoxide.

Methylenation of aldehydes.[1]   The reagent (**1**) prepared from this combination or from CH_2I_2–Zn–Al(CH$_3$)$_3$ (**8**, 339) effects selective methylenation of aldehydes. However, if the aldehyde group is first protected by reaction with $Ti[N(C_2H_5)_2]_4$, then methylenation of ketones can be effected with **1**.

Example:

$$\underset{}{CH_3\overset{\overset{\displaystyle O}{\|}}{C}(CH_2)_8CHO} \quad \xrightarrow[\underset{76\%}{\text{2) CH}_2I_2\text{–Zn–Ti(O-}i\text{-Pr)}_4}]{\text{1) Ti[N(C}_2H_5)_2]_4} \quad CH_3\overset{\overset{\displaystyle CH_2}{\|}}{C}(CH_2)_8CHO$$

[1] T. Okazoe, J. Hibino, K. Takai, and H. Nozaki, *Tetrahedron Letters*, **26**, 5581 (1985).

Diisobutylaluminum hydride.

Diastereoselective reduction of chiral β-keto sulfoxides (**11**, 291–292).   Chiral β-keto sulfoxides **1**, prepared by reaction of *p*-(tolylsulfinyl)methyllithium with esters, are reduced by DIBAH in THF diastereoselectively to (R,S)-**2**. In the presence of $ZnCl_2$, the opposite diastereoselectivity obtains. The paper includes a new method for conversion of these β-hydroxy sulfoxides into chiral epoxides.[1]

(R,R)-**2**, 95:5   (R)-**1**   (R,S)-**2**, 95:5

(R)-**3**   (S)-**3**

This reduction can be used to obtain either enantiomer of 4-substituted butenolides from the corresponding chiral β-keto sulfones (equation I).[2]

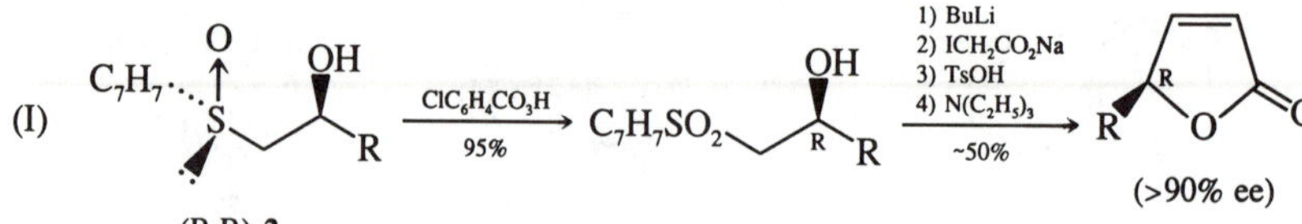

(I)

(R,R)-2

[1] G. Solladié, G. Demailly, and C. Greck, *Tetrahedron Letters*, **26**, 435 (1985); *idem, J. Org.*, **50**, 1552 (1985).
[2] G. Solladié, C. Fréchou, G. Demailly, and C. Greck, *J. Org.*, **51**, 1912 (1986).

Diisobutylaluminum hydride–Boron trifluoride etherate.

Reduction of γ-amino-α,β-unsaturated esters.[1]   These substrates when complexed with BF_3 etherate can be reduced selectively to γ-amino esters with only slight reduction of the ester group (equation I).

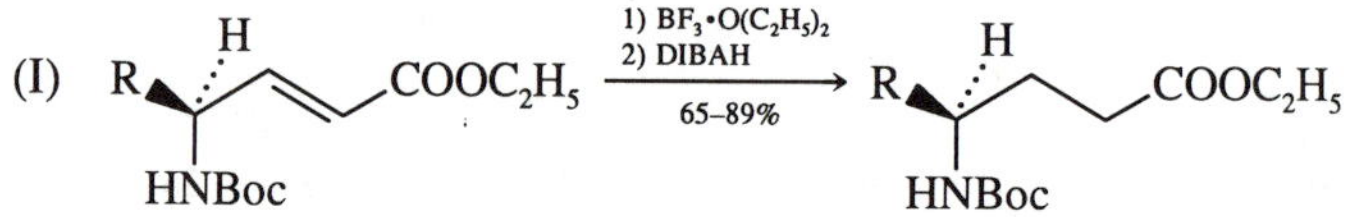

(I)

R = H, CH_3, $CH(CH_3)_2$, C_6H_5

[1] T. Moriwake, S. Hamano, D. Miki, S. Saito, and S. Torii, *Chem. Letters*, 815 (1986).

Diisobutylaluminum hydride–Tin(II) chloride.

Debromination.[1]   This reducing system reduces α-bromo ketones to ketones and *vic*-dibromoalkanes to alkenes (equations I and II).

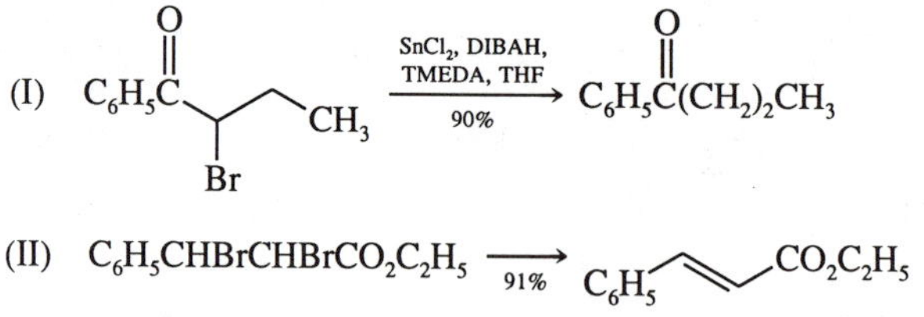

(I)

(II) $C_6H_5CHBrCHBrCO_2C_2H_5$ $\xrightarrow{91\%}$

Asymmetric reduction of ketones.[2]   A reducing agent prepared by treatment of a mixture of $SnCl_2$ and (S)-1-[1-methyl-2-pyrrolidinyl]methylpiperidine (**11**, 525, **12**, 490) with DIBAH effects asymmetric reduction of prochiral ketones in 60–80% ee (equation I).

$$\text{(I)} \quad R^1\overset{\overset{\displaystyle O}{\|}}{C}R^2 \xrightarrow[70\text{–}90\%]{\underset{1,\ CH_2Cl_2}{SnCl_2,\ DIBAH,}} R^1\overset{*}{C}HOHR^2$$

$$(60\text{–}80\% \ ee)$$

1

[1] T. Oriyama and T. Mukaiyama, *Chem. Letters*, 2069 (1984).
[2] *Idem, ibid.*, 2071 (1984).

Diisobutylaluminum phenylselenide, $i\text{-Bu}_2\text{AlSeC}_6\text{H}_5$ (**1**). This selenoaluminum reagent is prepared by addition of $(C_6H_5Se)_2$ (1 equiv.) to DIBAH (2 equiv.) in hexane.[1]

*Contra-***trans-***diaxial cleavage of epoxides.*[2] The reaction of the α,β-epoxy alcohol **2** with $NaSeC_6H_5$ provides the 1,3-diol **3** in 90% yield. Use of **1** results in the opposite regioselectivity to provide **4** as the major product.

[1] K. Maruoka, T. Miyazaki, M. Ando, Y. Matsumura, S. Sakane, K. Hattori, and H. Yamamoto, *Am. Soc.*, **105**, 2831 (1983).
[2] B. A. McKittrick and B. Ganem, *J. Org.*, **50**, 5897 (1985).

Diisopinocampheylborane.

Asymmetric hydroboration.[1] The key step in a synthesis of natural (+)-hirsutic acid-C (**1**), based on an earlier synthesis of racemic **1**, is an efficient asymmetric hydroboration of the *meso*-alkene **2**. Reaction of **2** with (+)-diisopinocampheylborane (90% ee) followed by oxidation provides the *exo*-alcohol **3** in 73% yield and in 92% optical purity. Ring expansion of the corresponding ketone with ethyl diazoacetate is not regioselective even in the presence of BF_3 etherate or $(C_2H_5)_3O^+$ BF_4^-, but does afford the desired α-keto ester in the presence of $SbCl_5$ (**8**, 500–501). Decarboxylation of the crude product gives (–)-**4** in 90% ee after chromatography.

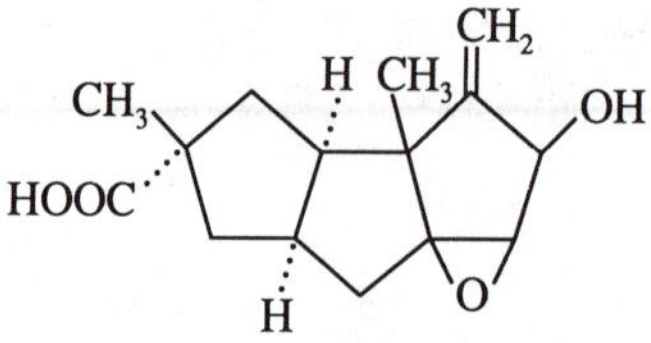

1

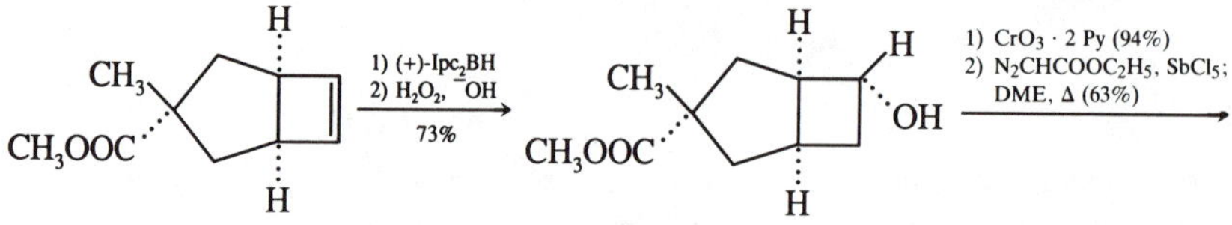

2   3 (92% ee)

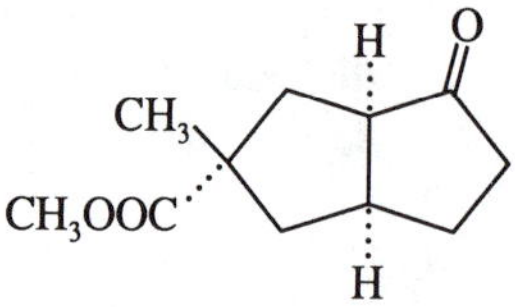

1) CrO₃ · 2 Py (94%)
2) N₂CHCOOC₂H₅, SbCl₅;
   DME, Δ (63%)

(−)-4 (90% ee)

[1] A. E. Greene, M.-J. Luche, and A. A. Serra, *J. Org.*, **50**, 3957 (1985).

3,3-Dimethoxy-1-trimethylstannylpropane, $(CH_3)_3Sn(CH_2)_2CH(OCH_3)_2$ (**1**). The acetal is obtained by reaction of the Grignard reagent from 3-bromo-1,1-dimethoxypropane with $(CH_3)_3SnCl$ (61% yield).

Cyclopentanes.[1]   The acetal serves as a bifunctional annelation reagent to form cyclopentanes from silyl enol ethers.

Example:

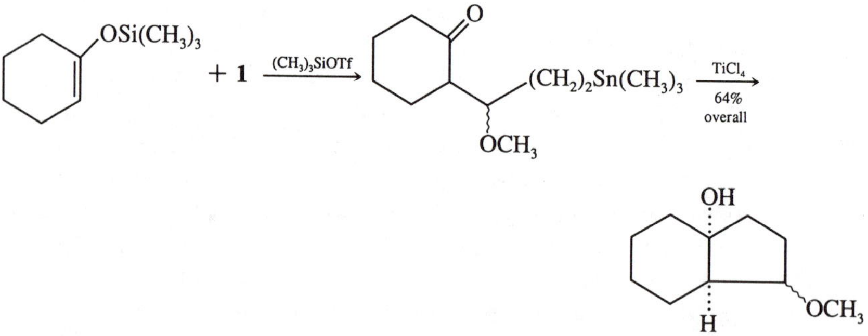

[1] T. V. Lee and K. A. Richardson, *Tetrahedron Letters*, **26**, 3629 (1985).

(3-Dimethylamino)propyltriphenylphosphonium   bromide, $(CH_3)_2N(CH_2)_3\overset{+}{P}$-$(C_6H_5)_3Br^-$ **(1)**. The reagent is prepared by consecutive reaction of $(C_6H_5)_3P$ with $Br(CH_2)_3Br$ and $HN(CH_3)_2$. It is converted into the ylide, $(CH_3)_2N(CH_2)_2$-$CH=P(C_6H_5)_3$ **(2)**, by $KN[Si(CH_3)_3]_2$ at 23°.

1,3-Dienes.[1]   The ylide (**2**) reacts even with hindered aldehydes to form (Z)-alkenes, which are convertible into 1,3-dienes.

Example:

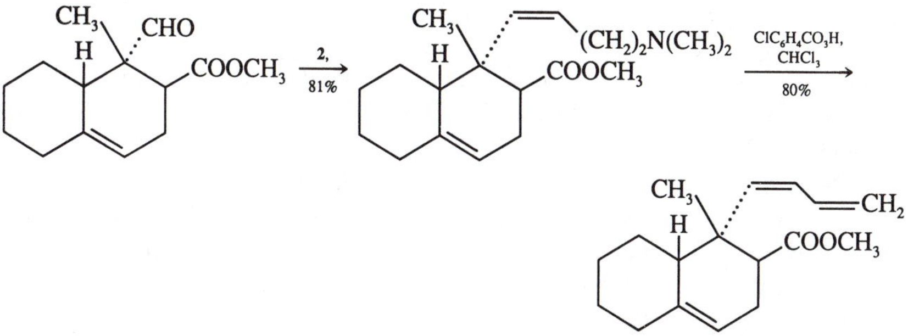

[1] E. J. Corey and M. C. Desai, *Tetrahedron Letters*, **26**, 5747 (1985).

(R,R)- and (S,S)-*trans*-2,5-dimethylborolanes (1).

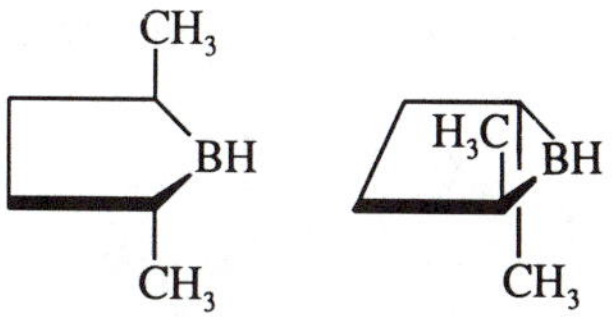

(R,R)-1   (S,S)-1

Asymmetric hydroboration.[1]   With the exception of 1,1-disubstituted ethylenes, a wide variety of alkenes including 1,1,2-trisubstituted alkenes undergo hydroboration with (R,R)- or (S,S)-**1** in 70–90% yield and with high enantioselectivity (97–100% ee).

Examples:

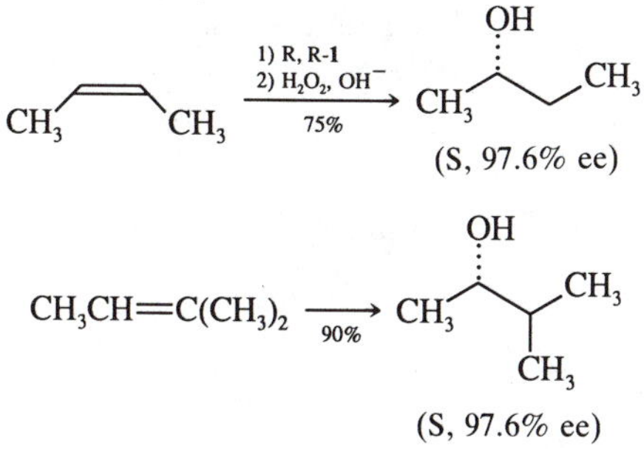

(S, 97.6% ee)

(S, 97.6% ee)

The synthesis of **1** starts with the reaction of dichloro(dimethylamino)borane with the Grignard reagent from 2,5-dibromohexane, and includes a resolution with (S)-prolinol.

[1] S. Masamune, B. M. Kim, J. S. Petersen, T. Sato, and S. J. Veenstra, and T. Imai, *Am. Soc.*, **107**, 4549 (1985).

Dimethyldioxirane,

$$\underset{CH_3 \quad CH_3}{\overset{O-O}{\diagdown\diagup}}$$

(1). This and related methylalkyldioxiranes have been prepared by reaction of potassium peroxomonosulfate with acetone (or 2-alkanones, **11**, 442). Dimethyldioxirane can be prepared *in situ*. Dimethyldioxirane on standing dimerizes to acetone diperoxide.

Epoxidation.[1]   This reagent converts *cis*- and *trans*-stilbene to the corresponding epoxides in 73% yield. It converts phenanthrene into the 9,10-oxide in 83% yield. The yield obtained with material prepared *in situ* is 60%. It converts triphenylphosphine into the oxide (quantitative yield) and methyl phenyl sulfide into the sulfoxide (94% yield).

The acetone–oxone system of Curci can be used for epoxidation of α,β-unsaturated acids if $NaHCO_3$ is present as a buffer to prevent Baeyer-Villiger oxidation of acetone; yields are in the range 60–92% (4 examples).[2]

$RNH_2 \rightarrow RNO_2$.[3]   Dimethyldioxirane oxidizes aliphatic or aromatic amines to nitro compounds in 85–97% yield, probably via the intermediates shown in equation (I).

$$(I) \quad RNH_2 + 1 \longrightarrow \left[RN\underset{H}{\overset{OH}{\diagup}} \overset{1}{\longrightarrow} RN(OH)_2 \overset{-H_2O}{\longrightarrow} RN{=}O \right] \overset{1}{\longrightarrow} RNO_2$$

[1] R. W. Murray and R. Jeyaraman, *J. Org.*, **50**, 2847 (1985).
[2] P. F. Corey and F. E. Ward, *ibid*, **51**, 1925 (1986).
[3] R. W. Murray, R. Jeyaraman, and L. Mohan, *Tetrahedron Letters*, **27**, 2335 (1986).

Dimethylformamide dimethyl acetal (1).

Mannich bases.   A nonacidic and regioselective route to Mannich bases (**3**) from ketones (or esters) involves reaction with DMF dimethyl acetal (**1**) at 110° to form an enaminone, which is reduced by $LiAlH_4$ to the Mannich base.[1]

Example:

2   3

Reagent **1** is preferred because of commercial availability, but it is less reactive than the amide acetal *t*-butoxybis(dimethylamino)methane **4** (Brederick's reagent). Thus **4** reacts with α-methylcyclohexanone to give the enaminone **2** in 99% yield. Trisdimethylaminomethane (**5**) is even more reactive than **4**, but is not as readily available.

[1] P. F. Schuda, C. B. Ebner, and T. M. Morgan, *Tetrahedron Letters*, **27**, 2567 (1986).

Dimethylformamide di-*t*-butyl acetal.

Preparation.[1]   Dialkyl acetals of DMF are generally prepared by transacetalization of the dimethyl or diethyl acetal, but this reaction is not useful in the case of the di-*t*-butyl acetal. Replacement of one methoxy group of DMF dimethyl acetal occurs on refluxing in *t*-butyl alcohol; conversion to the di-*t*-butyl acetal is effected in the presence of 2,4,6-tri-*t*-butylphenol (equation I).

$$(I) \quad (CH_3)_2NCH(OCH_3)_2 \xrightarrow[\Delta]{(CH_3)_3COH,} \left[(CH_3)_2NCH{<}^{OCH_3}_{OC(CH_3)_3} \right] \xrightarrow[50-60\%]{H^+, \Delta}$$

$$(CH_3)_2NCH[OC(CH_3)_3]_2$$
b.p. 53–54°/3 mm.

[1] E. Mohacsi, *Syn. Comm.*, **15**, 723 (1985).

Dimethyl(methylene)ammonium chloride.

Preparation[1]:

$$HCHO + HN(CH_3)_2 \xrightarrow{C_2H_5OH} C_2H_5OCH_2N(CH_3)_2 \xrightarrow[CH_3CN]{CH_3SiCl_3,}$$

$$H_2C{=}\overset{+}{N}(CH_3)_2 \ Cl^- + CH_3\overset{Cl}{\underset{Cl}{\overset{|}{\underset{|}{C}}}}{-}OC_2H_5$$
98%

[1] C. Rochin, O. Babot, J. Dunogues, and F. Doboudin, *Synthesis*, 228 (1986).

Dimethyl(methylthio)sulfonium tetrafluoroborate (1).

Intramolecular oxysulfenylation.[1]   Intramolecular oxysulfenylation (**11**, 205) of γ,δ-unsaturated alcohols or acids can be used for preparation of cyclic ethers or lactones, respectively. A base is not essential, but optimal yields are obtained in the presence of diisopropylethylamine (1.1 equiv.). Formation of five-membered rings is favored over formation of six-membered rings. The reaction is carried out at 25° and requires 1–3 days.

Example:

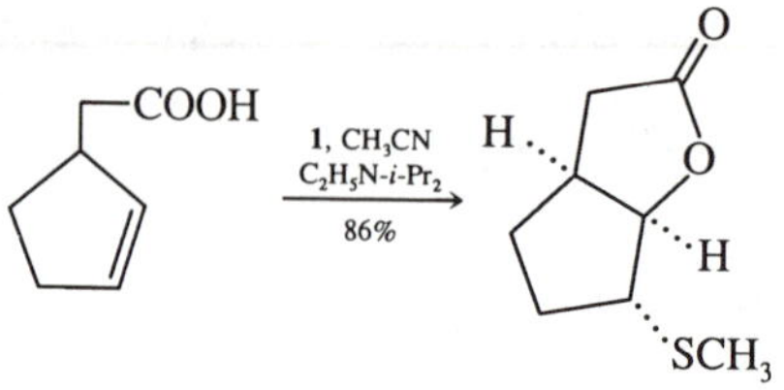

Addition of allyltins to thioketals.[2]   The reagent can convert dimethyl thio-ketals to a thionium ion, $>=\overset{+}{S}CH_3$, which can react with a nucleophile such as an allyltin.
Example:

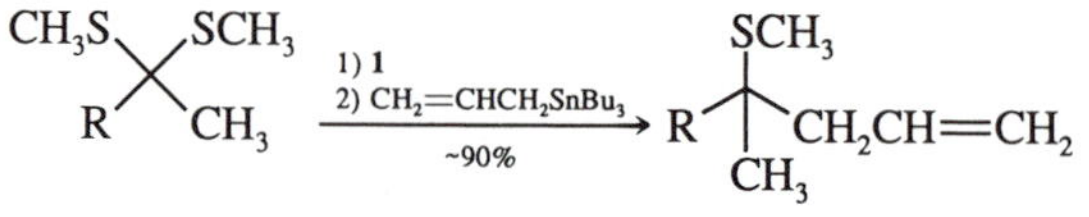

This reaction has been used for the macrocyclization of **2** to **3**.

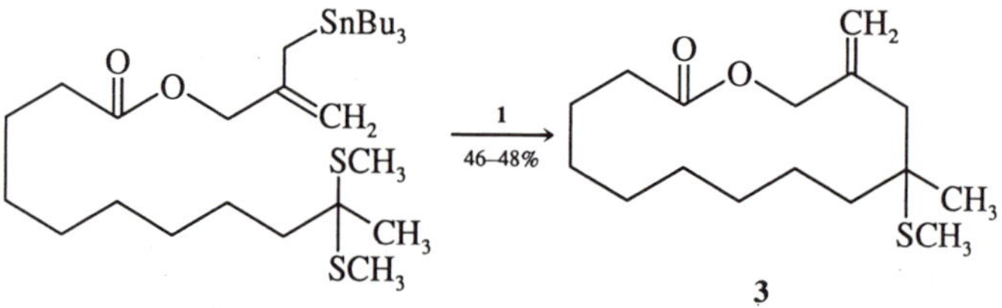

[1] G. J. O'Malley and M. P. Cava, *Tetrahedron Letters*, **26**, 6159 (1985).
[2] B. M. Trost and T. Sato, *Am. Soc.*, **107**, 719 (1985).

1,3-Dimethyl-2-oxohexahydropyrimidine (1,3-Dimethylpropyleneurea, DMPU).

Dipolar aprotic solvent.   Sandoz AG (Basel) reports that two investigations have shown no evidence of mutagenic activity of DMPU. It is recommended as a substitute for HMPT wherever possible.[1] It can generally replace HMPT in reactions of organolithium reagents.[2]

[1] *Chimia*, **39**, 148 (1985).
[2] M. Eyer and D. Seebach, *Am. Soc.*, **107**, 3601 (1985).

2,4-Dimethylpentane-2,4-diol chromate(VI) diester (1).

The diol is prepared by reaction of diacetone alcohol with 2 equiv. of CH_3Li (ether, $-70-0°$), 97% crude yield. The chromate ester (**1**) is obtained by reaction of the diol in CCl_4 with CrO_3; after about 10 minutes P_2O_5 is added and then the clear solution is stored for use.

1

Oxidation of sec-alcohols.[1]   In the presence of CH_3CO_3H, this chromium(VI) reagent can be used in catalytic amount (2 mole %) for oxidation of *sec*-alcohols to ketones in high yield. Under these conditions cholesterol is oxidized to Δ^5-3-cholestanone (84% yield). Primary alcohols can be oxidized to aldehydes (~80% yield). *t*-Butyl hydroperoxide or anhydrous H_2O_2 cannot replace the peracetic acid as the reoxidant.

[1] E. J. Corey, E.-P. Barrette, and P. A. Magriotis, *Tetrahedron Letters*, **26**, 5855 (1985).

Dimethylphenylsilane, $C_6H_5(CH_3)_2SiH$.

Stereoselective reduction of β-keto amides.[1]   α-Methyl-β-keto amides are reduced by this silane in combination with tris(diethylamino)sulfonium difluorotrimethylsilicate (**10**, 452–453) in DMPU to the *anti*-alcohol, but are reduced by this silane in TFA to the *syn*-alcohol.
  Example:

anti/syn = 98:2

syn/anti = 99:1

Reduction of oximes.[2]   Protected derivatives of oximes can be reduced by this silane in combination with an acid to substituted hydroxylamines.
  Examples:

$$C_6H_5CH{=}NOBzl \xrightarrow[75\%]{\substack{C_6H_5Me_2SiH,\\ TFA}} C_6H_5CH_2NHOBzl$$

(*syn/anti* = 99:1)

[1] M. Fujita and T. Hiyama, *Am. Soc.*, **107**, 8294 (1985); *Org. Syn.*, submitted (1986).
[2] M. Fujita, H. Oishi, and T. Hiyama, *Chem. Letters*, 837 (1986).

Dimethylsulfonium methylide. The reagent can be generated conveniently by re-action of NaOH with trimethylsulfonium chloride rather than NaH, required for generation from trimethylsulfonium iodide, the usual precursor.[1] An added advantage is that the chloride can be prepared readily by reaction of dimethyl sulfide with methyl chloroformate in 73% yield.[2]

[1] M. Rosenberger, W. Jackson, and G. Saucy, *Helv.*, **63**, 1665 (1980).
[2] B. Byrne and L. M. Lafleur Lawter, *Tetrahedron Letters*, **27**, 1233 (1986).

Dimethyl sulfoxide.

Dehydrostannation; α,β-enones.[1]   A novel route to α-methylene ketones involves cleavage of 1-alkyl-1-silyloxycyclopropanes with $SnCl_4$ in CH_2Cl_2 at 15° to provide β-trichlorostannyl ketones in 70–90% yield. These products undergo dehydrostannation when treated with excess DMSO in $CHCl_3$ at 60° to furnish α-methylene ketones in usually good yield. Pyridine and DMF are less useful than DMSO.

Example:

The paper reports application of this sequence to a 1-alkyl-2-silyloxycyclopropane (equation I) to furnish an α,β-enal.

[1] I. Ryu, S. Murai, and N. Sonoda, *J. Org.*, **51**, 2389 (1986).

Dimethyl tartrate.

Asymmetric bromination of ketals of alkyl aryl ketones.[1]   α-Bromination of the ketals of alkyl aryl ketones obtained with the dimethyl esters of (2R,3R)- or (2S,3S)-tartaric acid is highly diastereoselective. Hydrolysis of the products provides optically active 2-bromo ketones.

Example:

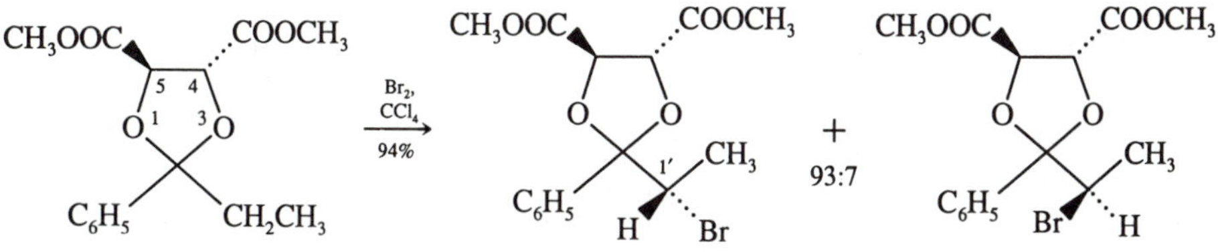

(4R, 5R)   →   (1′S)   +   (1′R)

Br₂, CCl₄
94%

93:7

¹ G. Castaldi, S. Cavicchioli, C. Giordano, and F. Uggeri, *Angew. Chem. Int. Ed.*, **25**, 259 (1986).

Diphenyl diselenide.

Alkenes from organomercurials.[1] Photoinitiated reaction of diphenyl dise-lenide with the organomercurial **1** provides a mixture of the corresponding α- and β-phenyl selenides, which undergo oxidative elimination to **2**. The reaction provides a key step in a total synthesis of K-76 (**3**), which counteracts the inflammatory

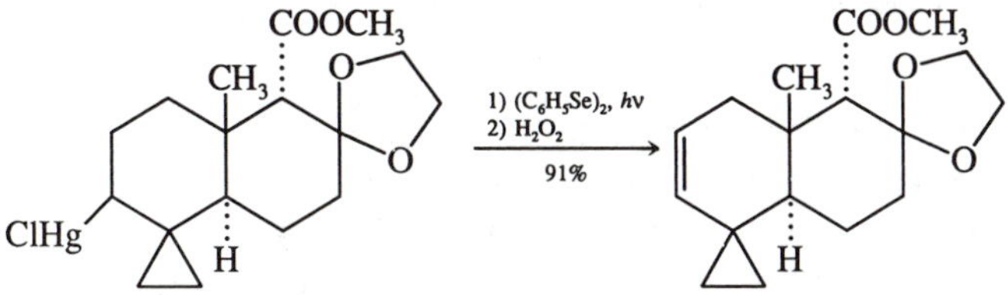

1) (C₆H₅Se)₂, hv
2) H₂O₂
91%

1   **2**

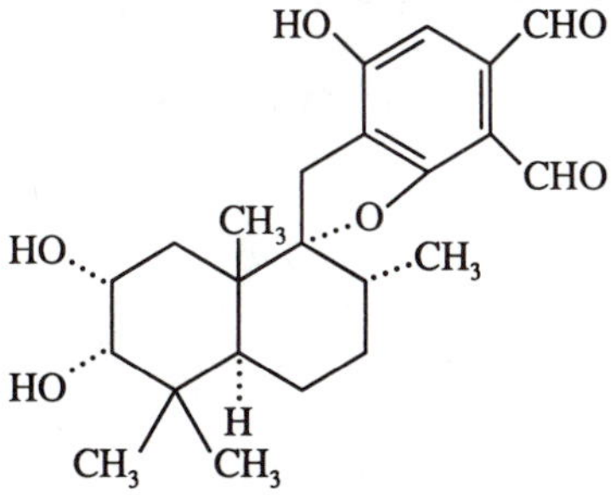

3

response to invading microorganisims. The cyclopropyl group was used in the synthesis in place of a *gem*-dimethyl group because this group does not prevent subsequent hydroxylation of the double bond with OsO_4.

¹ J. E. McMurry and M. D. Erion, *Am. Soc.*, **107**, 2712 (1985); M. D. Erion and J. E. McMurry, *Tetrahedron Letters*, **26**, 559 (1985).

Diphenyl-2-(1,3-dithianyl)phosphine oxide,

The reagent is prepared from 1,3-dithiane, BuLi, and $ClP(C_6H_5)_2$.

Ketene dithioketals.[1] The reagent undergoes Wittig-Horner reactions with aldehydes or ketones to give 2-alkylidene-1,3-dithianes in 80–100% yield. The corresponding P=S analog reacts with aldehydes in only moderate yield (25–65% yield).

[1] E. Juaristi, B. Gordillo, and L. Valle, *Tetrahedron*, **42**, 1963 (1986).

(Diphenylphosphine)lithium.

Deoxygenation of α-hydroxy ketones.[1] These ketones undergo dehydroxyl-ation when treated in THF with $(C_6H_5)_2PLi$ (excess) and, after a suitable time, acetic acid and CH_3I (1 equiv.) Benzoins are reduced in higher yield (75–85%) than acyloins (50–60%).

[1] A. Leone-Bay, *J. Org.*, **51**, 2378 (1986).

Diphenylsilane–Palladium(II) chloride–Triphenylphosphine.

Conjugate reduction of α,β-enals and -enones.[1] A catalyst composed of $PdCl_2$ and $P(C_6H_5)_3$ in the ratio 1:4 with a Lewis acid cocatalyst, $ZnCl_2$, effects this conjugate reduction with a hydridosilane, particularly $(C_6H_5)_2SiH_2$, in >90% yield. Almost any Pd(0) or Pd(II) catalyst in combination with triphenylphosphine can be used. Commercial $ZnCl_2$ is satisfactory, since small amounts of water are not harmful. Chloroform is the solvent of choice. α,β-Unsaturated esters, amides, and nitriles are not reduced. Saturated carbonyl groups do not undergo 1,2-reduction.

Example:

[1] E. Keinan and N. Greenspoon, *Tetrahedron Letters*, **26**, 1353 (1985); *idem, Org. Syn.*, submitted (1986).

Diphenylsilane–Tetrakis(triphenylphosphine)palladium(0)–Zinc chloride, $(C_6H_5)_2$-SiH_2—Pd(0)–$ZnCl_2$.

Conjugate reduction.   Silicon hydrides and a Pd(0) catalyst reductively cleave allylic acetates selectively (equation I).[1]

(I)

This system when catalyzed by $ZnCl_2$ can effect conjugate reduction of α,β-enones and -enals.[2]

[1] E. Keinan and N. Greenspoon, *J. Org.*, **48**, 3545 (1983).
[2] *Idem, Tetrahedron Letters*, **26**, 1353 (1985).

Diphosphorus tetraiodide.
$ArOCH(R^1)OR^2 \rightarrow ArOH$.[1]   Alkoxymethyl aryl ethers are cleaved by P_2I_4 in CH_2Cl_2 at 25° to phenols in 55–90% yield. This reagent is more useful for cleavage of SEM, MOM, and MEM ethers than Bu_4NF, BBr_3, or tris(diethylamino)sulfonium difluorotrimethylsilicate (TASF).

Aziridines.   2-Amino alcohols are converted into aziridines on reaction with P_2I_4 (equation I).[2]

(I)

[1] H. Saimoto, Y. Kusano, and T. Hiyama, *Tetrahedron Letters*, **27**, 1607 (1986).
[2] H. Suzuki and H. Tani, *Chem. Letters*, 2129 (1984).

Di-2-pyridyl thionocarbonate,

(**1**), m.p. 98–100°.

The pale-yellow reagent is prepared by reaction of thiophosgene with 2-hydroxypyridine and triethylamine in CH_2Cl_2 (85% yield).

Isothiocyanates and carbodiimides.[1]   The reagent converts alkyl and primary amines into isothiocyanates (equation I).

$$\text{(I)} \quad RNH_2 + \mathbf{1} \xrightarrow[85-95\%]{\overset{CH_2Cl_2}{25°}} RN{=}C{=}S + 2 \text{ PyOH}$$

Carbodiimides are obtained by reaction of **1** and DMAP with some N,N'-

disubstituted thioureas. The reaction fails or is slow with most N,N'-dialkylthio-ureas, but proceeds in high yield when one substituent is an aryl group.

Example:

$$C_6H_5HN-\overset{\overset{\displaystyle S}{\|}}{C}-NHC_6H_{11} \xrightarrow[87\%]{\substack{1,\ \text{DMAP,} \\ CH_3CN,\ 25°}} C_6H_5N{=}C{=}NC_6H_{11} + CS_2 + 2PyOH$$

[1] S. Kim and K. Y. Yi, *Tetrahedron Letters*, **26**, 1661 (1985).

E

OC_2H_5

(S)-2-Ethoxy-4-isopropyloxazoline, $CH(CH_3)_2$ **(1).** The reagent is obtained by alkylation of the corresponding oxazolidone (**11**, 379) with Meerwein's reagent.

Asymmetric alkylation of amines.[1] Reaction of **1** with tetrahydroisoquinoline affords the aminooxazoline **2**, which is alkylated with high asymmetric induction

at $-100°$. The diastereoselectivity decreases to only 60% at $-78°$. The oxazoline group is removed in high yield by hydrazinolysis.

[1] R. E. Gawley, G. Hart, M. Goicoechea-Pappas, and A. L. Smith, *J. Org.*, **51**, 3076 (1986).

3-Ethylbenzothiazolium bromide, [structure] Br⁻ **(1)**, m.p. 156–157°. The

salt is prepared by reaction of 1,3-benzothiazole with ethyl bromide at 70–80°; yield 47%.

Acyloin condensations. Formaldehyde in the presence of **1** undergoes self-condensation to form dihydroxyacetone (a triose) in high yield rather than the expected glycolaldehyde.[1] Surprisingly, the condensation of formaldehyde and another aldehyde catalyzed by **1** in the presence of $N(C_2H_5)_3$ results almost exclusively in a 1-hydroxy-2-one, $RCOCH_2OH$. Sodium cyanide, a known catalyst for benzoin condensations, is not effective for the condensation of formaldehyde and benzaldehyde in the presence of $N(C_2H_5)_3$.[2]

Example:

$$CH_3(CH_2)_2CHO + HCHO \xrightarrow[73\%]{\substack{1,\ N(C_2H_5)_3, \\ C_2H_5OH}} CH_3(CH_2)_2\overset{\displaystyle O}{\overset{\displaystyle \|}{C}}CH_2OH$$

[1] T. Matsumoto, and S. Inoue, *J.C.S. Chem. Comm.*, 171 (1983); T. Matsumoto, H. Yamamoto, and S. Inoue, *Am. Soc.*, **106**, 4829 (1984).

[2] T. Matsumoto, M. Ohishi, and S. Inoue, *J. Org.*, **50**, 603 (1985).

Ethyl (Z)-3-bromoacrylate, $BrCH{=}CHCOOC_2H_5$ **(1).**

Ethyl p-hydroxybenzoates.[1] The dianions (LDA) of 1,3-diketones react with **1** to give ethyl 5,7-diketo-3-octenoates in 40–55% yield by an addition–elimination process. The products on reaction with dilute sodium ethoxide undergo dehydrocyclization to afford ethyl p-hydroxybenzoates in 60–80% yield.

Examples:

[1] A. B. Smith, III, and S. N. Kilényi, *Tetrahedron Letters*, **26**, 4419 (1985).

Ethyl 3,3-diethoxyacrylate, $(C_2H_5O)_2C$=$CHCOOC_2H_5$ **(1).** Preparation.[1]

Ortho ester Claisen rearrangement.[2]   The [3,3]sigmatropic rearrangement of allyl vinyl alcohols prepared from this ester results in a regiospecific synthesis of diethyl allylmalonates.

Example:

[1] S. A. Glickman and A. C. Cope, *Am. Soc.*, **67**, 1017 (1945).
[2] S. Raucher, K.-W. Chi, and D. S. Jones, *Tetrahedron Letters*, **26**, 6261 (1985).

Ethyl N-(diphenylmethylene)-2-acetoxyglycinate (1), m.p. 62–65°.
Preparation:[1]

Amino acid synthesis.[2]   This glycine cation equivalent reacts with organocopper reagents, particularly $R_2Cu(CN)Li_2$, to form Schiff bases of amino acids.

Example:

[1] M. J. O'Donnell, W. D. Bennett, and R. L. Polt, *Tetrahedron Letters*, **26**, 695 (1985).
[2] M. J. O'Donnell and J.-B. Falmagne, *ibid.*, **26**, 699 (1985).

Ethyl diphenylphosphinite, $(C_6H_5)_2POC_2H_5$ **(1).** Supplier: Aldrich.

Peptides.   A new amide or peptide synthesis is based on the formation of iminophosphoranes, $R'N$=PR_3, from the reaction of azides with a tertiary phosphine. These phosphoranes react with carboxylic acids to form amides.[1] Ethyl diphenylphosphinite is more useful than a triarylphosphine because the by-product is hydrolyzed to diphenylphosphinic acid, which can be readily removed. The iminophosphorane **2**, derived from **1** and ethyl azidoacetate, reacts with CboGly-L-Phe-OH to give optically pure **3** in 70% yield.[2]

$$\underset{\underset{}{}}{C_2H_5OOCCH_2N_3} \xrightarrow{\overset{1}{CH_3COOC_2H_5}} \underset{2}{C_2H_5OOCCH_2N=P(C_6H_5)_2OC_2H_5} \xrightarrow[\underset{\text{overall}}{70\%}]{CboGly\text{-}L\text{-}Phe\text{-}OH}$$

$$\underset{3}{CboGly\text{-}L\text{-}Phe\text{-}GlyOC_2H_5}$$

[1] J. Garcia, F. Urpi, and J. Vilarrasa, *Tetrahedron Letters*, **25**, 4841 (1984).
[2] J. Zaloom, M. Calandra, and D. C. Roberts, *J. Org.*, **50**, 2601 (1985).

F

Ferric chloride.

Oxidative dimerization of allylic sulfones.[1] This oxidant is the reagent of choice for coupling of allylic sulfones to 1,6-disulfones by [3-3']coupling. Iodine effects coupling to 1,2-disulfones.

Example:

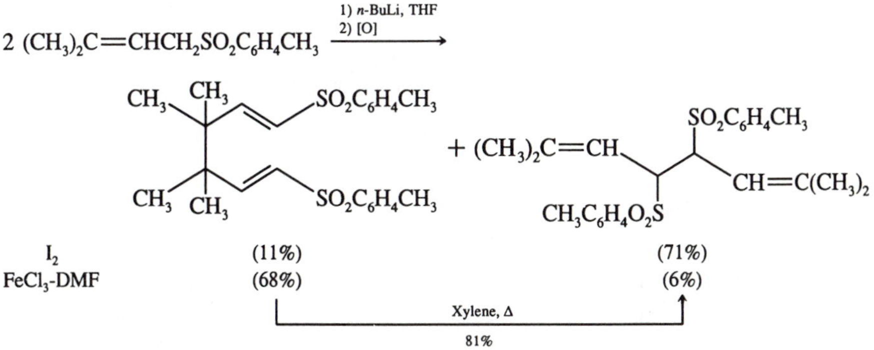

$$2\ (CH_3)_2C\!\!=\!\!CHCH_2SO_2C_6H_4CH_3 \xrightarrow[\text{2) [O]}]{\text{1) }n\text{-BuLi, THF}}$$

$$
\begin{array}{ccc}
 & \text{(11\%)} & \text{(71\%)} \\
\text{I}_2 & \text{(68\%)} & \text{(6\%)} \\
\text{FeCl}_3\text{-DMF} & & \\
\end{array}
$$

Xylene, Δ

81%

The 1,6-disulfones undergo Cope rearrangement to the 1,2-disulfones, which are obtained as single diastereomers.

Si-directed Nazarov cyclization.[2] This reaction proceeds with satisfactory stereoselectivity in the case of cyclohexenyl systems; only *cis* ring-fused products **1** and **2** are formed. The stereoselectivity is significantly influenced by the bulk of the R group in the six-membered ring (equation I).

Examples:

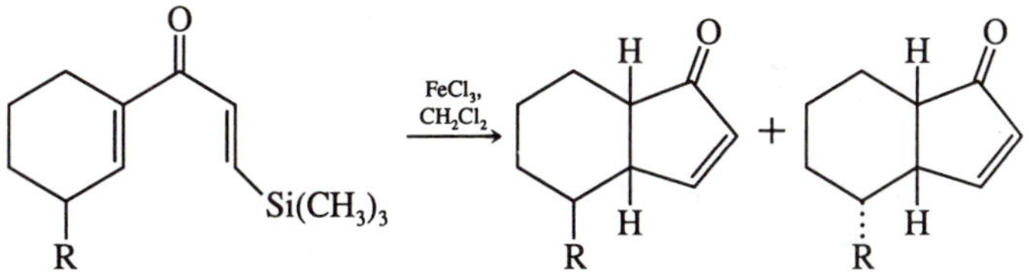

		1	**2**
R = CH$_3$	99%	78%	22%
R = C(CH$_3$)$_3$	63%	94%	6%

133

In contrast, stereocontrol by an R-substituent in cyclopentenyl systems is modest; but increasing the bulk on silicon favors formation of *cis*-isomers.

[1] G. Büchi and R. M. Freidinger, *Tetrahedron Letters*, **26**, 5923 (1985).
[2] S. E. Denmark, K. L. Habermas, G. A. Hite, and T. K. Jones, *Tetrahedron*, **42**, 2821 (1986).

Ferric chloride–Silica, $FeCl_3/SiO_2$. The pale, green powder can be prepared by stirring a mixture of SiO_2 with anhydrous $FeCl_3$ (8% of the weight of SiO_2).

   Ring enlargement of cyclobutanols.[1]   The reagent effects dehydration of secondary and tertiary alcohols. Dehydration of the tertiary cyclobutanol (**1**) results in dehydration and ring enlargement to give isolaurene (**2**) in quantitative yield.

[1] A. Fadel and J. Salaün, *Tetrahedron*, **41**, 413 (1985).

Ferric nitrate/K10 Bentonite (Clayfen). $\rangle C{=}S \to \rangle C{=}O$.[1] This transformation is possible with this reagent, but yields are high (60–100%) only with diaryl ketones.

[1] S. Chalais, A. Cornelis, P. Laszlo, and A. Mathy, *Tetrahedron Letters*, **26**, 2327 (1985).

9-Fluorenylmethyl pentafluorophenyl carbonate.

(**1**), m.p. 84–86°

The reagent is prepared by reaction of 9-fluorenylmethyl chloroformate (**3**, 145, **4**, 237, **8**, 230) with pentafluorophenol and triethylamine in ether (86% yield).

   Protection of amines.[1]   The reagent can be used to prepare N-fluorenylmethoxycarbonyl (Fmoc) amino acids (70–99% yield).

Example:

$$\underset{H_2NCHCOOH}{\overset{CH_3}{|}} + 1 \xrightarrow[93\%]{\underset{(CH_3)_2C=O}{NaHCO_3,}} \underset{FmocN-CHCOOH}{\overset{H \quad CH_3}{|\quad\;|}} + HOC_6F_5$$

[1] I. Schön and L. Kisfaludy, *Synthesis*, 303 (1986).

Fluorine.

α-Fluoro aldehydes and ketones.[1] Reaction of silyl enol ethers with 5% F_2 in N_2 in $FCCl_3$[2] at $-78°$ provides α-fluoro aldehydes or ketones in 60–80% yield. The by-product is the volatile $(CH_3)_3SiF$. The advantage of this fluorination is that glass vessels and tubing can be used. Trifluoromethyl hypofluorite (**10**, 420–421) has been used for this purpose, but this reagent is a toxic gas.

[1] S. T. Purrington, N. V. Lazaridis, and C. L. Bumgardner, *Tetrahedron Letters*, **27**, 2715 (1986).
[2] Available from Air Products Co.

Fluorine–Acetonitrile.

Epoxidation. Alkenes are epoxidized when treated with a reagent formed from dilute F_2 (10% in N_2) in acetonitrile containing some water. Both CH_3CN and H_2O are essential for formation of the oxidant, which is probably not peroxyacetimidic acid (**1**, 470; **8**, 387). The new reagent effects epoxidation of alkenes rapidly, even at $-15°$. When used in excess it also effects epoxidation of α,β-enones in high yield.[1]

Example:

$$C_6H_5CH=CHCOCH_3 \xrightarrow[85\%]{\underset{CH_3CN}{F_2, H_2O,}} \underset{C_6H_5CH-CHCOCH_3}{\overset{O}{\diagup\;\diagdown}}$$

[1] S. Rozen and M. Brand, *Angew. Chem. Int. Ed.*, **25**, 554 (1986).

Fluoromethyl phenyl sulfone, $C_6H_5SO_2CH_2F$ (**1**).

Preparation:

$$\underset{C_6H_5SCH_3}{\overset{O}{\overset{\|}{}}} \xrightarrow[85\%]{DAST} C_6H_5SCH_2F \xrightarrow[93\%]{m\text{-}ClC_6H_4CO_3H} 1$$

Vinyl fluorides.[1] Fluoro(phenylsulfonyl)methyllithium (**2**), obtained by reaction of **1** with BuLi in THF at $-78°$, reacts with carbonyl compounds to form fluorohydrins (**3**), which are dehydrated to (E)-α-fluoro-α,β-unsaturated sulfones (**4**) in 67–92% yield with CH_3SO_2Cl or orthophosphoric acid. Conversion of **4** into vinyl fluorides (**5**) is effected with aluminum amalgam (~90% yield). The vinyl fluorides are obtained as a 1:1 mixture of (E)- and (Z)-isomers.

Example:

$$\underset{\textbf{2}}{C_6H_5SO_2\overset{\overset{\displaystyle Li}{|}}{C}HF} + C_6H_5CHO \longrightarrow \underset{\textbf{3}}{C_6H_5\overset{\overset{\displaystyle OH}{|}}{C}H\underset{\underset{\displaystyle F}{|}}{C}HSO_2C_6H_5} \xrightarrow[\substack{67\% \\ \text{overall}}]{\substack{CH_3SO_2Cl, \\ N(C_2H_5)_3}}$$

$$\underset{\textbf{4}}{C_6H_5CH=C\overset{\displaystyle F}{\underset{\displaystyle SO_2C_6H_5}{\big\langle}}} \xrightarrow[90\%]{Al/Hg} \underset{\textbf{5 (E/Z = 1:1)}}{C_6H_5CH=CHF}$$

[1] M. Inbasekaran, N. P. Peet, J. R. McCarthy, and M. E. LeTourneau, *J.C.S. Chem. Comm.*, 678 (1985).

Formaldehyde.

α-Methylene ketones. The hydroxymethylation of allyl β-keto carboxylates (**1**) with HCHO and K_2CO_3 as base proceeds in almost quantitative yield. Treatment of the corresponding acetates (**2**) with $Pd_2(dba)_3 \cdot CHCl_3—P(C_6H_5)_3$ induces decarboxylation–deacetoxylation to provide α-methylene ketones (**3**) and allyl acetate.[1]

Example:

$$\underset{\textbf{1}}{CH_3\overset{\overset{\displaystyle O}{\|}}{C}\underset{\underset{\displaystyle CO_2CH_2CH=CH_2}{|}}{C}H(CH_2)_4CH_3} \xrightarrow[\text{2) Ac}_2\text{O, Py}]{\text{1) HCHO, K}_2\text{CO}_3} \underset{\textbf{2}}{CH_3\overset{\overset{\displaystyle O}{\|}}{C}—\overset{\overset{\displaystyle CH_2OAc}{|}}{\underset{\underset{\displaystyle CO_2CH_2CH=CH_2}{|}}{C}}—(CH_2)_4CH_3}$$

$$\substack{88\% \\ \text{overall}} \Big\downarrow \substack{Pd_2(dba)_3 \cdot CHCl_3 \\ P(C_6H_5)_3, CH_3CN, 20°}$$

$$\underset{\textbf{3}}{CH_3\overset{\overset{\displaystyle O}{\|}}{C}\underset{\underset{\displaystyle CH_2}{\|}}{C}(CH_2)_4CH_3}$$

[1] J. Tsuji, M. Nisar, and I. Minami, *Tetrahedron Letters*, **27**, 2483 (1986).

Formic acid.

Detritylation.[1]   Trityl ethers are cleaved by reaction with formic acid in ether in high yield in 7–45 minutes. Under these conditions isopropylidene and benzylidene acetals and *t*-butyldimethylsilyl ethers are not affected, but tetrahydropyranyl ethers are partially cleaved.

[1] M. Bessodes, D. Komiotis, and K. Antonakis, *Tetrahedron Letters*, **27**, 579 (1986).

Formylmethyltriphenylarsonium bromide $[(C_6H_5)_3As^+CH_2CHO]Br^-$ (**1**). The salt (m.p. 160–161°) is prepared by reaction of $(C_6H_5)_3As$ and bromoacetaldehyde in CH_3CN (90% yield).

(E)-α,β-Enals.[1]   The ylide derived from **1** (K_2CO_3) reacts with aldehydes in THF/ether to give (E)-α,β-enals in 80–95% yield.

Example:

[1] Y. Huang, L. Shi, and J. Yang, *Tetrahedron Letters*, **26**, 6447 (1985).

G

Grignard reagents.

1,4-*Bis(bromomagnesio)pentane* (1); 1,5-*bis(bromomagnesio)hexane* (2). The reaction of α,ω-diprimary di-Grignard reagents with esters results in 1-(ω-hydroxy-alkyl)cycloalkanols; this reaction is particularly useful when cyclopentanols or cyclohexanols can be formed.[1] A similar reaction of the primary-secondary bis-Grignard reagent **1** with esters is highly stereoselective and results mainly in *trans*-2-methyl-1-substituted cyclopentanols. A similar stereoselectivity obtains in reactions of **2** with esters, but yields are lower.[2]

$$RCOOC_2H_5 + BrMg(CH_2)_3\overset{\underset{\textstyle |}{CH_3}}{C}HMgBr \longrightarrow$$

1

R = CH₃	77%	6:94
R = C₆H₅	78%	11:89

$$RCOOC_2H_5 + BrMg(CH_2)_4\overset{\underset{\textstyle |}{CH_3}}{C}HMgBr \longrightarrow$$

2

R = CH₃	53%	25:75
R = C₆H₅	54%	23:77

Acyloins.[3] Grignard reagents of the type R^1CH_2MgX react with the O-trimethylsilyl ethers of aldehyde cyanohydrins to form products that are hydrolyzed by acids to acyloins in 75–85% yield.

Example:

$$\underset{\underset{\textstyle OSi(CH_3)_3}{|}}{Pr}CHCN + ArCH_2MgCl \longrightarrow \underset{\underset{\textstyle OSi(CH_3)_3}{|}}{Pr}CH\overset{\overset{\textstyle NMgCl}{||}}{C}CH_2Ar \xrightarrow[\sim75\%]{H_3O^+} \underset{\underset{\textstyle OH}{|}}{Pr}CH\overset{\overset{\textstyle O}{||}}{C}CH_2Ar$$

Conjugate addition to nitroarenes.[4] Grignard reagents undergo facile conjugate addition to nitrobenzene (equation I). In fact only aldehydes can compete successfully with nitroarenes in this reaction.

Methallylmagnesium chloride (1).[5]   The reaction of **1** with the α-oxoketene dithioacetal **2** provides a 1,2-adduct **3** that on treatment with HBF₄ is converted into the aryl sulfide **4**.

This reaction provides a synthesis of *trans*-calamene (**7**) from the ketene dithioacetal (**5**) of *l*-menthone.

[1] P. Canonne, D. Bélanger, and G. Lemay, *Tetrahedron Letters*, **22**, 4995 (1981).
[2] P. Canonne and M. Bernatchez, *J. Org.*, **51**, 2147 (1986).
[3] M. Gill, M. J. Kiefel, and D. A. Lally, *Tetrahedron Letters*, **27**, 1933 (1986).
[4] G. Bartoli, M. Bosco, and R. Dalpozzo, *ibid.*, **26**, 115 (1985).
[5] R. K. Dieter and Y. J. Lin, *ibid.*, **26**, 39 (1985).

H

Hexamethyldisilazane.

Primary amides.[1] Acyl chlorides react with hexamethyldisilazane in CH_2Cl_2 at room temperature to give, after acid hydrolysis, primary amides in 60–90% yield from the carboxylic acid.

$$RCOOH \longrightarrow RCOCl \xrightarrow[\substack{60-90\%}]{\substack{1)\ HN[Si(CH_3)_3]_2,\ CH_2Cl_2 \\ 2)\ CH_3OH,\ H_2SO_4}} RCONH_2$$

[1] R. Pellegata, A. Italia, M. Villa, G. Palmisano, and G. Lesma, *Synthesis*, 517 (1985).

Hexamethyldisilazane–Chlorotrimethylsilane.

Monosilylation of anilines. The usual method of silylation of anilines, BuLi, $ClSi(CH_3)_3$, is inconvenient on a large scale and not widely applicable because of ready hydrolysis of some silylanilines. A new method involves reaction of the aniline with $HN[Si(CH_3)_3]_2$ (3 equiv.) and catalytic amounts of $ClSi(CH_3)_3$ and LiI. After 95% silylation, the $ISi(CH_3)_3$ is quenched with cyclohexene oxide and the product isolated by distillation. Yields are 80–97%.

The dilithium derivatives of 2-alkyl-N-trimethylsilylanilines condense with esters to give C_2-substituted indoles in generally satisfactory yield.[1]

Example:

This protocol was used for synthesis of cinchonamine (**1**) as outlined in equation (I).

1

[1] A. B. Smith III, M. Visnick, J. N. Haseltine, and P. A. Sprengeler, *Tetrahedron*, **42**, 2957 (1986).

Hexamethylditin, $(CH_3)_3Sn—Sn(CH_3)_3$ (**1**), m.p. 23°, b.p. 182°.

Vinyllithiums.[1]   In the presence of $Pd[P(C_6H_5)_3]_4$ and LiCl, vinyl triflates couple with **1** to form vinyltrimethyltin reagents in generally high yield and with retention of the stereochemistry. The vinyltrimethyltins react with CH_3Li to form the corresponding vinyllithiums and $Sn(CH_3)_4$. Since the kinetic and thermodynamic enol triflates of unsymmetrical ketones can be obtained selectively by reaction of lithium enolates with N-phenyl trifluoromethanesulfonimide (**12**, 395), this new coupling reaction provides a regioselective route to vinyllithiums.

[1] W. D. Wulff, G. A. Peterson, W. E. Bauta, K.-S. Chan, K. L. Faron, S. R. Gilbertson, R. W. Kaesler, D. C. Yang, and C. K. Murray, *J. Org.*, **51**, 277 (1986).

Hexamethylphosphoric triamide (HMPT).

Stereoselective Michael additions.   In the absence of strong steric effects, the stereochemistry of Michael addition of amide enolates depends on the enolate geometry, with (Z)-enolates giving mainly *anti*-adducts and (E)-enolates giving mainly *syn*-adducts.[1] Ester enolates show higher stereoselectivity than amide enolates, as shown by the (E)- and (Z)-enolates of *t*-butyl propionate (**1**). The (E)-

enolate is obtained by use of LDA in THF at $-78°$, and the (Z)-enolate is obtained by use of THF and HMPT (Ireland technique, **7**, 209–210).[2]

Example:

$$ t\text{-BuC}\!\!\!\overset{\displaystyle O}{\|}\!\!\!\diagdown\!\!\diagup\!\!\diagdown_{C_2H_5} \; + \; t\text{-BuOC}\!\!\overset{\displaystyle OLi}{|}\!\!=\!\!CHCH_3 \; \longrightarrow $$

1

E/Z = 12:88 THF, HMPT
E/Z = 94:6 THF

syn   anti

5:95
91:9

Regioselective alkylation of a methyl ketone.[3]   Even though the kinetic enolate of 2-heptanone consists of a mixture of terminal and internal enolates in the ratio 87:13, benzylation in DME results in preferential internal alkylation. Regioselective benzylation at the terminal position can be enhanced by addition of various ligands such as benzo-14-crown-4 and DMF, but HMPT is the most effective ligand, resulting in a ratio of terminal to internal benzylation of 11:1. The three ligands also increase the rate of alkylation. The same effect, but less marked, is observed in alkylation with the less reactive electrophile butyl iodide.

Cyclodehydration of 1,4- and 1,5-diols.[4]   These diols are converted into tetrahydrofurans and tetrahydropyrans, respectively, when heated with HMPT (0.3 equiv.).

Example:

$$ \xrightarrow[87\%]{HMPT,\ \Delta} $$

[1] C. H. Heathcock, M. A. Henderson, D. A. Oare, and M. A. Sanner, *J. Org.*, **50**, 3019 (1985).
[2] C. H. Heathcock and D. A. Oare, *ibid.*, **50**, 3022 (1985).
[3] C. L. Liotta and T. C. Caruso, *Tetrahedron Letters*, **26**, 1599 (1985).
[4] J. Diab, M. Abou-Assali, C. Gervais, and D. Anker, *ibid.*, **26**, 1501 (1985).

Hydrazine.

Reduction of nitro groups.[1]   Graphite catalyzes reduction of aromatic and aliphatic nitro compounds by hydrazine in refluxing ethanol (2 hours, 85–98% yield).[2]

[1] A. Furst, R. C. Berlo, and S. Hooton, *Chem. Rev.*, **65**, 51 (1965).
[2] B. H. Han, D. H. Shin, and S. Y. Cho, *Tetrahedron Letters*, **26**, 6233 (1985).

Hydridotetrakis(triphenylphosphine)rhodium(I).

Isomerization of monoepoxides of 1,3-dienes.[1]   In the presence of this catalyst, monoepoxides of a 1,3-diene isomerize in benzene at 105° to α,β-unsaturated aldehydes or ketones with high (E)-selectivity.

Example:

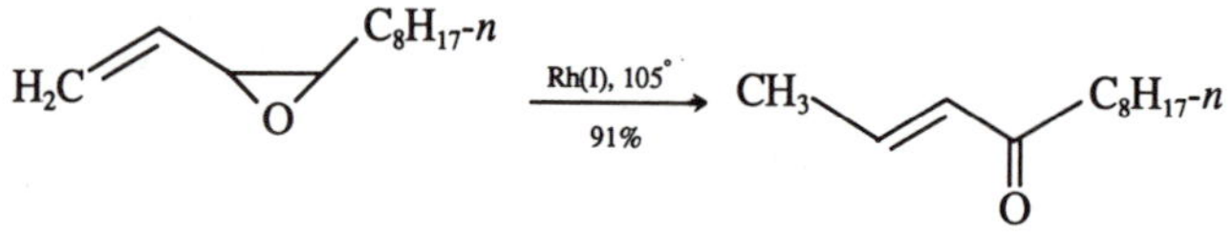

(E/Z = 10:1)

The reaction provides the key step in a synthesis of *ar*-turmerone (**2**), equation (I).

(I)

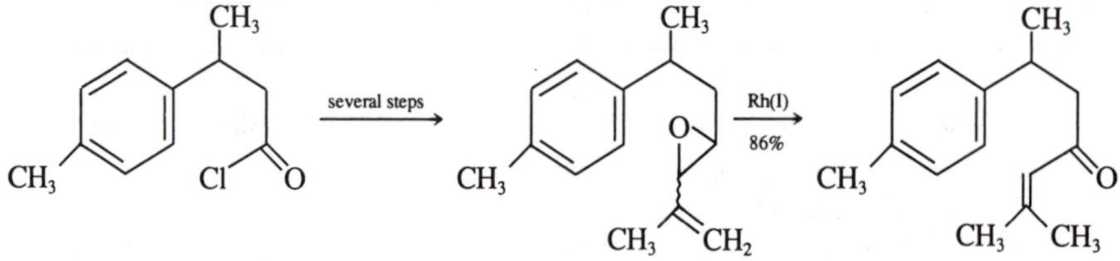

The α-alkylidene-γ-butyrolactone **3** is isomerized by this method to the butenolide **4**.

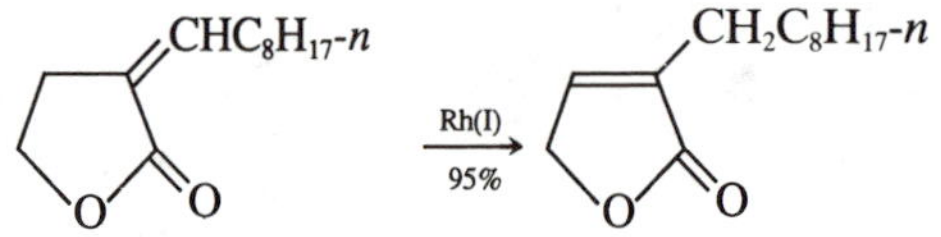

[1] S. Sato, I. Matsuda, and Y. Izumi, *Tetrahedron Letters*, **26**, 1527 (1985).

Hydrogen hexachloroplatinate(IV)–Copper(II) chloride, H_2PtCl_6–$CuCl_2$.

Oxygenation of alcohols.[1]   Primary and secondary alcohols are converted into the corresponding aldehydes or ketones by O_2, visible light, and H_2PtCl_6—$CuCl_2$ in the ratio 1:2. Yields are almost quantitative, but rates are reduced by steric factors.

[1] R. E. Cameron and A. B. Bocarsly, *Am. Soc.*, **107**, 6116 (1985).

Hydrogen peroxide.

Selenoxide and telluride elimination.[1]   A recent study of the oxidation of *sec*-alkyl phenyl selenides confirms that selenoxide elimination normally results in *trans*-alkenes with high selectivity regardless of the nature or amount of the oxidant (H_2O_2, $ClC_6H_4CO_3H$, $NaIO_4$). In contrast, oxidation of the corresponding *sec*-alkyl phenyl tellurides is much slower and the stereochemistry of the resulting alkene is dependent on the amount of oxidant employed. As the concentration of the oxidant is increased, the proportion of the *cis*-isomer is increased. In any case, the yield of alkene is lower than that obtained from the corresponding selenide. One possible reason for the differing behavior may be that telluroxides are easily oxidized further to tellurones.

[1] S. Uemura, Y. Hirai, K. Ohe, and N. Sugita, *J.C.S. Chem. Comm.*, 1037 (1985).

Hydrogen peroxide–Sodium tungstate.

Epoxidation of unsaturated alcohols.[1]   One advantage of H_2O_2 epoxidation catalyzed by Na_2WO_4 is that the reaction can be effected in water or water-alcohol mixtures. However, glycols are obtained unless the medium is buffered [NaOAc or $(CH_3)_3NO$] at about pH 4.5. Under these conditions, isolated double bonds react very slowly at 25°.

Allylic and *cis*-homoallylic alcohols are epoxidized readily, but *trans*-homoallylic and bishomoallylic alcohols react slowly, if at all. The stereoselectivity in the epoxidation of acyclic allylic alcohols is the same as and is comparable to that observed with *t*-BuOOH/VO(acac)$_2$. The stereoselectivity in epoxidation of acyclic homoallylic alcohols is also the same but lower than that obtained with *t*-BuOOH/VO(acac)$_2$. Epoxidation of cyclic allylic alcohols proceeds more slowly and in lower yield than that of acyclic allylic alcohols.

Cleavage of **vic**-*diols to carboxylic acids.*[2]   In the presence of catalytic amounts of Na_2WO_4·$2H_2O$ and H_3PO_4 and at a pH of 2, H_2O_2 (40%) oxidizes a variety of water-soluble 1,2-diols to carboxylic acids at 90° via an α-ketol. Addition of a phase-transfer catalyst permits oxidation of water-insoluble diols.

[1] D. Prat and R. Lett, *Tetrahedron Letters*, **27**, 707 (1986); D. Prat, B. Delpech, and R. Lett, *ibid.*, **27**, 711 (1986).
[2] C. Venturello and M. Ricci, *J. Org.*, **51**, 1599 (1986).

1-(2-Hydroxyethyl)-4,6-diphenylpyridine-2-thione,

(1)

The reagent is prepared in 67% overall yield from 4,6-diphenyl-2-pyrone by thionation (P_4S_{10}) followed by reaction with ethanolamine.

RX → RSH.[1]   The reagent converts primary iodides or benzyl bromides directly into thiols. Primary chlorides or bromides undergo the same reaction in acetonitrile in the presence of $(C_2H_5)_4NI$ (equation I).

(I)   RX + 1 $\xrightarrow[60-90\%]{}$ RSH +

[1] P. Molina, M. Alajarin, M. J. Vilaplana, and A. R. Katritzky, *Tetrahedron Letters*, **26**, 469 (1985).

(S)-2-(1-Hydroxy-1-methylethyl)pyrrolidine (1).

Asymmetric addition of Grignard reagents to α,β-unsaturated amides.[1]   In the presence of DBU, Grignard reagents undergo asymmetric conjugate addition to α,β-unsaturated amides **2** derived from **1** or (S)-prolinol in yields of 50–89% ee.

Optical yields are markedly reduced on additions to the methyl ether of **2** because of decreased chelation.

[1] K. Soai, H. Machida, and A. Ookawa, *J.C.S. Chem. Comm.*, 469 (1985).

$$CH_2$$
1-Hydroxy-3-trimethylsilylmethyl-3-butene, $(CH_3)_3SiCH_2CCH_2CH_2OH$ (**1**). The reagent is prepared by lithiation and silylation of 3-methyl-3-butene-1-ol (38% yield).

Indolizidines; quinolizidines.[1]   Mitsunobu condensation of **1** with succinimide provides the imide **2**, which can be converted as shown into the indolizidine (**4**). A similar sequence but starting with glutarimide provides the quinolizidine (**5**).

[1] J.-C. Gramain and R. Remuson, *Tetrahedron Letters*, **26**, 327 (1985).

I

Iodine.

Oxidative coupling of dianions (**12**, 278). The oxidative coupling of dianions of carboxylic acids has been used for synthesis of enterolactone (**6**), a lignan urinary metabolite.[1] Thus the dianion **2**, generated from **1** with LDA, when treated with I_2 (0.5 equiv.) couples to **3**, obtained as a mixture of two isomers in a 4.5:1 ratio. This product is converted in 89% overall yield to the dimethyl ether (**5**) of enterolactone (**6**).

Coupling of unsymmetrical dianions is also possible. Thus the acylsulfonamide dianion **7** couples with sodium iodoacetate to give **8** in high yield. On reduction and acidification, **8** is converted into the butyrolactone **9** in 75% yield.[2]

$$C_6H_5CH_2CONHSO_2C_6H_5 \xrightarrow[\text{THF, } -78-0°]{n\text{-BuLi}} \overset{Li^+ \quad Li^+}{C_6H_5\underset{|}{C}HCONSO_2C_6H_5} \xrightarrow[96\%]{ICH_2COONa}$$

7

8

$$\xrightarrow[\text{2) H}_3\text{O}^+]{\substack{\text{1) BH}_3 \cdot \text{S(CH}_3)_2 \\ \text{THF}}} \quad \underset{75\%}{} $$

9

Iodolactamization; γ-lactams.   γ,δ-Unsaturated amides can be converted into γ-lactams by treatment with trimethylsilyl triflate (triethylamine, DMAP) to give an imidate, which is cyclized by iodine to an iodo-γ-lactam in 50–70% yield (equation I). δ-Lactams can be prepared by this methodology, but yields are low because

(I)

of partial reversion to the unsaturated amide on chromatography. Some other lactams (and the yields) prepared are formulated.[3]

(58%, *trans/cis* = 11:1)   (63%)   (35%)

[1] J. L. Belletire and S. L. Fremont, *Tetrahedron Letters*, **27**, 127 (1986).
[2] J. L. Belletire and E. G. Spletzer, *ibid.*, **27**, 131 (1986).
[3] S. Knapp and K. E. Rodriques, *ibid.*, **26**, 1803 (1985).

Iodine–Mercury(II) oxide.

Ring expansion via the hypoiodite reaction.[1]   Irradiation of **1**, formed by photocycloaddition of β-naphthol with acrylonitrile, in the presence of HgO and iodine results in **2** in 40% yield.

Medium-sized lactones.[2] A new method for synthesis of lactones that is particularly useful in the case of medium sized ones involves the hypoiodite reaction with a catacondensed lactol such as **1**, which results in cleavage to the ten-membered isomeric iodo lactone **2**. Phoracantholide I (**3**) is obtained on reductive removal of iodine.

[1] H. Suginome, C. F. Liu, M. Tokuda, and A. Furusaki, *J.C.S. Perkin I*, 327 (1985).
[2] H. Suginome and S. Yamada, *Tetrahedron Letters*, **26**, 3715 (1985).

(Iodomethyl)trimethyltin, $(CH_3)_3SnCH_2I$.

Stereospecific Wittig rearrangement.[1] Wittig rearrangement of the optically active (Z)-allylic ether **1** (**9**, 475 for a related reaction) gives (R,E)-**2** with complete chirality transfer. Rearrangement of (E)-**1** results in two products, again with complete chirality transfer.[1]

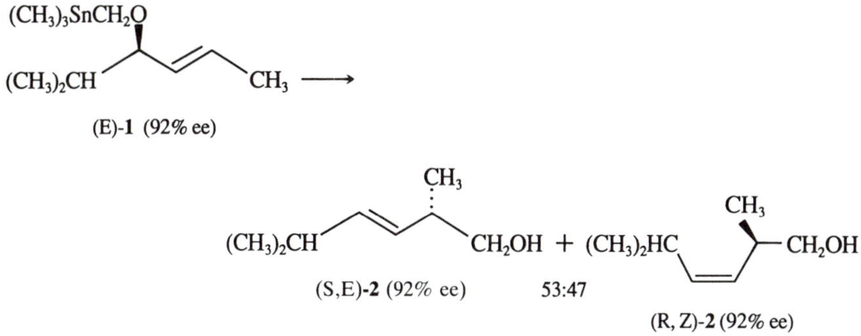

$(CH_3)_3SnCH_2O$

$(CH_3)_2CH$ — CH=CH — CH_3 ⟶

(E)-1 (92% ee)

CH_3

$(CH_3)_2CH$ — CH=CH — CH_2OH + $(CH_3)_2HC$ — C=C — CH_2OH (CH_3)

(S,E)-2 (92% ee)   53:47

(R, Z)-2 (92% ee)

This stereospecific rearrangement is a key step in a synthesis of optically active 26-hydroxycholesterol and 25,28-dihydroxy-7,8-dihydroergosterol.

[1] M. M. Midland and Y. C. Kwon, *Tetrahedron Letters*, **26**, 5013, 5017, 5021 (1985).

Iodosylbenzene.

Vinyliodonium tetrafluoroborates.[1]   These salts are obtained with retention of configuration by reaction of vinylsilanes with iodosylbenzene activated with triethyloxonium tetrafluoroborate (equation I).

(I)

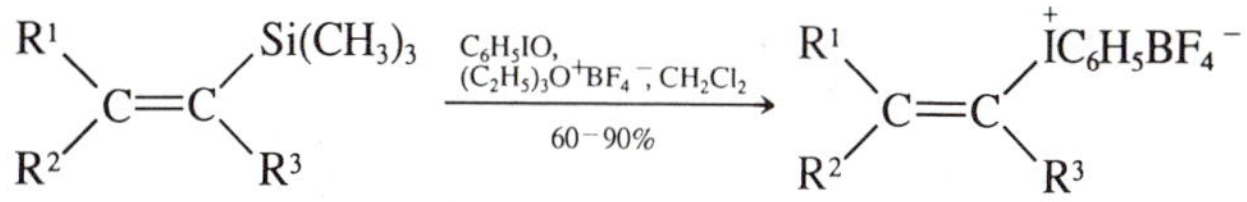

R^1 R^2 C=C $Si(CH_3)_3$ R^3 $\xrightarrow[60-90\%]{\substack{C_6H_5IO, \\ (C_2H_5)_3O^+BF_4^-, CH_2Cl_2}}$ R^1 R^2 C=C $\overset{+}{I}C_6H_5BF_4^-$ R^3

The salts react with various nucleophiles with loss of iodobenzene. Example:

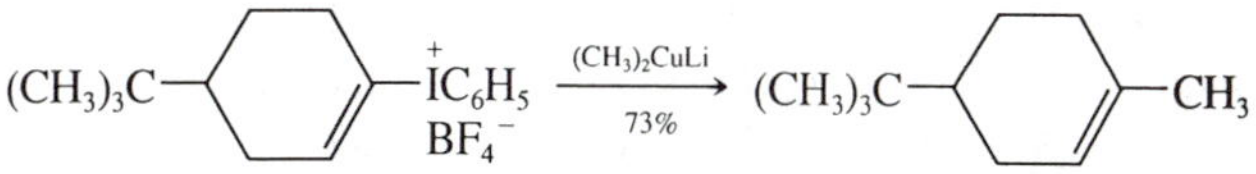

$(CH_3)_3C$ —⟨ ⟩— $\overset{+}{I}C_6H_5$ BF_4^- $\xrightarrow[73\%]{(CH_3)_2CuLi}$ $(CH_3)_3C$ —⟨ ⟩— CH_3

1,4-Diarylbutane-1,4-diones.[2]   Silyl enol ethers of aryl methyl ketones couple in the presence of $C_6H_5IO–BF_3$ etherate (1:3) in CH_2Cl_2 or ether to give 1,4-diarylbutane-1,4-diones in moderate yield.

[1] M. Ochiai, K. Sumi, Y. Nagao, and E. Fujita, *Tetrahedron Letters*, **26**, 2351 (1985).
[2] R. M. Moriarty, O. Prakash, and M. P. Duncan, *J.C.S. Chem. Comm.*, 420 (1985).

Ion-exchange resins.

Hydrolysis of acetals.[1]   Amberlyst-15 is an excellent catalyst for hydrolysis of acetals or ketals in aqueous acetone. It is particularly valuable for hydrolysis of acetals of readily epimerizable aldehydes or ketones.

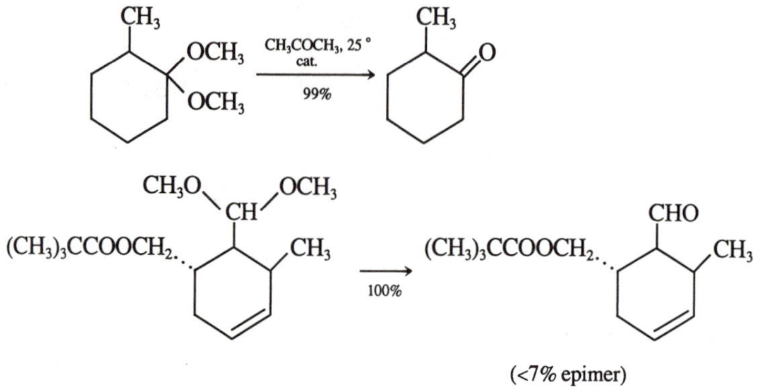

(<7% epimer)

[1] G. M. Coppola, *Synthesis*, 1021 (1984).

Iron carbonyl.

Indole synthesis.[1]   Indoles are formed by deoxygenation of *o*-nitrostyrenes with carbon monoxide catalyzed by $Fe(CO)_5$ to give a nitrene, which undergoes intramolecular cyclization. $Ru_3(CO)_{12}$ and $Rh_6(CO)_{16}$ show comparable selectivity.

Example:

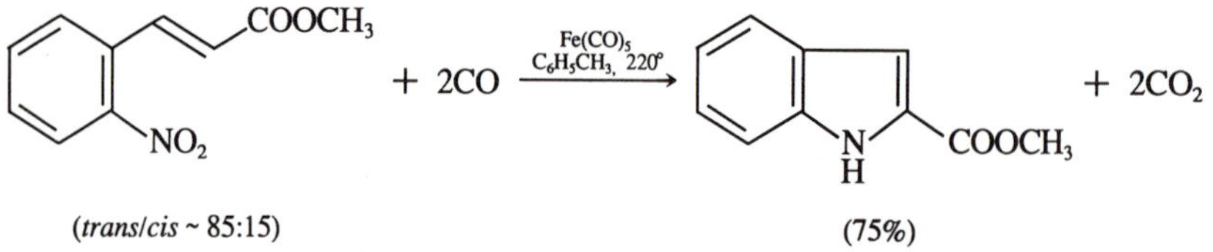

(*trans/cis* ~ 85:15)   (75%)

Dicarbonylation of strained bonds.[2]   The acyltetracarbonylferracycle (**2**), obtained on irradiation of quadricyclane (**1**) in the presence of excess $Fe(CO)_5$, is oxidized to the α-diketone **3** in 81% yield by $CuCl_2$ in aqueous CH_3CN.

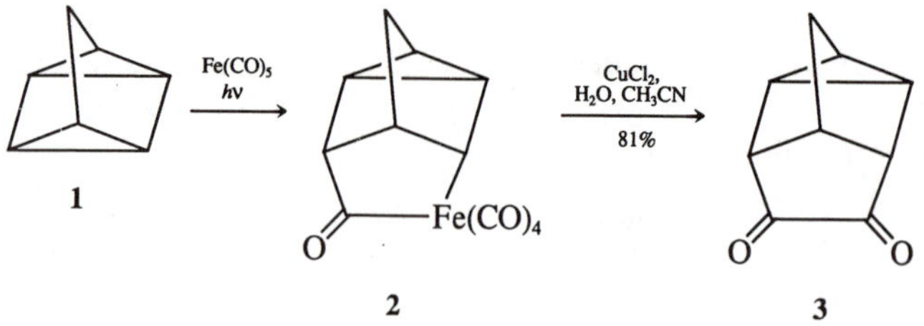

[1] C. Crotti, S. Cenini, B. Rindone, S. Tollari, and F. Demartin, *J.C.S. Chem. Comm.*, 784 (1986).
[2] R. Yamaguchi, S. Tokita, Y. Takeda, and M. Kawanisi, *ibid.*, 1285 (1985).

2,3-O-Isopropylidene-2,3-dihydroxy-1,4-bis(diphenylphosphine)butane (DIOP).

Asymmetric hydrosilylation.[1] Hydrosilylation of the 3,4-dihydro-(2*H*)-pyrroles (**1**) with $(C_6H_5)_2SiH_2$ catalyzed by $[Rh(COD)Cl]_2$ complexed with (+)- or (−)DIOP can proceed with as much as 64% ee. (S)-Enantiomers are formed preferably with (+)-DIOP; (R)-enantiomers are favored with (−)-DIOP.
Example:

[1] R. Becker, H. Brunner, S. Mahboobi, and W. Wiegrebe, *Angew. Chem. Intl. Ed.*, **24**, 995 (1985).

Isopropylsulfonyl chloride, $(CH_3)_2CHSO_2Cl$. Mol. wt. 142.61, b.p. 175°/19 mm. Supplier: Aldrich.

Dehydroxylation.[1] Deoxygenation of the mesylate or tosylate of the alcohol **1**, an intermediate in a synthesis of (+)-hirsutene (**2**), with $Li(C_2H_5)_3BH$ results

only in cleavage to the alcohol. The desired deoxygenation was effected by reaction with isopropylsulfonyl chloride (93% yield) and $Li(C_2H_5)_3BH$ reduction of the isopropylsulfonate (72% yield).

[1] D. H. Hua, G. Sinai-Zingde, and S. Venkataraman, *Am. Soc.*, **107**, 4088 (1985).

K

Ketenylidenetriphenylphosphorane (1).

 α,β-*Enones*.[1]   Grignard reagents react with this phosphorane to form an adduct that is hydrolyzed to an acyl ylide (**2**), which can be hydrolyzed to a methyl ketone (**3**) or allowed to react with an aldehyde to give an α,β-enone (**4**) (equation I).

$$(I) \quad (C_6H_5)_3P{=}C{=}C{=}O \xrightarrow[\text{60-95\%}]{\substack{R^1MgBr, \\ THF}} (C_6H_5)_3P{=}CHCOR^1 \xrightarrow{H_2O} CH_3COR^1$$

1   **2**   **3**

35-55% | R²CHO

$$\underset{H}{\overset{R^2}{>}}C{=}C\underset{COR^1}{\overset{H}{<}}$$

4

 α,β-*Unsaturated esters and lactones*.[2]   α,β-Unsaturated esters are obtained directly in 80–95% yield by reaction of an alcohol, an aldehyde, and **1** in refluxing toluene (equation II).

$$(II) \quad \textbf{1} \; + \; ROH \longrightarrow [(C_6H_5)_3P{=}CHCOOR] \xrightarrow[\text{94\%}]{BuCHO} \underset{H}{\overset{Bu}{>}}C{=}C\underset{COOR}{\overset{H}{<}}$$

Generation of the α,β-unsaturated ester in the presence of a diene results in the Diels-Alder adduct.

Example:

$$BuOH + BuCHO + \textbf{1} + \underset{CH_3}{\overset{CH_3}{>}}\hspace{-2pt}C\hspace{-2pt}\underset{CH_2}{\overset{CH_2}{<}} \xrightarrow[\text{51\%}]{120°}$$

[1] H. J. Bestmann, M. Schmidt, and R. Schobert, *Angew. Chem. Int. Ed.*, **24**, 405 (1985).
[2] H. J. Bestmann and R. Schobert, *ibid.*, **24**, 790 (1985).

L

Lead tetraacetate.

Fragmentation of γ-hydroxyalkylstannanes.[1] These compounds are cleaved stereoselectively by Pb(OAc)$_4$ (1 equiv.) in refluxing benzene to unsaturated carbonyl compounds.

Examples:

Cleavage of 1-(trimethylsilyloxy)bicyclo[n.1.0]alkanes (**8**, 269–270). Lead tetraacetate fragmentation of the *exo-* and *endo*-methyl substituted silyl cyclopropyl ethers (**1**) is essentially stereospecific.[2] Thus *exo*-**1** is fragmented to the (E)-alkenoate **2** exclusively and *endo*-**1** is converted into (Z)-**2** exclusively.

Oxidation of a phenol to an acetoxy enone.[3]   The key step in a recent synthesis of anthracyclinones is the oxidative dearomatization of the A ring of **1** to the enone **2** in 50–55% yield. The product was converted in several steps into the aglycone SM-173B (**3**).

[1] K. Nakatani and S. Isoe, *Tetrahedron Letters*, **25**, 5335 (1984).
[2] G. M. Rubottom, E. C. Beedle, C.-W. Kim, and R. C. Mott, *Am. Soc.*, **107**, 4230 (1985).
[3] A. S. Kende and S. Johnson, *J. Org.*, **50**, 727 (1985).

Lead tetraacetate–Diphenyl disulfide.

Hydroxysulfenylation.   Hydroxysulfenylation of cyclohexene is possible with diphenyl disulfide in CH_2Cl_2—TFA under air in the absence of a metal salt, but the reaction requires several days at 25°. With $Pb(OAc)_4$ reaction occurs in 30 minutes at 0°. Actually $Mn_3O(OAc)_7$ is as efficient as $Pb(OAc)_4$, particularly in reactions with dibenzyl disulfide. The trifluoroacetoxy sulfides can be converted into acetamido sulfides by reaction with acetonitrile containing conc. H_2SO_4 (Ritter reaction, equation I).[1]

2,2'-Dipyridyl disulfide and bis(2-aminophenyl) disulfide do not undergo hydroxysulfenylation using $Pb(OAc)_4$ or in the absence of a metal salt, but each reacts satisfactorily in the presence of catalytic amounts of copper(II) acetate.[2]

[1] A. Bewick, J. M. Mellor, and W. M. Owton, *J.C.S. Perkin I*, 1039 (1985).
[2] A. Bewick, J. M. Mellor, D. Milano, and W. M. Owton, *ibid.*, 1045 (1985).

Lead tetraacetate–Manganese(II) acetate.

$—CH_2OH \rightarrow —CHO$. This combination in the ratio 1:0.1–0.5 effects oxidation of primary alcohols in refluxing benzene to aldehydes as the major product.[1]

[1] M. L. Mihailović, S. Konstantinović, and R. Vukićević, *Tetrahedron Letters*, **27**, 2287 (1986).

N-Lithioethylenediamine (LEDA).

Rearrangement of allylic alcohols to aldehydes.[1] Primary allylic alcohols can isomerize to aldehydes on treatment with LEDA in ethylenediamine (EDA) at 95–130° or with lithium 3-aminopropylamide (**6**, 476).

Example:

[1] H. M. R. Hoffmann, A. Köver, and D. Pauluth, *J.C.S. Chem. Comm.*, 812 (1985).

Lithium.

Alka-2,3-dienes.[1] The chlorohydrins **1** react with BuLi and then CH_3I to form O-methyl chlorohydrins (**2**), which can be isolated if desired; *in situ* lithiation of **2** provides allenes (**3**).

Lithiation of imines and acetylene.[2] Imines and acetylenes can be lithiated directly by reaction with metallic lithium in THF at 10° with phenanthrene as the hydrogen acceptor (it is converted into 1,2,3,4,5,6,7,8-octahydrophenanthrene).

Example:

$$C_5H_{11}C\equiv CH \xrightarrow[\substack{75\%}]{\substack{1)\ Li,\ ArH,\ THF \\ 2)\ CH_3(CH_2)_3I}} C_5H_{11}C\equiv C(CH_2)_3CH_3$$

[1] J. Barluenga, J. R. Fernandez, and M. Yus, *J.C.S. Chem. Comm.*, 203 (1985); *idem, J.C.S. Perkin I*, 447 (1985).
[2] E. A. Mistryukov and I. K. Korshevetz, *Synthesis*, 947 (1984).

Lithium–Ammonia.

Desulfuration/α-alkylation of β-keto sulfones.[1]   Reductive alkylation of sulfones is routinely carried out by α-alkylation of the enolate followed by desulfuration (Al/Hg). It can also be effected in one pot by desulfuration with lithium in liquid ammonia, which generates an enolate that can be alkylated. The yield is markedly enhanced by conversion of the lithium enolate to an alkylstannyl enolate by addition of Bu₃SnCl with HMPT as cosolvent.

Example:

[1] M. J. Kurth and M. J. O'Brien, *J. Org.*, **50**, 3846 (1985).

Lithium–Ethylamine.

Chiral allylic alcohols.[1]   Desulfuration of the β-hydroxy sulfoxides **1** with Raney nickel (**11,** 292) proceeds with simultaneous reduction of the double bond, but can be effected selectively with lithium in ethylamine at −78° to give optically active allylic alcohols (**2**).

[1] G. Solladié, G. Demailly, and C. Greck, *J. Org.*, **50**, 1552 (1985).

Lithium aluminum hydride–Bis(cyclopentadienyl)nickel.

Desulfuration.[1]   The combination of Cp₂Ni and LiAlH₄ (1:1) effects desulfuration of thiols, sulfides, and thioacetals in 40–85% yield, without reduction of carbonyl, ester, and C=C groups.

[1] M.-C. Chan, K.-M. Cheng, M. K. Li, and T.-Y. Luh, *J.C.S. Chem. Comm.*, 1610 (1985).

Lithium aluminum hydride–Cerium(III) chloride.

Reduction of RX.[1]   Alkyl halides, even alkyl fluorides, are reduced in high yield by LiAlH$_4$/CeCl$_3$ (3:1) in refluxing DME or THF. The actual reagent may be a low-valent cerium compound, since direct hydride attack is not involved. Under the same conditions phosphine oxides are reduced to phosphines.

[1] T. Imamoto, T. Takeyama, and T. Kusumoto, *Chem. Letters,* 1491 (1985).

Lithium aluminum hydride–Hexamethylphosphoric triamide.

Reduction of ketoximes.[1]   Ketoximes are reduced by LiAlH$_4$ to a mixture of primary and secondary amines. In contrast, reduction with LiAlH$_4$–HMPT in the molar ratio 1:10 in refluxing THF (130°, 3 hours) results in ketones. HMPT is believed to prevent further reduction of the imine intermediate and to facilitate hydrolysis. This method is not useful for reversion of aldoximes to aldehydes because of dehydration to nitriles.

[1] S. S. Wang and C. N. Sukenik, *J. Org.,* **50**, 5448 (1985).

Lithium aluminum hydride–Sodium methoxide.

syn-1,3-*Amino alcohols*.[1]   The mixture of *anti-* and *syn*-isomers (**1**) of O-benzyloximes of β-hydroxy ketones is reduced by LiAlH$_4$ in the presence of NaOCH$_3$ (**4,** 295) or KOCH$_3$ stereoselectively to *syn*-1,3-amino alcohols.

1 (*anti/syn* = 48:52)   2 (*syn/anti* = 96:4)

[1] K. Narasaka, S. Yamazaki, and Y. Ukaji, *Chem. Letters*, 2065 (1984); *Bull. Chem. Soc. Japan,* **59**, 525 (1986).

Lithium amides, chiral. Koga *et al.*[1] have prepared a series of lithium amides of the type in which one carbon atom adjacent to the nitrogen is chiral and bears a bulky group (phenyl, naphthalene, *t*-butyl). Highest enantioselective deprotonation

$$(CH_3)_2CHN\overset{R}{C}HCH_2X$$
Li*

of prochiral 4-alkylcyclohexanones (2) is observed with 1 in which X is N-methylpiperazyl and forms a five-membered chelate structure (equation I). Asymmetric induction is also highly dependent on the bulk of the alkyl group in the cyclohex-

$$1, X = -N \underset{\underset{}{}}{\overbrace{\hspace{2cm}}} NCH_3$$

anone. In some cases, enantioselectivity is enhanced by addition of HMPT, probably because aggregation is suppressed.

Reactions of other chiral lithium amides have been reported previously (**10**, 245, **12**, 318, 421).

2a, R = C(CH₃)₃	51%	(R)**3a**, 97% ee
2b, R = CH(CH₃)₂	85%	**3b**, 66% ee
2c, R = CH₃	68%	**3c**, 50% ee

[1] R. Shirai, M. Tanaka, and K. Koga, *Am. Soc.*, **108**, 543 (1986).

Lithium 3-aminopropylamide, $LiNH(CH_2)_3NH_2$ (**1**). Lithium metal dissolves in 1,3-diaminopropane at 25° and reacts to form the amide **1** at 70°.

Isomerization of triple bonds.[1]   This base is easier to prepare than potassium 3-aminopropylamide, but is less effective for isomerization of triple bonds to the terminal position. However when used together with potassium *t*-butoxide, isomerization of triple bonds occurs in 1–4 hours at 25°. Isomerization of acetylenic acids results in terminal alkynoic acids and 3,5-dienoic acids. However, acetylenic alcohols (primary or secondary) can be isomerized to ω-hydroxy-1-alkynes in high yield.

Example:

$$HOCH_2C \equiv C(CH_2)_6CH_3 \xrightarrow[\substack{H_2N(CH_2)_3NH_2 \\ 86-88\%}]{1, \ KOC(CH_3)_3} HO(CH_2)_8C \equiv CH$$

[1] S. R. Abrams, *Can. J. Chem.*, **62**, 1333 (1984); idem, *J. Org.*, **49**, 3587 (1984); S. R. Abrams and A. C. Shaw, *Org. Syn.* submitted (1985).

Lithium bis(dimethylphenylsilyl)cuprate ($C_6H_5Me_2Si)_2CuLi$ (1).

Allylsilanes. The cuprate (1) derived from $C_6H_5(CH_3)_2SiLi$ reacts with secondary allyl acetates and urethanes in $THF/O(C_2H_5)_2$/pentane (but not in THF alone) to give allylsilanes with fair to good regioselectivity (*cf.* **10**, 163). (E)-Allylic acetates react to give mainly the allylsilane having the silyl group at the less hindered end of the allyl group, whereas (Z)-allylic acetates react mainly with allylic shift and with more regiocontrol. Yields can be increased by addition of triphenylphosphine.[1]

Examples:

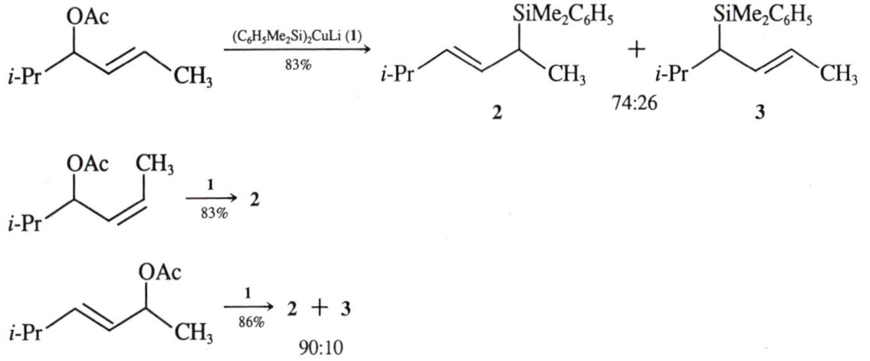

Conjugate addition of **1** to α,β-unsaturated esters, ketones, or aldehydes followed by reaction with an alkyl iodide (or bromide) proceeds by stereoselective attack *anti* to the silyl group, apparently because of electronic effects.[2] Protonation of the β-silyl enolate of **5** proceeds in the same sense to give the opposite diastereomer (Chart I).

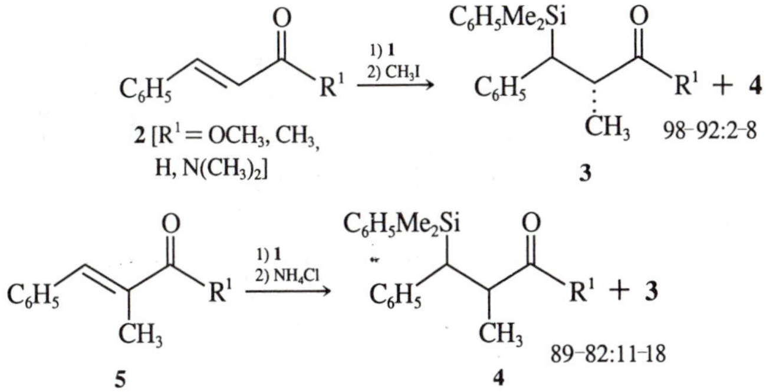

Chart I

This methodology can be extended to alkylation of β-silyl-α-alkyl enolates; the diastereoselectivity is dependent on the size of the α-alkyl substituent, decreasing as the size of the alkyl group is increased (equation I).

(I)

[1] I. Fleming and A. P. Thomas, *J.C.S. Chem. Comm.*, 411 (1985).
[2] I. Fleming, J. H. M. Hill, D. Parker, and D. Waterson, *ibid.*, 318 (1985).

Lithium 4,4'-di-*t*-butylbiphenylide (1).

Reduction of pyranosyl chlorides to glycals (10, 240).[1]   Reduction of the pyranosyl chloride **2** with lithium–ammonia results in a modest yield of the expected glycal **3** (**8**, 282–283) and the tetrahydropyran **4** in a 1:1 ratio. This reduction was

Li/NH₃	50%	1:1
C₁₀H₈Na, THF	32%	>50:1
1, THF	82%	>50:1

then examined in detail with a model furanosyl chloride. Sodium naphthalenide is superior in that dehydrohalogenation is suppressed, but the overall yield is low. Finally **1** was found to be the most useful reagent for the desired conversion of both pyranosyl and furanosyl chlorides to the corresponding glycals.

[1] R. E. Ireland, D. W. Norbeck, G. S. Mandel, and N. S. Mandel, *Am. Soc.*, **107**, 3285 (1985).

Lithium diisopropylamide.

γ-Alkylation of α,β-enones.[1]   Dienolates of α,β-enones undergo alkylation at the α-position. However, conversion of the α'-phenylsulfonyl derivative (**1**) of (E)-3-pentene-2-one to the trianion followed by alkylation results in **2** and **3**.

2-Azaallyl anion cycloadditions.   Imines bearing one or more aryl groups are converted by LDA into 2-azaallyl anions. These anions undergo cycloaddition not only with activated alkenes,[2] but can also undergo intramolecular cycloaddition with a double bond to form *cis*-fused bicyclic pyrrolidines.[3]

Example:

α-Alkylation of α-halo ketones.[4]   The anions of α-bromo or α-chloro ketones (**1**) are too reactive to be useful for direct alkylation. However, α-haloimines (**2**) are deprotonated by LDA (THF, 0°) to give relatively stable carbanions that can be alkylated to provide α-alkyl-α-chloroimines (**3**), precursors to α-alkyl-α-halo ketones (**4**).

Example:

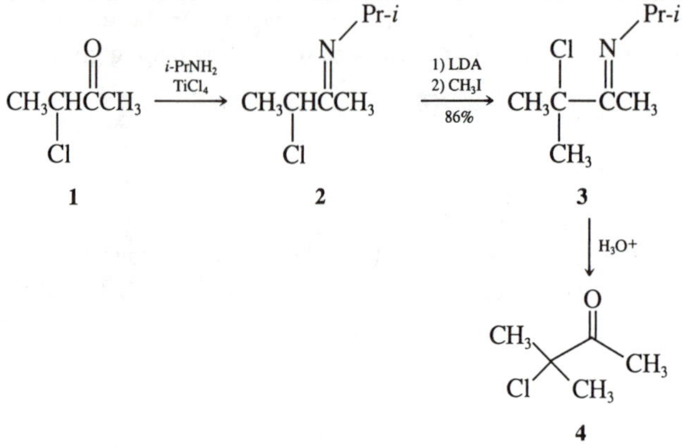

1   **2**   **3**

4

[1] P. T. Lansbury, G. E. Bebernitz, S. C. Maynard, and C. J. Spagnuolo, *Tetrahedron Letters,* **26**, 169 (1985).
[2] T. Kauffmann, *Angew. Chem. Int. Ed.,* **13**, 627 (1974).
[3] W. H. Pearson, M. A. Walters, and K. D. Oswell, *Am. Soc.,* **108**, 2769 (1986).
[4] N. De Kimpe, P. Sulmon, and N. Schamp, *Angew. Chem. Int. Ed.,* **24**, 881 (1985).

Lithium diisopropylamide–Potassium *t*-butoxide.

Cleavage of homoallyl ethers; 2,4-dienols.[1]   3,6-Dihydro-2H-pyrans are cleaved to (Z)-2,4-pentadienols by this combination of bases in 35–75% yield.
Example:

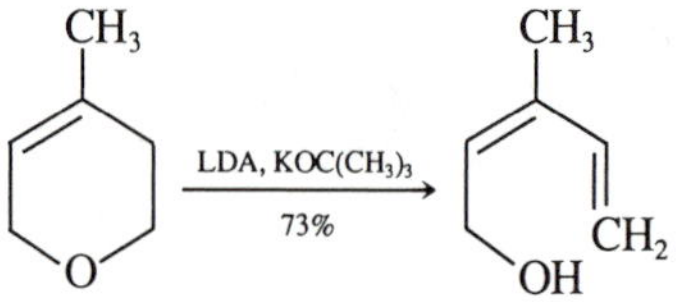

The same cleavage is observed with acyclic homoallyl ethers.
Example:

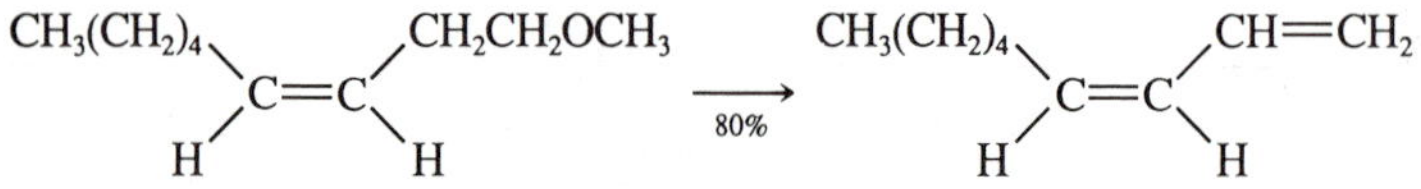

[1] C. Margot and M. Schlosser, *Tetrahedron Letters,* **26**, 1035 (1985).

Lithium 1-(dimethylamino)naphthalenide (LDMAN).

Reductive lithiation of allyl phenyl sulfides.[1] This reagent is particularly useful for preparation of allyllithium reagents at temperatures at which the anions are stable. Moreover, regioselectivity in reactions can be achieved by conversion to allyltitanium(IV) complexes by metal exchange with Ti(O-*i*-Pr)$_4$. Thus the unsymmetrical anion formed from the allyl sulfide **1** with LDMAN reacts with crotonaldehyde to give a mixture of 1,2- and 1,4-adducts. The 1,2-adduct **2** can be obtained in high yield as two diastereomers (9:1) by use of the allyltitanium complex (equation I).

(I)

1) LDMAN, THF, −60°
2) Ti(O-*i*-Pr)$_4$
3) CH$_3$CH═CHCHO

92%

1

2 (9:1)

[1] T. Cohen and B.-S. Guo, *Tetrahedron,* **42,** 2803 (1986).

Lithium hexamethyldisilazide.

α-Silyl ketones.[1] Reaction of a primary α-bromo ketone with LiN[Si(CH$_3$)$_3$]$_2$ (**1**) followed by chlorotrimethylsilane or chlorotriethylsilane results in a trialkyl-silyloxyvinyl bromide (**a**), which is not isolated but treated with BuLi (2 equiv.) to effect O–C silyl migration (equation I).

(I) $(CH_3)_3CCCH_2Br$ $\xrightarrow[\text{2) R}_3\text{SiCl}]{\text{1) 1, −78°}}$ $\left[(CH_3)_3CC\text{═}CHBr \right]$ $\xrightarrow[\text{67−68\%}]{\text{BuLi}}$ $(CH_3)_3CCCH_2SiR_3$

a

Reaction of secondary α-bromo ketones in this sequence can result in an acetylene because of Peterson olefination.

[1] P. Sampson and D. F. Wiemer, *J.C.S. Chem. Comm.,* 1746 (1985).

Lithium hydride,[1] LiH. Commercially available LiH is essentially inert as a metallation reagent. A highly reactive form of LiH can be prepared by hydrogenation of butyllithium in the presence of TMEDA (1–1.2 equiv.) in hexane at 30–35°. No reaction is observed in the absence of TMEDA. This material metallates dibenzyl ketone almost instantly at 25°.

Extremely reactive NaH and KH are obtained by hydrogenation of butylsodium or butylpotassium, respectively, in the presence of TMEDA. Caution: These hydrides are pyrophoric. They are more reactive than the LiH described above.

[1] P. A. A. Klusener, L. Brandsma, H. D. Verkruijsse, P. v. R. Schleyer, T. Friedl, and R. Pi, *Angew. Chem. Int. Ed.*, **25**, 465 (1986).

Lithium *o*-lithiophenoxide,

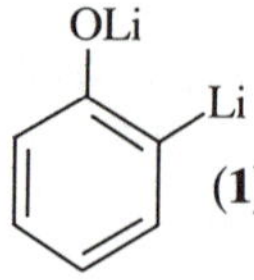

$$\text{OLi}$$

(**1**). This dianion can be obtained by reaction of phenol with *t*-BuLi (2.8 equiv.) in tetrahydropyran (THP) at 25°. The dianion reacts with various electrophiles to give *o*-substituted phenols in 40–67% yield.[1]

Example:

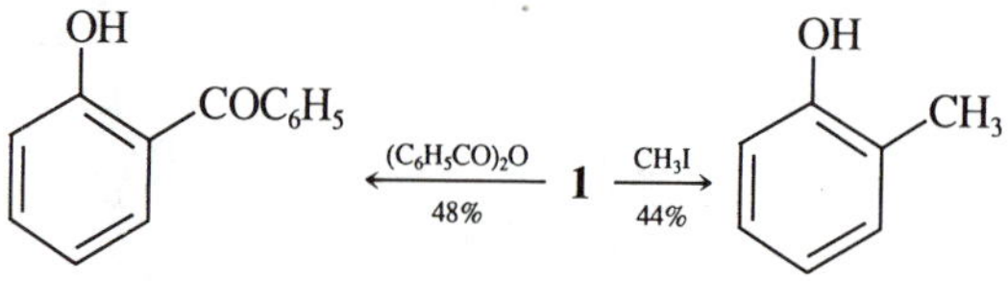

[1] G. H. Posner and K. A. Canella, *Am. Soc.*, **107**, 2571 (1985).

Lithium methoxyacetylide (1).
 Generation *in situ*[1]:

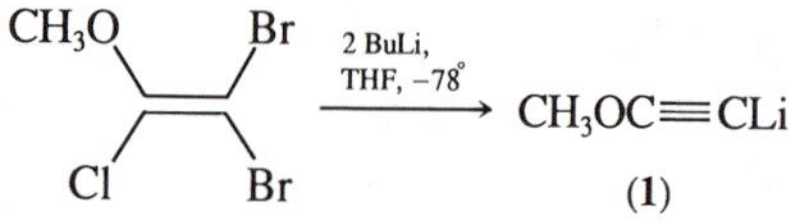

α,β-*Unsaturated esters*. The reagent reacts with aldehydes or ketones to provide α-acetylenic alcohols, which undergo acid catalyzed rearrangement to α,β-unsaturated esters (Meyer-Schuster rearrangement).
 Examples:

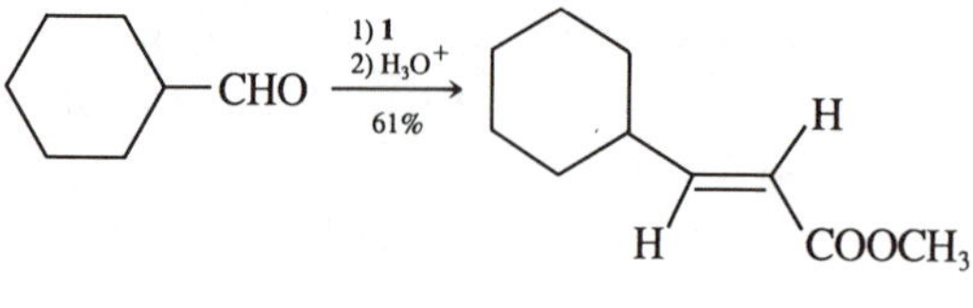

(E/Z = 2:1)

The overall yields are similar to those obtained via a Wittig-Horner reaction, but the method is more (E)-selective.[1]

[1] R. H. Smithers, *Syn. Comm.*, **15**, 81 (1985).

Lithium 2,2,6,6-tetramethylpiperidide.

Dehydrohalogenation.[1]   This lithium dialkylamide shows a greater preference for the Hoffmann product in the dehydrohalogenation of 2-bromobutane than less hindered bases of this type. The preference is increased by addition of 12-crown-4 (equation I).

(I) $CH_3CH_2CHBrCH_3$ $\xrightarrow[\substack{80\% \\ + \text{ crown ether } 100\%}]{\substack{\text{LiTMP} \\ \text{THF, } 0°}}$ $CH_3CH_2CH=CH_2 + CH_3\ CH=CH\ CH_3$

85:15	(E/Z=1:1)
99:1	(E/Z=1:1)

[1] I. E. Kopka, M. A. Nowak, and M. W. Rathke, *Syn. Comm.*, **16**, 27 (1986).

Lithium tri-*sec*-butylborohydride.

Optically pure 1,2-diols.[1]   The acyllactones **1,** obtained by reaction of a Grignard reagent with the acid chloride derived from (R)-(–)- or S-(+)-glutamic acid, are reduced by lithium tri-*sec*-butylborohydride almost exclusively to *syn*-alcohols (**2**), regardless of the nature of the R group. In contrast, reduction of **1** with sodium

2 (*syn/anti* 92–99:8–1)

borohydride or zinc borohydride proceeds with lower diastereoselectivity in favor of the *anti*-isomer (*anti/syn* ~70:30).

The syn alcohol **3**, obtained from (R)-(−)-glutamic acid was used for an enantiospecific synthesis of (+)-exobrevicomin (**4**), the aggregation pheromone of the female Western pine beetle.

Stereoselective acyclic enolization.[2]   This reaction can sometimes be achieved by conjugate reduction of acyclic α,β-enones by lithium tri-*sec*-butylborohydride (**1**). When an enone reacts with **1** mainly by 1,2-reduction Li/NH$_3$ can be used for conjugate reduction. In at least one case reduction with **1** is more stereoselective than that with Li/NH$_3$. The stereoselectivity in general can be correlated with conformational preferences of enones.[3]

Examples:

---

[1] M. Larchevêque and J. Lalande, *J.C.S. Chem. Comm.*, 83 (1985).
[2] A. R. Chamberlin and S. H. Reich, *Am. Soc.*, **107**, 1440 (1985).
[3] H.-J. Oelichmann, D. Bougeard, and B. Schroder, *Angew. Chem. Suppl.*, 1404 (1982).

Lithium triethylborohydride.

Desulfonation of allylic sulfones.[1]   These sulfones undergo desulfonation with this borohydride in the presence of [1,3-bis(diphenylphosphine)propane]-dichloropalladium(II), PdCl$_2$(dppp). This method is superior to use of NaBH$_4$

catalyzed by Pd[P(C$_6$H$_5$)$_3$]$_4$ previously used (**11,** 512–513). The reaction was used in a synthesis of squalene (**3**) from farnesyl *p*-tolyl sulfone (**1**).

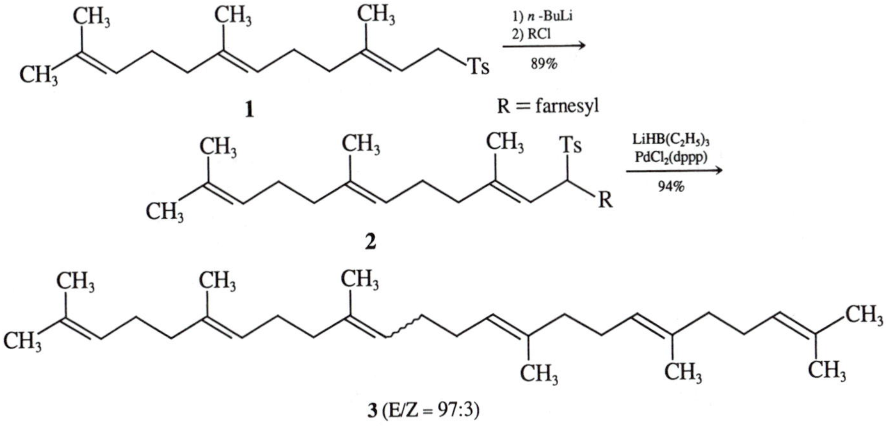

1

R = farnesyl

2

3 (E/Z = 97:3)

[1] M. Mohri, H. Kinoshita, K. Inomata, and H. Kotaki, *Chem. Letters,* 451 (1985).

M

Magnesium.

gem-Debromination. An efficient synthesis of penicillanic acid S,S-dioxide (sublactam, **3**), a clinically useful β-lactamase inhibitor, involves diazotization/bromination of 6-aminopenicillanic acid S,S-dioxide (**1**) using NaNO$_2$, Br$_2$, HBr, CH$_3$OH in CH$_2$Cl$_2$ to give **2** in 90% yield. Debromination of **2** has been carried out by hydrogenolysis using Pd/C, but is effected more efficiently with magnesium metal in aqueous ethyl acetate at pH 3.5.[1]

Reduction of bis(trialkyltin) oxides to R$_6$Sn$_2$.[2] This reduction can be effected with Mg powder activated by a few drops of BrCH$_2$CH$_2$Br (or I$_2$) in the presence of a small amount of THF (82% yield). Other metals, Ti,K, and Na, can be used but yields are in the range 60–75%.

[1] J. C. Kapur and H. P. Fasel, *Tetrahedron Letters*, **26**, 3875 (1985).
[2] B. Jousseaume, E. Chanson, and M. Pereyre, *Organometallics*, **5**, 1271 (1986).

Magnesium–Methanol.

Desulfonation.[1] Magnesium, activated by 10 washes with dilute aqueous HCl followed by 5 washes with distilled water, effects desulfonation in dry CH$_3$OH at 50° in satisfactory yield. For example, phenyl β-phenethyl sulfone is converted into ethylbenzene in 68% yield. It is also useful for desulfonation of 1,1- and 1,2-disulfones.

Reduction of α,β-unsaturated esters.[2]   These esters are reduced to the saturated esters by Mg in CH_3OH in almost quantitative yield. Even double bonds conjugated with an aromatic ring are reduced.

Example:

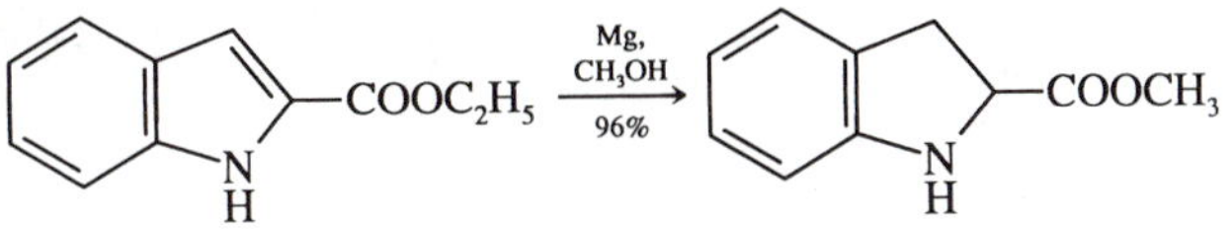

[1] A. C. Brown and L. A. Carpino, *J. Org.*, **50**, 1749 (1985).
[2] I. K. Youn, G. H. Yon, and C. S. Pak, *Tetrahedron Letters*, **27**, 2409 (1986).

Magnesium iodide.

Vinyl iodides.[1]   Vinyl triflates are converted into vinyl iodides on reaction with MgI_2 and $N(C_2H_5)_3$ in CS_2 or cyclohexane (65–85% yield).

[1] A. G. Martinez, R. M. Alvarez, A. G. Fraile, L. R. Subramanian, and M. Hanack, *Synthesis*, 222 (1986).

Manganese(III) acetate.

γ-Lactones.[1]   Alkenes can be converted into α-substituted γ-lactones by reaction with cyanoacetic acid or malonic acids in the presence of $Mn_3O(OAc)_7$ (1–2 equiv.).

Examples:

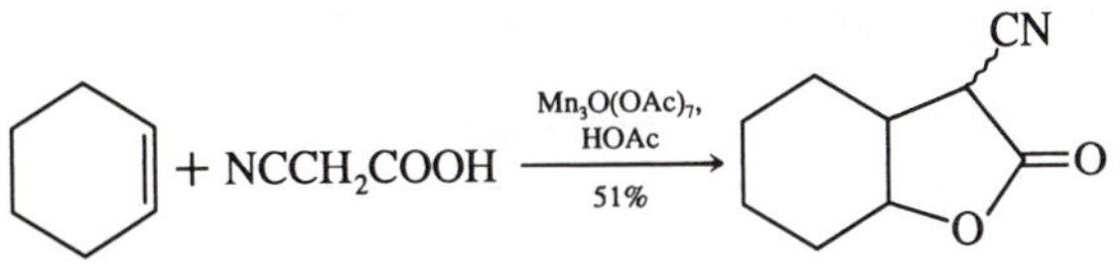

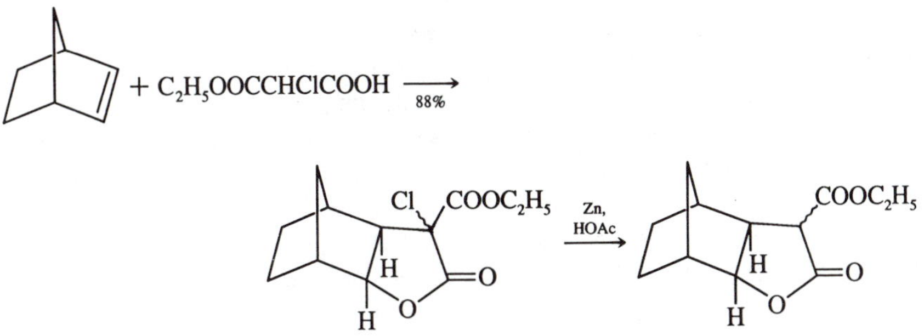

[1] E. J. Corey and A. W. Gross, *Tetrahedron Letters*, **26**, 4291 (1985).

Meldrum's acid.

Stereoselective intramolecular Diels-Alder reactions.[1]   The reaction of Meld-rum's acid (**1**) with (R)-citronellal (**2**) in the presence of ethylenediammonium acetate (EDDA)[2] at 15–20° results in the tricyclic dihydropyran **3** as the major product with an optical purity of >98%. The product evidently results from an intramolecular hetero-Diels-Alder addition. It can be converted by acid into the optically pure α-methoxycarbonyllactone **4**.

3 (>98% de)   4

[1] L.-F. Tietze and G. v. Kiedrowski, *Tetrahedron Letters*, **22**, 219 (1981); L. F. Tietze, G. v. Kiedrowski, and K.-G. Fahlbusch, *Org. Syn.* submitted (1985).
[2] EDDA, m.p. 114°, is prepared by reaction of ethylenediamine with acetic acid at 4°.

Menthol.

Chiral 1,2-cycloalkanedicarboxylic acids.[1]   This ring system can be obtained by coupling 1,ω-dihalides or -ditosylates with the dienolate of *d*- or *l*-menthyl succinate (**1**). The dienolate generated with lithium 2,2,6,6-tetramethylpiperidide is mainly the (E,E)-isomer, whereas deprotonation with LDA and HMPT generates mainly the (Z,Z)-isomer. Thus reaction of **1** (R = *l*-menthyl) with LiTMP in THF at −78° with CH$_2$BrCl results in (S,S)-(−)-dimenthyl cyclopropane-*trans*-1,2-dicarboxylate (**2**) in 58% chemical yield and 99% de. A similar reaction of **1**, R = *d*-menthyl, results in (R,R)-**2** in 60% yield and 80% de.

This reaction can be extended to preparation of the corresponding cyclobu-tane-, cyclopentane-, and cyclohexane-1,2-dicarboxylic esters in 65–92% de.

1, R = *l*-menthyl   (S,S)-2 (99% de,
                                   trans/cis > 95:1)

***Diastereoselective Diels-Alder reactions.*[2]**   The (S)- and (R)-3-(2-pyridylsul-
finyl)acrylates (1) are obtained by addition of 2-mercaptopyridine to (+)-menthyl
propiolate followed by oxidation and fractional crystallization of the diastereoiso-
meric sulfoxides. In the presence of $(C_2H_5)_2AlCl$, both undergo highly diastereoselective Diels-Alder reactions with furan to give mainly *endo*-adducts without
epimerization of the sulfoxide group. The products are converted as shown into
(+)- and (−)-*endo*-2-hydroxymethyl-7-oxabicyclo[2.2.1]heptane (2).

[1] A. Misumi, K. Iwanaga, K. Furuta, and H. Yamamoto, *Am. Soc.*, **107**, 3343 (1985).
[2] H. Takayama, A. Iyobe, and T. Koizumi, *J.C.S. Chem. Comm.*, 771 (1986).

(S)-(−)-Menthyl *p*-toluenesulfinate (1).
    ***Chiral α-methylene-γ-lactones.*[1]**   (R)-(+)-Alkyl *p*-tolyl sulfoxides (2), read-
ily obtainable in almost quantitative yield from (1),[2] on lithiation (LiTMP) and
reaction with lithium α-bromomethylacrylate (3) are converted into α-methylene-
γ-sulfinyl carboxylic acids (4), which can be separated by chromatography or crystallization. Reduction of optically pure 4 provides γ-tolylthio acids [(S)-5], which
on methylation and treatment with potassium *t*-butoxide are converted into (4R)-
α-methylene-γ-lactones (6), with inversion of chirality.

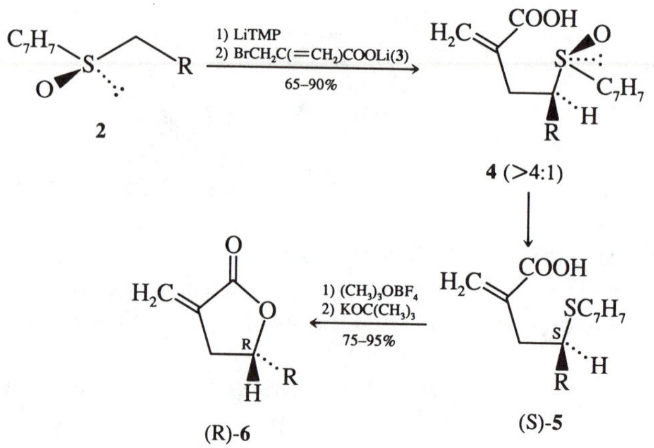

2

4 (>4:1)

(R)-6

(S)-5

[1] P. Bravo, G. Resnati, and F. Viani, *Tetrahedron Letters*, **26**, 2913 (1985).
[2] J. Drabowicz, B. Bujnicki, and M. Mikolajczyk, *J. Org.*, **47**, 3325 (1982).

Mercury.

Propargylic and allenic organomercurials.[1] Mercury reacts with 1-iodo-2-alkynes and 3-iodo-1-alkynes to provide propargylic and allenic mercurials, respectively (equations I and II). These products react with acyl chlorides in the presence of $AlCl_3$ (1 equiv.) with rearrangement to give allenic and propargylic ketones. The latter ketones are unstable, and rearrange in the presence of alumina to allenic ketones.

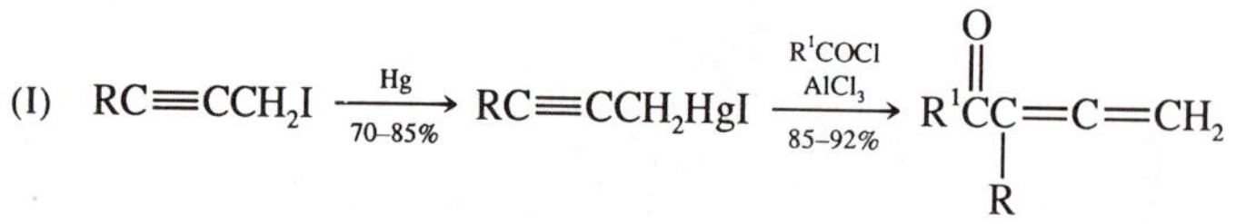

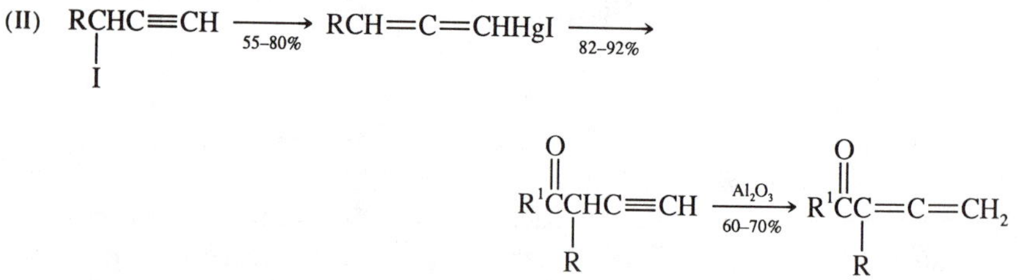

[1] R. C. Larock and M.-S. Chow, *Tetrahedron Letters*, **25**, 2727 (1984); R. C. Larock, M.-S. Chow, and S. J. Smith, *J. Org.*, **51**, 2623 (1986).

Mercury(II) chloride.

Cyclization of silyl enol ethers of ε-acetylenic ketones or aldehydes. Reaction of these derivatives of ynones or -als with $HgCl_2$ (1.1 equiv) and hexamethyldisilazane (0.2 equiv.) as an acid scavenger in CH_2Cl_2 at 30° followed by acidification with aqueous HCl and NaI (2 equiv. each) results in cyclic β,γ-enones. Thermal cyclization of these substrates requires high temperatures, which lead to decomposition and rearrangements.[1]

Example:

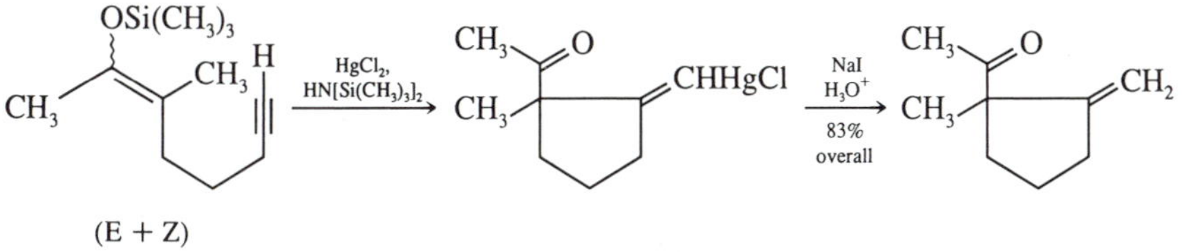

(E + Z)

---

[1] J. Drouin, M.-A. Boaventura, and J.-M. Conia, *Am. Soc.*, **107**, 1726 (1985); J. M. Conia and P. LePerchec, *Synthesis*, 1 (1975).

Mercury(II) trifluoroacetate.

Aminomercuration; aminoalditols.[1] The synthesis of the natural aza-alditol 1-deoxynojirimycin (**4**) from methyl α-D-glucopyranoside involves the reductive ring-opening (Zn) and reductive amination ($NaBH_3CN$) of **1** to give **2** in 91% yield. Aminomercuration of **2** results mainly in the bromomercurial **3**, which was converted into the aza-sugar **4** by reductive oxygenation and deprotection.

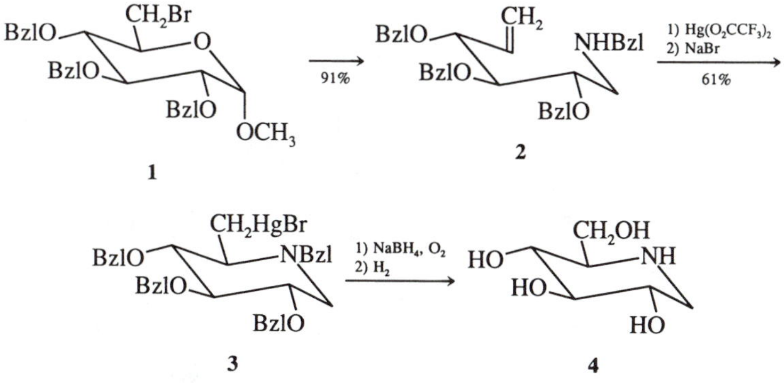

---

[1] R. C. Bernotas and B. Ganem, *Tetrahedron Letters*, **26**, 1123 (1985).

Methanesulfonyl chloride.

Intramolecular cyclization of allylic stannanes.[1]   The pyrrolizidine alkaloid isoretronecanol (3) can be synthesized efficiently by cyclization of the mesylate of the hydroxylactam 1 to give 2 stereoselectively. Oxidative cleavage followed by reduction of the keto group gives 3.

1   2 (endo/exo = 74:1)

[1] G. E. Keck and E. J. Enholm, *Tetrahedron Letters*, **26**, 3311 (1985).

Methanesulfonyl chloride–4-Dimethylaminopyridine.

An effective reagent for dehydration (1) is prepared by stirring MsCl (2.5 equiv), DMAP (1.25 equiv.), and H_2O (1 equiv.) in CH_2Cl_2 (13 ml.) for 2–3 days at 25°. The water is essential for hydrolysis of MsCl to MsOH and HCl.

Dehydration.[1]   Usual methods for dehydration of the hydroxy aldehyde 2 are unsatisfactory, but 1 effects the desired conversion to 3 in 90% yield. The product is a precursor to 6-protoilludene (4), a natural product considered to be an intermediate in the biosynthesis of humulenes.

2   3

4

[1] J. Furukawa, N. Morisaki, H. Kobayashi, S. Iwasaki, S. Nozoe, and S. Okuda, *Chem. Pharm. Bull.*, **33**, 440 (1985).

Methoxyallene.

Methylenomycins.[1]   The cyclopentannelation reaction (**12**, 310) has been modified to provide a general synthesis of methylenomycins. Thus, the adduct (**2**) of α-lithio-α-(methoxymethyl)allene (**1**) with 3-methyl-3-butene-2-one cyclizes to methylenomycin B (**3**) in the presence of trifluoroacetic anhydride and 2,6-lutidine.

The only limitation noted in the synthesis of related methylenomycins is that only the adducts of (E)-enones undergo cyclization.

[1] M. A. Tius, D. P. Astrab, A. H. Fauq, J.-B. Ousset, and S. Trehan, *Am. Soc.*, **108**, 3438 (1986).

Methoxyamine.

Ketone synthesis.[1]   Alkyllithium or Grignard compounds (2 equiv.) react with O-methyloximes of aldehydes to form ketones in moderate to high yield.

Example:

[1] S. Itsuno, K. Miyazaki, and K. Ito, *Tetrahedron Letters*, **27**, 3033 (1986).

1-Methoxy-1,3-bis(trimethylsilyloxy)-1,3-butadiene (1).

Diastereoselective aldol condensations.[1]   The reaction of (S)-(–)-2-benzyl-oxyhexanal (2) complexed with $TiCl_4$ with the bistrimethylsilyl enol ether (1) of methyl acetoacetate gives almost exclusively the *syn*-aldol adduct (3). This reaction

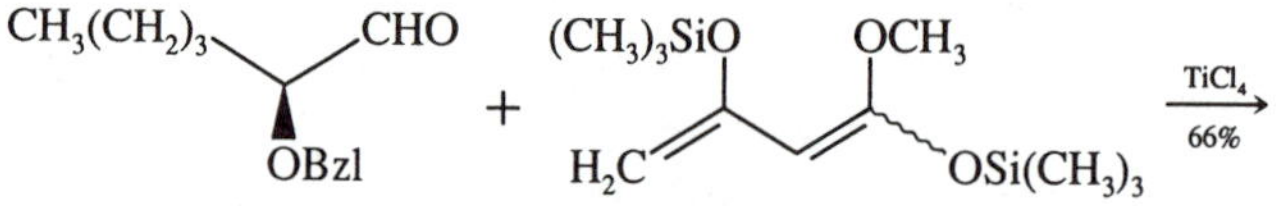

(S)-2 (98% ee)   1

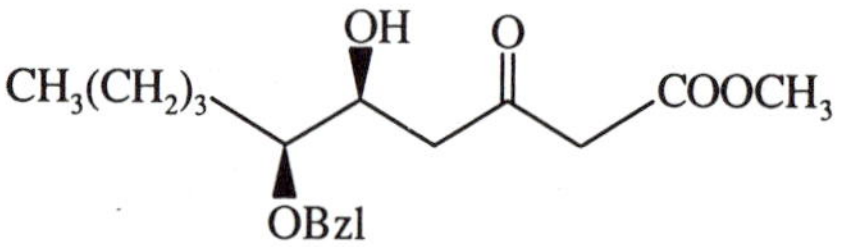

syn-3 (98% de)

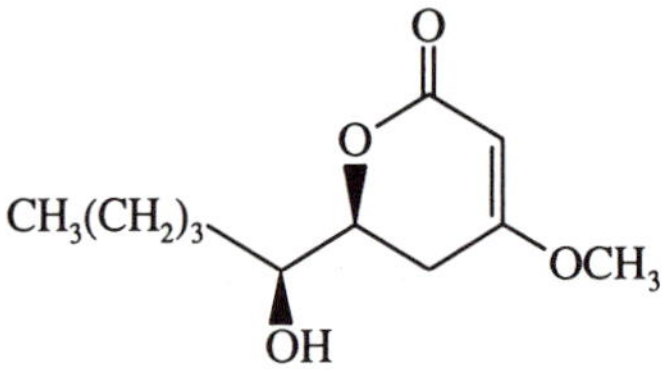

4

provides a synthesis of optically active (–)-pestalotin (4). In contrast, reaction of the lithium enolate of methyl acetoacetate with (R)-2 gives *anti*-3 with only modest selectivity (64:36).

[1] H. Hagiwara, K. Kimura, and H. Uda, *J.C.S. Chem. Comm.*, 860 (1986).

2-Methoxy-1,3-butadiene.

Anthracyclinones.[1]   The anthraquinone 1, available from the commercial dye-stuff 1,4,5-trihydroxy-9,10-anthraquinone, reacts regiospecifically because of the halo substituent with 2-methoxybutadiene to give 2. The product can be converted by ketalization and oxidation into the quinone 3, a known precursor to anthra-cyclinones.

[1] D. W. Cameron, G. I. Feutrill, P. G. Griffiths, and B. K. Merrett, *Tetrahedron Letters*, **27**, 2421 (1986).

1-Methoxy-1-butene-3-yne, **(1).** Supplier: Aldrich.

1,4-*Bis(trimethylsilyl)*-1,3-*butadiyne* (2). This unstable butadiyne is readily prepared from **1** (equation I).[1]

$$(I) \quad 1 \xrightarrow[-\text{LiOCH}_3]{\substack{3 \text{ BuLi,} \\ \text{THF, } -25°}} [\text{LiC}{\equiv}\text{CC}{\equiv}\text{CLi}] \xrightarrow[80\%]{\text{ClSi(CH}_3)_3} (\text{CH}_3)_3\text{SiC}{\equiv}\text{CC}{\equiv}\text{CSi(CH}_3)_3$$
$$\textbf{2} \text{ (m.p. } 107°)$$

(Z)-1-*Methoxy*-4-(*trimethylsilyl*)-1-*butene*-3-yne (3)[1] (b.p. 77–79°/12 mm) is prepared by silylation of **1**. Lithiation of **3** at −72° provides the anion **a,** which reacts with various electrophiles to provide methoxy enynes, (equation I). At higher temperatures **a** is converted into **b,** which is useful for preparation of butadiyne derivatives.

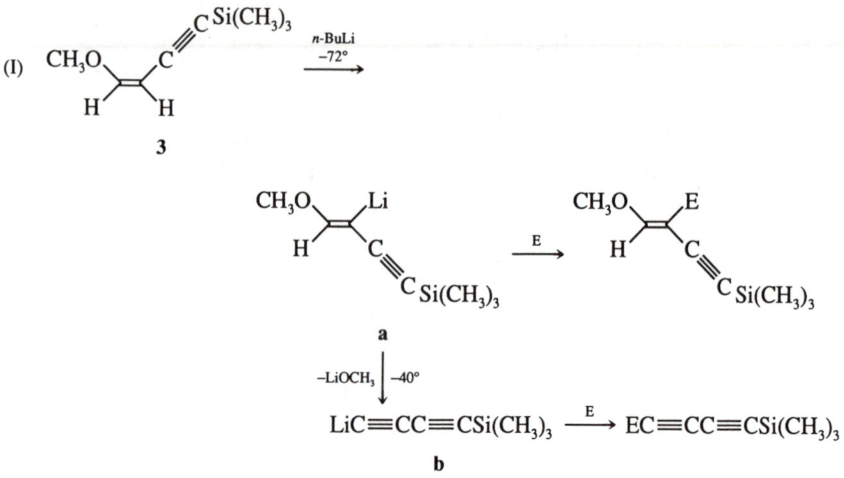
(I)
CH_3O
$Si(CH_3)_3$
n-BuLi
−72°
3
Li
E
a
−LiOCH$_3$
−40°
LiC≡CC≡CSi(CH$_3$)$_3$
EC≡CC≡CSi(CH$_3$)$_3$
b

[1] G. Zweifel and S. Rajagopalan, *Am. Soc.*, **107**, 700 (1985).

(S)-(+)-2-Methoxymethylpyrrolidine (1).

Chiral α-ketoenamines.[1] The α-ketoenamine **2** is obtained by reaction of 1,2-cyclohexanedione with **1** in benzene in the presence of molecular sieves. Grignard reagents add to **2** to provide (R)-**3** in high optical purity. Surprisingly, alkyllithiums add to **2** to give (S)-**3**, but usually in lower optical yields.

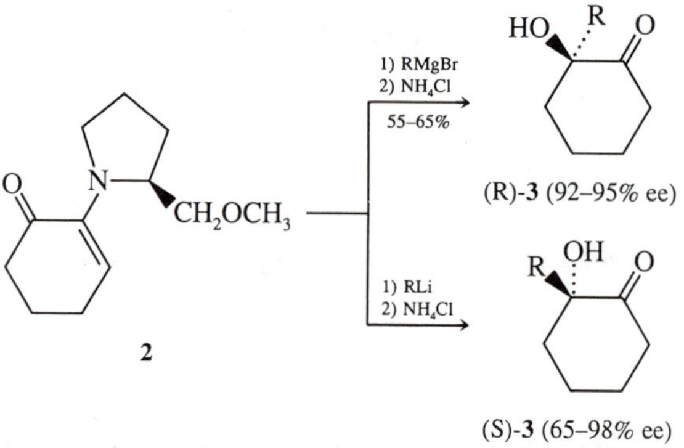
CH_2OCH_3
2
1) RMgBr
2) NH_4Cl
55–65%
1) RLi
2) NH_4Cl
(R)-3 (92–95% ee)
(S)-3 (65–98% ee)

[1] T. Fujisawa, M. Watanabe, and T. Sato, *Chem. Letters*, 2055 (1984).

1-Methoxy-2-methyl-3-trimethylsilyloxy-1,3-pentadiene (1).

Imino Diels-Alder reactions.[1] The cycloaddition of dihydro-β-carbolines with polyoxygenated dienes provides heterocycles of the yohimbine type. Thus reaction of **1** and **2** catalyzed by $ZnCl_2$ gives **3** and **4** in the ratio 4:1 (58%).

3, $R^1 = CH_3$, $R^2 = H$
4, $R^1 = H$, $R^2 = CH_3$

[1] S. Danishefsky, M. E. Langer, and C. Vogel, *Tetrahedron Letters*, **26**, 5983 (1985).

p-Methoxyphenol, $HOC_6H_4OCH_3$-p (**1**).

Protection of primary alcohols.[1] p-Anisyl ethers are readily prepared from primary alcohols by the Mitsunobu reaction [$P(C_6H_5)_3$; DEAD]. The ethers are stable to 3 N HCl or 3 N NaOH at 100°, to Jones or PCC oxidation, and to $LiAlH_4$. Deprotection is effected in 85–95% yield by oxidation with CAN in aqueous CH_3CN.

[1] T. Fukuyama, A. A. Laird, and L. M. Hotchkiss, *Tetrahedron Letters*, **26**, 6291 (1985).

Methoxy(phenylthio)methyllithium (1).

Aldehydes from alkylboronic esters, $RB(OR^1)_2$. This reagent reacts with alkylboronic esters (**2**) to provide an ate complex (**a**). Addition of $HgCl_2$ induces alkyl migration from boron to carbon to give **3**, which is oxidized to a homologated aldehyde **4**.[1]

This sequence has now been used to obtain optically pure aldehydes by preparation of optically active boronic esters from diisopinocampheylborane or monoisopinocampheylborane.[2]

Example:

>99% ee   >99% ee

[1] H. C. Brown and T. Imai, *Am. Soc.*, **105**, 6285 (1983).
[2] H. C. Brown, T. Imai, M. C. Desai, and B. Singaram, *ibid.*, **107**, 4980 (1985).

Methoxy(phenylthio)trimethylsilylmethane (1).

Acylsilanes. Acylsilanes are obtained by alkylation of the anion of **1** followed by periodate oxidation.[1]

Example:

β-Trimethylsilyl-α,β-enones or enals.[2] These compounds can be obtained by the same sequence from 1-methoxy-3-phenylthio-3-trimethylsilyl-1-propene, **2**.

2

[1] T. Mandai, M. Yamaguchi, Y. Nakayama, J. Otera, and M. Kawada, *Tetrahedron Letters*, **26**, 2675 (1985).
[2] T. Mandai, H. Arase, J. Otera, and M. Kawada, *ibid.*, **26**, 2677 (1985).

4-Methoxy-2,2,6,6-tetramethyl-1-oxopiperidinium chloride (1).
Preparation:

1, m.p. 123°dec.

Oxidation of primary and secondary alcohols.[1] Aliphatic primary and secondary alcohols are oxidized rapidly and in high yield by 1. Benzylic and primary allylic alcohols are also oxidized in high yield. Benzoin is oxidized slowly (17 hours) and in low yield.

γ-*and* δ-*Lactones.* Primary 1,4- and 1,5-diols are oxidized by 1 (2 equiv.) to γ-and δ-lactones, respectively in good yield, but only traces of lactones are formed on oxidation of 1,3- or 1,6-primary diols. 2,3-Butanediol is oxidized by 1 (1 equiv.) to acetoin in 74% yield. 1,4-Pentanediol (2) is oxidized by 1 (3 equiv.) to the γ-valerolactone 3 (83% yield) because of selective oxidation of the primary hydroxyl group.[2]

Silver carbonate on Celite has been the preferred reagent for this oxidation (3, 247), but 10–20 equivalents of oxidant is required. In addition the regioselectivity is highly dependent on the solvent. Thus 2 is oxidized by Fetizon's reagent in CH_2Cl_2/C_6H_6 to a 1:1 mixture of 3 and 4-ketopentanol-1 (4), but in $CHCl_3$ to 3 and 4 in the ratio 9:1.

[1] T. M. Miyazawa, T. Endo, S. Shiihashi, and M. Okawara, *J. Org.*, **50**, 1332 (1985).
[2] T. Miyazawa and T. Endo, *ibid.*, **50**, 3930 (1985).

Methyl acrylate.
Three component [2 + 2 + 2]cycloadditions.[1] Lithium enolates of ketones react with methyl acrylate (2 equiv.) in THF at −78° to form cyclohexanols (equation I). The reaction involves two sequential Michael additions and an aldol con-

densation. The best yields are obtained with methyl acrylate, but α-substituted acrylates can be used. Lithium enolates of both cyclic and acyclic ketones can be used. This reaction is useful for regiospecific synthesis of aromatics.

Example:

[1] G. H. Posner, S.-B. Lu, E. Asirvatham, E. F. Silversmith, and E. M. Shulman, *Am. Soc.*, **108**, 511 (1986).

N-Methyl-*o*-aminothiophenol

(1), b.p. 84°/2 mm. The reagent is obtained on reduction of benzothiazole with $LiAlH_4$.

Protection of aldehydes and ketones.[1] The reagent reacts with aldehydes or ketones in refluxing ethanol (4–24 hours) to form 3-methylbenzothiazolines (**2**) in ~70–90% yield. Aldehydes usually react so much more readily than ketones that selective protection is possible. Deprotection is effected in 90–98% yield by treatment with $AgNO_3$ in aqueous CH_3CN at pH 7 or by reaction with $HgCl_2$ in refluxing aqueous CH_3CN.

2

[1] H. Chikashita, N. Ishimoto, S. Komazawa, and K. Itoh, *Heterocycles*, **23**, 2509 (1985).

α-Methylbenzylamine, $C_6H_5CH(CH_3)NH_2$ **(1).**

Chiral 2,2-disubstituted cycloalkanones.[1] The imine **2** prepared from racemic 2-methylcyclohexanone and (S)-(−)-**1**, reacts with methyl vinyl ketone to form an adduct that is hydrolyzed to the (R)-(+)-diketone **3** in 91% ee with recovery of **1** in almost quantitative yield. The reaction is described as a "deracemizing alkylation."

3 (91% ee)

2

The same process was used to obtain **4** from **2** and methyl acrylate, and to obtain **5** and **6** from the imine of 2-methylcyclopentanone.

4   **5**   **6**

Chiral α-methyl-α-amino acids.[2] The reaction of phenylacetone with (S)-(−)-**1** in the presence of NaCN in HOAc affords one (R,S)-diastereomer (**2**), which is converted to the amide **3** on hydrolysis. Hydrogenation of **3** provides (R)-(+)-2-methyl-3-phenylalanine (**4**) in >98% ee.

(S)-1

2 (α_D −125°)

89% | H_2SO_4

(R)-4 (>98% ee)

1) H_2, Pd/C (95%)
2) HBr (89%)

3 (α_D −36°)

[1] M. Pfau, G. Revial, A. Guingant, and J. d'Angelo, *Am. Soc.*, **107**, 273 (1985).
[2] P. K. Subramanian and R. W. Woodard, *Syn. Comm.*, **16**, 337 (1986).

Methyl N-benzyloxycarbonyl-2-chloroglycinate $\overset{\overset{\displaystyle Cl}{|}}{\underset{\underset{\displaystyle NHCbo}{|}}{HC}}$—COOCH₃ **(1),** m.p. 70°. The reagent is prepared by reaction of methyl N-benzyloxycarbonyl-α-methoxyglycinate with PCl₅ in CCl₄ (82.5% yield).[1]

β, γ-*Amino acids*.[2]   Vinyl Grignard reagents react with **1** in THF at −70° to give N-protected β, γ-unsaturated amino acids in 55–65% yield.

Example:

$$CH_2\!\!=\!\!CHMgBr + 1 \xrightarrow[\underset{65\%}{}]{\overset{THF,}{-70°}} CH_2\!\!=\!\!CH\underset{\underset{\displaystyle NHCbo}{|}}{C}HCOOCH_3$$

[1] Z. Berntein and D. Ben-Ishai, *Tetrahedron*, **33**, 881 (1977).
[2] A. L. Castelhano, S. Horne, R. Billedeau, and A. Krantz, *Tetrahedron Letters*, **27**, 2435 (1986).

Methyl bis(trifluoroethoxy)phosphinylacetate, $(CF_3CH_2O)_2\overset{\overset{\displaystyle O}{\|}}{P}CH_2COOCH_3$ **(1).**

(Z)-α,β-*Unsaturated esters*.[1]   A key step in a total synthesis of N-acetylneuraminic acid **(5)**, a component of complex carbohydrates, is the Wittig-Horner reaction of the aldehyde **2**, obtained in several steps from the cyclocondensation of an oxygenated diene with an aldehyde, with the Still reagent **(1)** to provide the (Z)-unsaturated ester **3**. This product undergoes stereospecific dihydroxylation on reaction with OsO₄ in pyridine to give **4**. Remaining steps to **5** include oxidative

2, TBS=SiMe₂-t-Bu   3

4   5

degradation (RuO_2—$NaIO_4$) of the furan group to a carboxyl group, migration of the benzoyl group from C_5 to C_4, and introduction of the amino group at C_5.

[1] S. J. Danishefsky and M. P. DeNinno, *J. Org.*, **51**, 2615 (1986).

N-Methyl-N,O-bis(trimethylsilyl)hydroxylamine, $CH_3N\overset{\displaystyle Si(CH_3)_3}{\underset{\displaystyle OSi(CH_3)_3}{}}$ **(1).**

This reagent is obtained in 52% yield by reaction of $CH_3NHOH\cdot HCl$ with $ClSi(CH_3)_3$ (2 equiv.) and $N(C_2H_5)_3$ (excess) in ether at 25°.

N-*Methylnitrones*.[1] This reagent reacts with aldehydes and ketones to form a hemiaminal (**2**) that decomposes at 50° to an N-methylnitrone (**3**) in generally high yield. With deactivated substrates, addition of trimethylsilyl triflate is useful.

Sequential [3 + 2] cycloaddition can be carried out *in situ*.
Examples:

[1] J. A. Robl and J. R. Hwu, *J. Org.*, **50**, 5913 (1985), *idem, Org. Syn.*, submitted (1986).

Methyl (E)-4-(*t*-butyldimethyl-silyloxy)-3-methyl-2-butenoate, $R_3SiOCH_2\overset{\displaystyle}{\underset{\displaystyle CH_3}{}}C=C\overset{\displaystyle H}{\underset{\displaystyle COOCH_3}{}}$ **(1).**

The dimethylacrylate **1** is readily available from dimethylacrylic acid.

β-*Methylene-γ-butyrolactones*.[1] Alkylation of β-methylene-γ-butyrolactone itself is impracticable, but substituted β-methylene-γ-butyrolactones can be prepared by deprotonation and alkylation of **1** to afford **2**, which cyclizes to a β-methylene-γ-butyrolactone (**3**) on treatment with 40% aqueous HF in acetonitrile (**9**, 238–239).

Example:

[1] A. E. Greene, F. Coelho, J.-P. Déprés, and T. J. Brocksom, *J. Org.*, **50**, 1973 (1985).

Methyllithium.

Regiospecific synthesis of naphthols.[1] The *o*-allylbenzamide **1**, obtained by directed *o*-metallation of tertiary amides (**12**, 97–99, and previous references cited), on treatment with CH$_3$Li (superior to LDA, *sec*-BuLi) cyclizes to the naphthol **2**. This reaction provides an attractive route to naphthoquinones (equation I).

A similar cyclization of the *o*-crotylbenzamide **3** provides 2-methyl-1-naphthol in 50% yield.

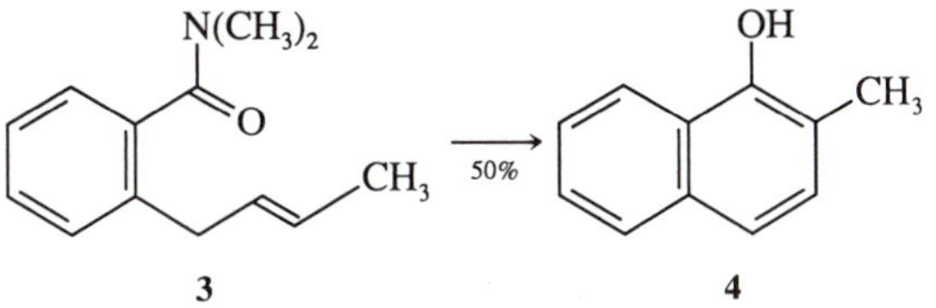

3   **4**

(E)-Allylsilanes.[2]   These silanes can be prepared regio- and stereoselectively by hydroalumination of (chloromethyl)dimethylsilyl-1-alkynes (**1**) to give the *cis*-adduct **2**. Reaction of CH₃Li (3 equiv.) with **2** results in (E)-allyltrimethylsilanes (**3**).

Example:

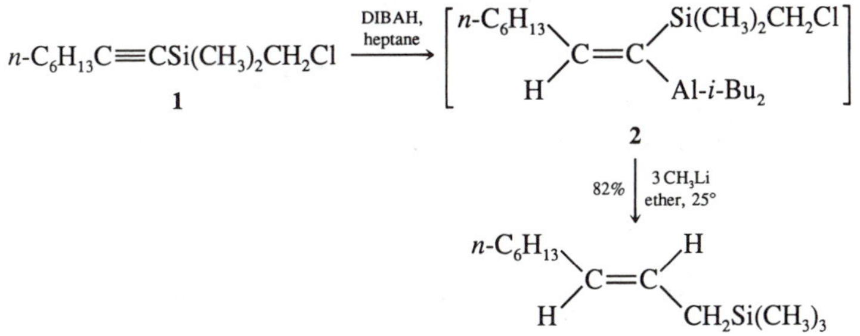

[1] M. P. Sibi, J. W. Dankwardt, and V. Snieckus. *J. Org.*, **51**, 271 (1986).
[2] H. Shiragami, T. Kawamoto, K. Utimoto, and H. Nozaki, *Tetrahedron Letters*, **27**, 589 (1986).

Methylmagnesium N-cyclohexylisopropylamide $CH_3MgN\begin{smallmatrix} C_6H_{11}\text{-}c \\ CH(CH_3)_2 \end{smallmatrix}$ (MMA, **1**).

The amide is prepared by reaction of lithium N-cyclohexylisopropylamide with methylmagnesium bromide.

Isomerization of epoxides to allylic alcohols.   The isomerization of (−)-vernolic acid (**1**) to (+)-coriolic acid (**2**) is best effected with this magnesium amide in toluene at 0°.[1]

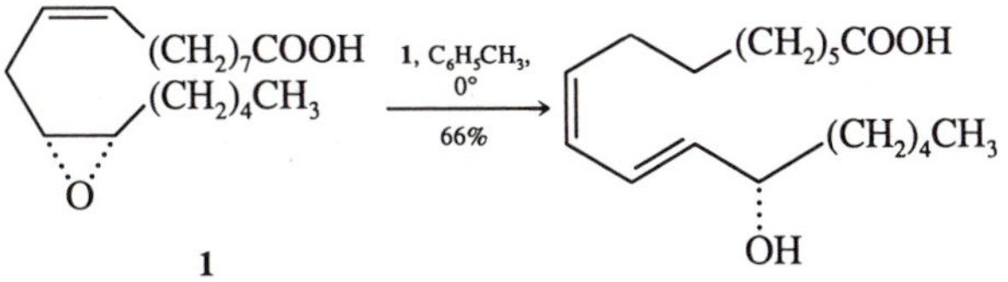

The reaction of the amide with epoxides can also result in nucleophilic addition of a methyl group.[2]

Examples:

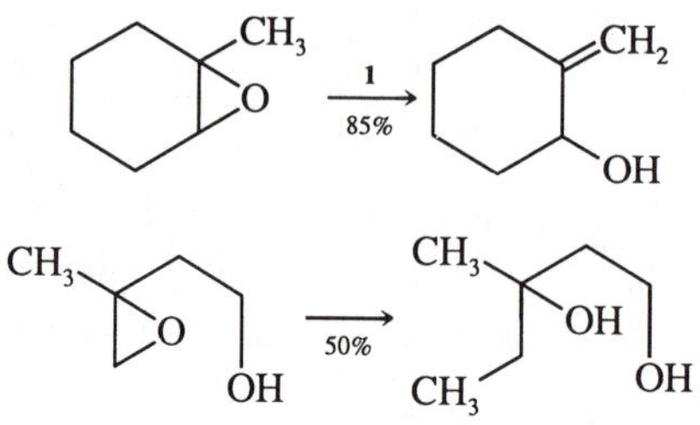

[1] C. A. Moustakis, D. K. Weerasinghe, P. Mosset, J. R. Falck, and C. Mioskowski, *Tetrahedron Letters*, **27**, 303 (1986).
[2] P. Mosset, S. Manna, J. Viala, and J. R. Falck, *ibid.*, **27**, 299 (1986).

2-Methyl-3-(phenylthio)-1,3-butadiene (1).

Preparation:

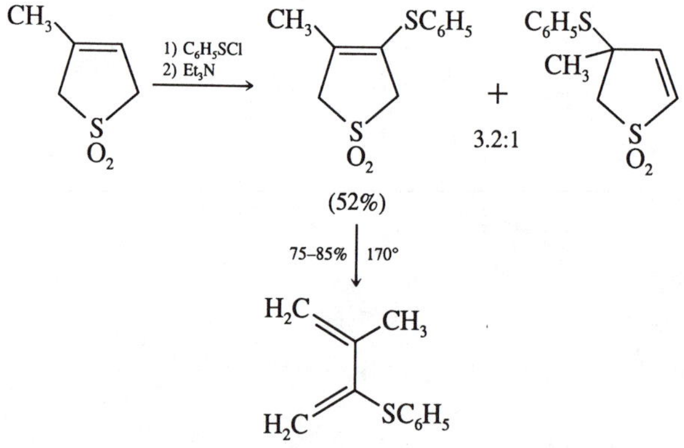

Regioselective Diels-Alder reactions.[1]   The phenylthio group controls the regioselectivity of the reaction of **1** with monosubstituted dienophiles.

Example:

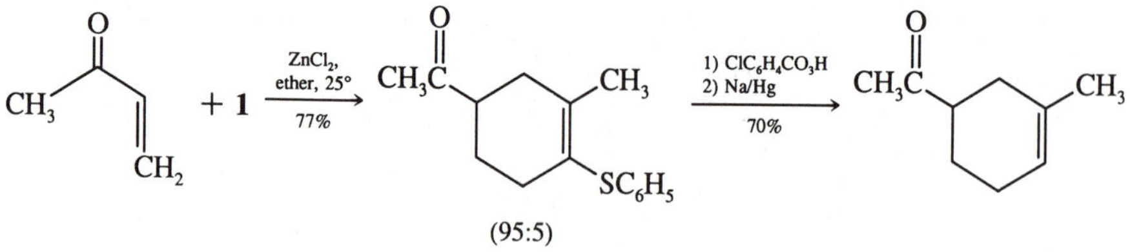

[1] P. J. Proteau and P. B. Hopkins, *J. Org.*, **50**, 141 (1985).

3-Methyl-3-(phenylthio)-1-butene, CH_2=$CHC(CH_3)_2SC_6H_5$ **(1).** This sulfide (b.p. 51°/0.5 mm.) is obtained by reaction of C_6H_5SH and isoprene in ether catalyzed by sulfuric acid. It rearranges when heated to the primary allylic isomer.[1]

Radical prenylation.[2]   The allylstannane corresponding to **1** cannot be used for radical prenylation because of facile isomerization. However, prenylation of alkyl halides can be effected by irradiation of **1** (3 equiv.) and hexabutylditin (1.5 equiv.) Presumably **1** reacts with $Bu_3Sn\cdot$ to form an allylstannane. The yields of this process are not as high as those observed with allyltributyltin (**11**, 15–16).

Examples:

$$HO(CH_2)_3Br + 1 \xrightarrow[69\%]{\overset{(Bu_3Sn)_2}{h\nu}} HO(CH_2)_4CH=C(CH_3)_2$$

A similar reaction can be carried out with 3-(phenylthio)-1-butene, CH_2=$CHCH(CH_3)SC_6H_5$; in this reaction a mixture of (Z)- and (E)-isomers is formed, with the latter predominating.

[1] S. N. Lewis. J. J. Miller, and S. Winstein, *J. Org.*, **37**, 1478 (1972).
[2] G. E. Keck and J. H. Byers, *ibid.*, **50**, 5442 (1985).

N,N-Methylphenylaminotributylphosphonium iodide (1).
Murahashi alkylation.   Murahashi alkylation (**8**, 346–347) of the optically active allylic alcohol **2** with butyllithium results in almost exclusive (99%) γ-alkylation to give a mixture of (S)-(E)-**3** and (R)-(Z)-**3** with only slight loss of chirality and with predominant *syn*-stereochemistry.[1] This *syn*-stereochemistry is opposite

to that observed in alkylation of a cyclic allylic alcohol, 5-methyl-2-cyclohexenyl alcohol, where *anti*-γ-alkylation prevails.[2]

[1] H. L. Goering and C. C. Tseng, *J. Org.*, **50**, 1597 (1985).
[2] H. L. Goering and S. S. Kantner, *ibid.*, **46**, 2144 (1981).

Methylthiomethyllithium, CH_3SCH_2Li (1).

Spiro epoxidation of α,β-enones.[1]   The conversion of ketones to spiro epoxides is usually carried out with dimethylsulfonium methylide, but this reaction can proceed in low and variable yields when extended to α,β-enones. A generally useful route to these vinyl spiro epoxides involves addition of **1**, methylation, and ring closure with base, as illustrated for cyclohexenone.

[1] S. P. Tanis, M. C. McMills, and P. M. Herrinton, *J. Org.*, **50**, 5887 (1985).

Methylthiomethyl p-tolyl sulfone (1).

Synthesis of aldehydes and ketones.[1]   The carbanion of **1** can be generated by aqueous NaOH and can be alkylated with use of methyltrioctylammonium chloride as phase-transfer catalyst. A second alkylation is possible using NaH in DMF as base. The dialkylated product (**3**) can be hydrolyzed to a ketone by HCl or H_2SO_4 in CH_3OH (equation I).

Surprisingly, the monoalkylated products (**2**) resist acid-catalyzed hydrolysis, but can be converted into aldehydes by irradiation in dioxane-water with a low-pressure Hg arc lamp, usually in the presence of $NaHCO_3$. Yields are in the range 25–70%.

[1] K. Ogura, K. Ohtsuki, M. Nakamura, N. Yahata, K. Takahashi, and H. Iida, *Tetrahedron Letters*, **26**, 2455 (1985).

Methyl (trifluoromethylsulfonyl)methyl sulfone $CF_3SO_2CH_2\overset{\alpha}{\underset{}{\overset{O}{\underset{O}{\|}}}}\overset{\alpha'}{SCH_3}$ (mesyltriflone,

1). The sulfone is obtained by reaction of dimethyl sulfone with ethylmagnesium bromide and then with triflyl fluoride (CF_3SO_2F, 3 M Co.), m.p. 115–116, yield 54%.

Alkene synthesis.[1] This sulfone can be subjected to successive alkylation in the order shown in equation (I) to give a tetraalkylated derivative, but the fourth alkylation requires elevated temperatures and an activated reagent such as benzyl bromide. The alkylated derivatives on treatment with base undergo a Ramberg-Bäcklund reaction to give alkenes.

(I) **1** $\xrightarrow[\begin{subarray}{l}1)\ BuLi,\ R^1X\\2)\ BuLi,\ R^2X\\3)\ BuLi,\ R^3X\\4)\ BuLi,\ R^4X\end{subarray}]{}$ $CF_3SO_2\overset{R^1\ R^4}{\underset{}{C}}\overset{O}{\underset{O}{\overset{\|}{\underset{\|}{S}}}}CH\overset{R^2}{\underset{R^3}{}}\xrightarrow{\text{base}} R^1R^4C{=}CR^2R^3 + SO_2 + CF_3SO_2H$

This general alkene synthesis has been applied to a synthesis of cyclopentenones, such as 2,3-dimethylcyclopentenone (equation II).[2]

(II) **1** $\xrightarrow[95\%]{\begin{subarray}{l}3\ BuLi,\\2\ CH_3I\end{subarray}}$ $CF_3SO_2\underset{CH_3}{\overset{O\ \ O}{\underset{}{C}HS}CH_2CH_3}$ $\xrightarrow[CH_2=CHCHO]{BuLi,}$

$\xrightarrow[30\%]{\text{Jones oxid.}}$

$\xleftarrow[100\%]{K_2CO_3}$

[1] J. B. Hendrickson, G. J. Boudreaux, and P. S. Palumbo, *Tetrahedron Letters*, **25**, 4617 (1984).

[2] J. B. Hendrickson and P. S. Palumbo, *ibid.*, **26**, 2849 (1985); *idem, J. Org.*, **50**, 2110 (1985).

Methyl vinyl ketone.

1,5-Dicarbonyl compounds.[1] The reaction of enol silyl ethers with methyl vinyl ketone catalyzed by BF_3 etherate results in 1,5-dicarbonyl compounds. Almost quantitative yields can be obtained, even from hindered ketones, by addition of an alcohol or even, to a less extent, of water.

Example:

$$+ CH_3OH \qquad . \quad 84\%$$
$$+ C_6H_5CH(OH)CH_3 \qquad 100\%$$

[1] P. Duhamel, L. Hennequin, N. Poirier, and J.-M. Poirier, *Tetrahedron Letters*, **26**, 6201 (1985).

Molybdenum carbonyl.

Functionalization of cycloheptadienyl-$\overset{+}{M}o(CO)_2CpPF_6^-$ (1).[1] This complex (**1**), prepared as shown, reacts with carbon nucleophiles to give products of stereoselective allylic substitution.

Reaction of a monosubstituted complex with a second nucleophile can also proceed stereo- and regioselectively with preferential attack at the other allylic position to give *cis*-disubstituted complexes.

Example:

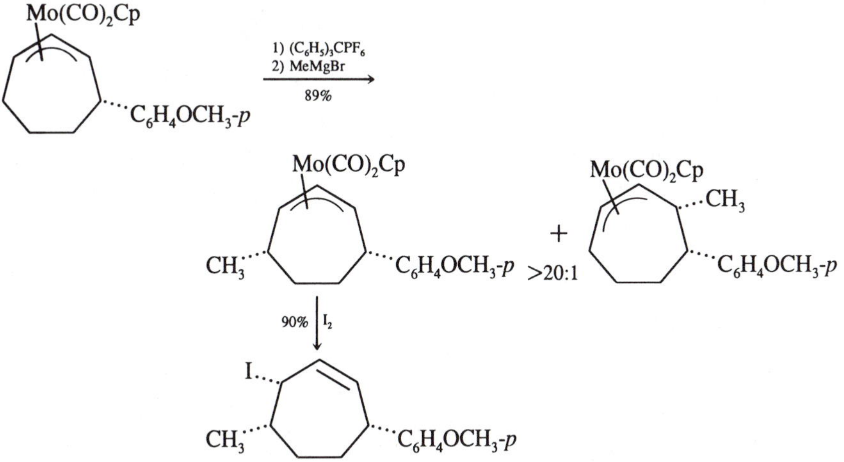

Mo(CO)₂Cp

1) (C₆H₅)₃CPF₆
2) MeMgBr
⟶
89%

Mo(CO)₂Cp

CH₃⋯ ⋯C₆H₄OCH₃-p >20:1

+

Mo(CO)₂Cp
⋯CH₃

⋯C₆H₄OCH₃-p

90% | I₂

I⋯

CH₃⋯ ⋯C₆H₄OCH₃-p

***Reductive cleavage of isoxazoles.*[2]** Isoxazoles (**1**) are reductively cleaved to β-amino-α,β-unsaturated aldehydes or ketones when refluxed in moist acetonitrile containing $Mo(CO)_6$ (0.2–1.0 equiv.).

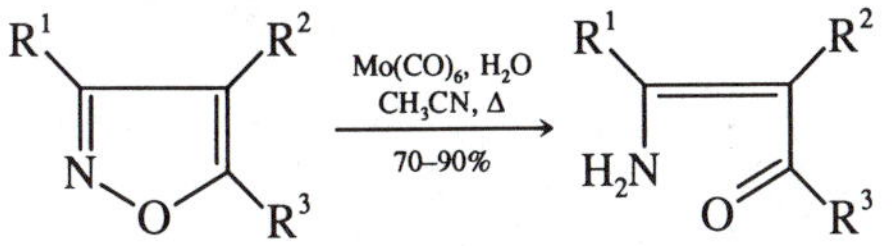

$$R^1 \quad\quad R^2$$
N–O R³

Mo(CO)₆, H₂O
CH₃CN, Δ
⟶
70–90%

$$R^1 \quad\quad R^2$$
H₂N   O R³

1, R^1, R^2, R^3 =
H, CH₃, C₆H₅

[1] A. J. Pearson and M. N. I. Khan, *J. Org.*, **50**, 5276 (1985).
[2] M. Nitta and T. Kobayashi, *J.C.S. Perkin I*, 1401 (1985).

2-(Morpholino)acrylonitrile $CH_2=C$ CN N O   (**1**), m.p. 64°.

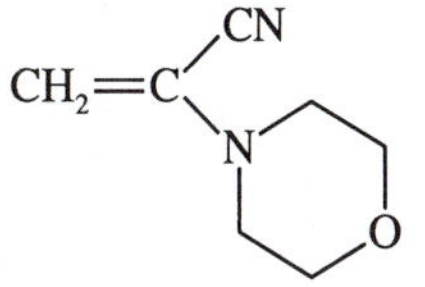

Preparation:

ClCH₂CHO

1) HN O, HCl
2) NaCN
⟶

$$\left[ClCH_2CH-N \quad O \right]$$
CN

NaOH
⟶
54%
1

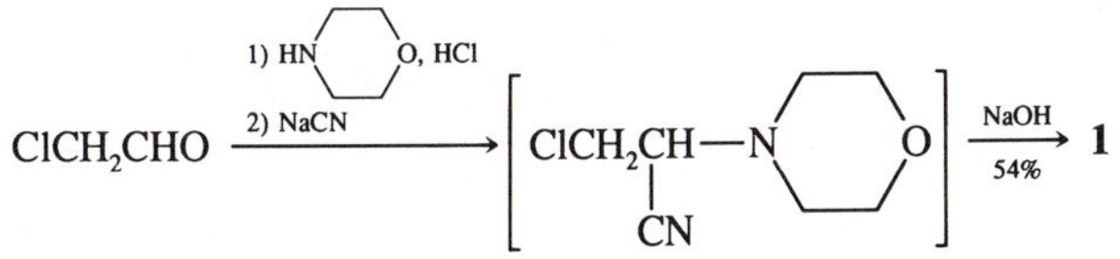

Diels-Alder reactions; cyclohexenones.   The reagent reacts with 1,3-dienes (excess) at 160° to form a mixture of 1:1 adducts in 72–92% yield. These can be hydrolyzed to cyclohexenones.[1]

Example:

[1] J.-L. Boucher and L. Stella, *Tetrahedron*, **41**, 875 (1985).

N

(CH₃)₂B   B(CH₃)₂

1,8-Naphthalenediylbis(dimethylborane),   (1),   air-sensitive

orange oil. The borane is obtained in 43% yield (pure) by reaction of 1,8-dilithio-naphthalene with $(CH_3)_2BOC_2H_5$ (BF_3 etherate catalysis).

Hydride sponge. This compound abstracts hydride from triorganoborohy-drides and Cp_2ZrClH.[1] It also forms an adduct with F^- but interacts only weakly with Cl^- and Br^-.

[1] H. E. Katz, *Am. Soc.*, **107**, 1420 (1985).

Nickel. An activated nickel(0) can be prepared by reduction of $NiCl_2$ with lithium in the presence of naphthalene.

Reformatsky-type reactions.[1] This Ni(0) can effect selective addition of halo-acetonitriles to aldehydes.

Example:

$$BrCH_2CN \ + \ C_3H_7 \diagdown \diagup CHO \xrightarrow[76\%]{\substack{Ni(0), \\ glyme, 85°}} C_3H_7 \diagdown \diagup CH(OH)CH_2CN$$

[1] S. Inaba and R. D. Rieke, *Tetrahedron Letters*, **26**, 155 (1985).

Nickel boride (1).
Reduction of derivatives of allylic alcohols. Nickel boride can effect reduc-tion of allylic alcohols to alkenes, but yields are generally improved by reduction of the acetates, benzoates, or trifluoroacetates.[1] Reduction of allylic benzyl ethers to alkenes is effected in higher yield with Raney nickel. Methyl ethers are not reduced by either reagent. The trimethylsilyl ethers of allylic alcohols are reduced to alkenes by nickel boride in diglyme.[2]

Example:

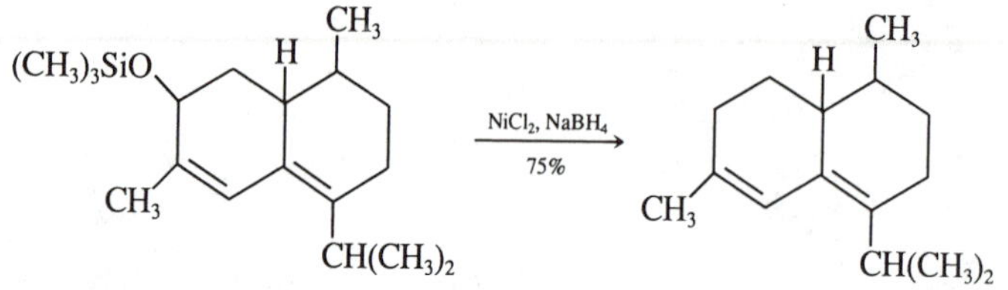
(CH3)3SiO
H
CH3
CH3
CH(CH3)2
NiCl2, NaBH4
75%
H
CH3
CH3
CH(CH3)2

Dehalogenation.[3] α-Bromo or α-chloro ketones are reduced by nickel boride in DMF to ketones in 70–95% yield. *vic*-Dibromides are reduced to alkenes in 80–90% yield.

(Z)-α,β-Unsaturated aldehydes.[4] A nickel boride catalyst similar to P-2 nickel boride is obtained by reaction of $NiCl_2$ and excess $NaBH_4$ in C_2H_5OH. It effects selective hydrogenation of α,β-alkynal acetals to the (Z)-α,β-alkenal acetals.

Example:

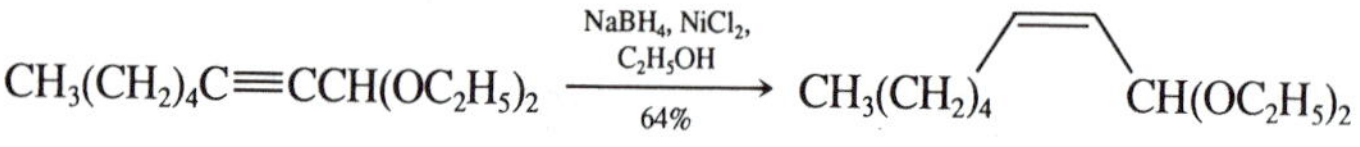
CH3(CH2)4C≡CCH(OC2H5)2
NaBH4, NiCl2,
C2H5OH
64%
CH3(CH2)4
CH(OC2H5)2

Z/E = 97:1

[1] D. N. Sarma and R. P. Sharma, *Tetrahedron Letters*, **26**, 2581 (1985).
[2] Idem, *ibid.*, **26**, 371 (1985).
[3] J. C. Sarma, M. Borbaruah, and R. P. Sharma, *ibid.*, **26**, 4657 (1985).
[4] B. Byrne, L. M. Lafleur Lawter, and K. J. Wengenroth, *J. Org.*, **51**, 2607 (1986).

Nickel carbonyl.

2,3,5-Trisubstituted-2-cyclopentenones.[1] A short synthesis of methylenomycin B (**3**) involves reaction of a mixture of allyl chloride and 2-butyn-1-ol with

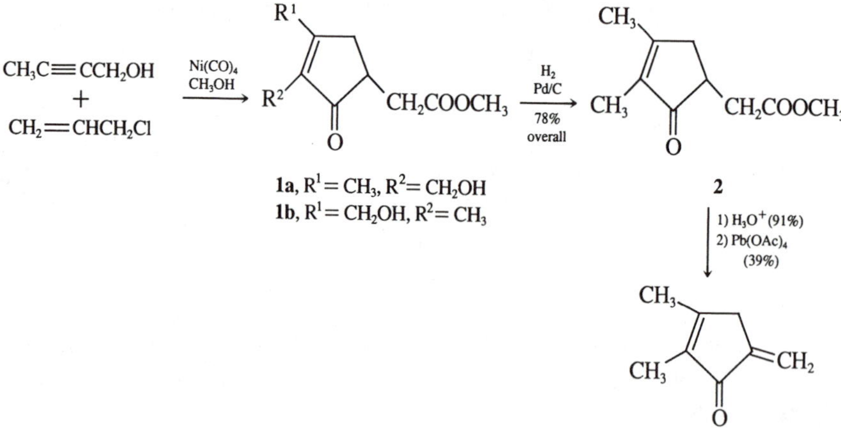
CH3C≡CCH2OH
+
CH2=CHCH2Cl
Ni(CO)4
CH3OH
R1
R2
CH2COOCH3
O
H2
Pd/C
78%
overall
CH3
CH3
CH2COOCH3
O
1) H3O+ (91%)
2) Pb(OAc)4
(39%)
CH3
CH3
CH2
O

1a, $R^1 = CH_3$, $R^2 = CH_2OH$
1b, $R^1 = CH_2OH$, $R^2 = CH_3$

2

3

Ni(CO)$_4$ in CH$_3$OH to give a mixture of **1a** and **1b** which was converted into the cyclopentenone **2**. Final steps involved hydrolysis and oxidative decarboxylation.

π-Allylnickel halides. Billington[2] has reviewed the preparation of these complexes from allylic halides using Ni(CO)$_4$ or Ni(COD)$_2$, and their use in synthesis, mainly of natural products (54 references). These complexes react with a wide range of both aliphatic and aryl bromides or iodides as well as aldehydes, ketones, epoxides, and quinones. One advantage is that both allyl ligands react. They do not react with acid chlorides, esters, ethers, nitriles, or acetals.

[1] F. Camps, J. Coll, J. M. Moretó, and J. Torras, *Tetrahedron Letters*, **26**, 6397 (1985).
[2] D. C. Billington, *Chem. Soc. Reviews*, **14**, 93 (1985).

o-**Nitrobenzyl alcohol,** *o*-NO$_2$C$_6$H$_4$CH$_2$OH (**1**), m.p. 70–72°.

Protection of aldehydes and ketones.[1] Bis-*o*-nitrobenzyl acetals or ketals are removable in 85–95% yield on irradiation at 350 nm in benzene. The acetals or ketals are easily prepared from **1** by an exchange reaction using 2,2-dimethoxy-propane (**1**, 268–269) catalyzed by an arenesulfonic acid. In the case of hindered ketones (17-keto steroids), the glycol *o*-NO$_2$C$_6$H$_4$CH(OH)CH$_2$OH (**2**) can be used.

[1] D. Gravel, S. Murray, and G. Ladouceur, *J.C.S. Chem. Comm.*, 1828 (1985).

Nitromethane.

Anthracyclinones.[1] In a short synthesis of these quinones, nitromethane is used to introduce the C$_{10}$-group simultaneously with the C$_9$-hydroxyl group (equation I).

(I)

R^1 = H, OCH$_3$
R^2 = H, OH
R^3 = H, CH$_3$

[1] K. Krohn and W. Priyono, *Angew. Chem. Int. Ed.*, **25**, 339 (1986).

2-Nitropropane.

gem-Dimethylcyclopropanation.[1]   The potassium salt (1) of 2-nitropropane, prepared with KOH or $KOC(CH_3)_3$ in DMSO, reacts with a variety of alkenes, particularly those with two electron-withdrawing groups, to form *gem*-dimethyl-cyclopropanes.

Example:

(E, 100%)

This cyclopropanation can be extended to anions of allylic, benzylic, and tertiary nitro compounds.

Example:

[1] N. Ono, T. Yanai, I. Hamamoto, A. Kamimura, and A. Kaji, *J. Org.*, **50**, 2806 (1985).

Norephedrine.

Aldol reactions.[1]   The chiral oxazolidine (1), formed from 3-pentanone and (−)-norephedrine, after conversion to the tin azaenolate reacts with aldehydes to give predominantly *anti*-aldols (2) in >90% ee. Reduction of the carbonyl group of the *anti*-aldol 2 provides (3S,4R)-4-methyl-3-heptanol (3) in 95% ee.

anti-**2**   syn-**2**

1) Ac$_2$O, N(C$_2$H$_5$)$_3$
2) NaBH$_4$
3) MsCl, N(C$_2$H$_5$)$_3$
4) LiAlH$_4$

3 (95% ee)

[1] K. Narasaka, *Pure and Appl. Chem.*, **57**, 1883 (1985).

O

Organoaluminum reagents.

$Bu_3SnAl(C_2H_5)_2$ (**12**, 339–341). Reaction of allylic phosphates with this reagent (**1**) and a Pd(0) catalyst affords allyltin compounds, which react with aldehydes to produce homoallylic alcohols in 65–85% yield. The reaction involves predominant inversion of stereochemistry.[1]

Example:

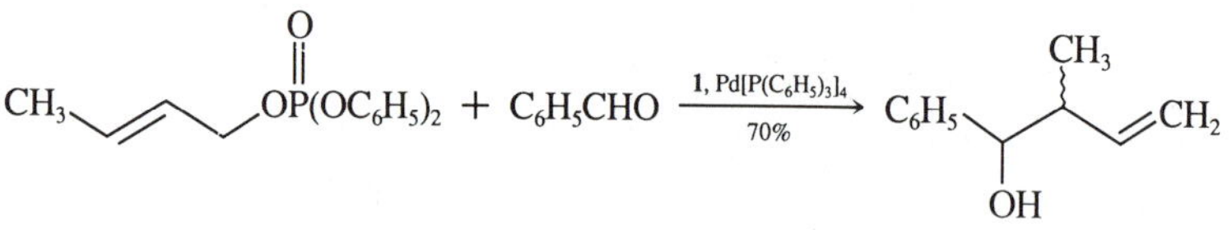

(*anti*/*syn* = 59:41)

Imines.[2] Bis(dichloroaluminum)phenylimide (**1**) is useful for conversion of ketones as well as aldehydes to imines because the by-product is a stable dialuminoxane (equation I). In addition the reagent converts acid chlorides into amides. The reagent is particularly useful for preparation of the anil (**3**) of chalcone (**2**).

(I) 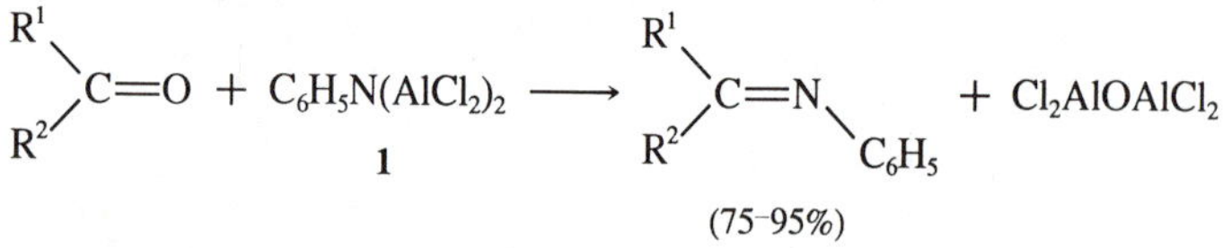

(II) 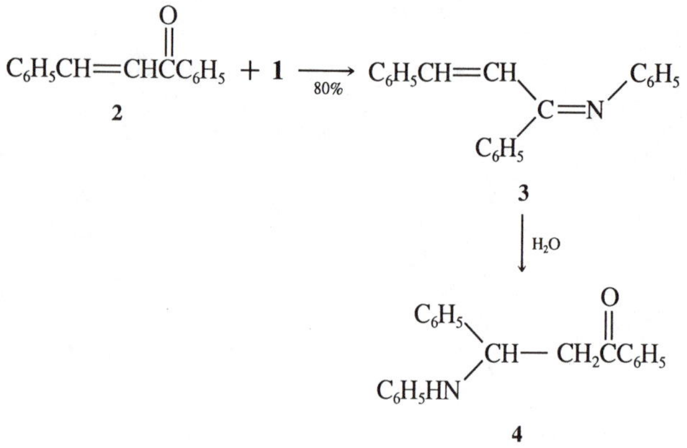

Methylaluminum bis(2,6-di-t-butyl-4-methylphenoxide) (*MAD*, 1); *methylaluminum bis(2,4,6-tri-t-butylphenoxide)* (*MAT*, 2). The complexes formed from 4-*t*-butylcyclohexanone and **1** or **2** (3 equiv.) react with methyllithium or CH_3MgBr

1, R = CH_3 (MAD)
2, R = $C(CH_3)_3$ (MAT)

with almost exclusive axial methylation (equation I). Complexation of 2- and 3-methylcyclohexanone with **1** and **2** also favors axial attack of RLi and RMgBr to afford equatorial alcohols. However, acyclic ketones and esters do not undergo

CH_3Li		R = CH_3	21:79
CH_3Li + MAD	84%	R = CH_3	99:1
CH_3Li + MAT	92%	R = CH_3	99.5:0.5
BuMgBr	58%	R = Bu	44:56
BuMgBr + MAD	67%	R = Bu	100:0

MAD-mediated alkylation. α-Substituted aldehydes complexed with **1** or **2** undergo *anti*-selective alkylation with Grignard reagents. In general **2** is more effective than **1**.[3]

Example:

C_2H_5MgBr	78%	84:16
C_2H_5MgBr + MAT	98%	20:80
BuMgBr + MAT	98%	33:67

C-Glycosides.[4]   A variety of organoaluminum reagents react with furanosyl or pyranosyl fluorides to form C-glycosides in 68–93% yield.

Examples:

α-F	$Al(C_2H_5)_3$	79%	$R = C_2H_5$, α/β = >20:1
β-F	$Al(C_2H_5)_3$	76%	$R = C_2H_5$, α/β = >20:1
α-F	$i\text{-}Bu_2AlCH{=}CHC_6H_{13}$	76%	$R = CH{=}CHC_6H_{13}$, α/β = >20:1

This reaction also effects stereoselective alkylation of the 6-fluoro-1,6-anhydro-glucose derivative (**1**).

Substitution of sulfones.[5]   Organoaluminum reagents, particularly vinyl- or alkynylalanes, undergo Lewis acid catalyzed substitution reactions with sulfones. Thus the vinylalane **2** reacts with the allylic sulfone **1** in the presence of $AlCl_3$ to

give mainly **3** (9:1). Substitution of the sulfone always occurs at the less hindered side of an allyl system, but not necessarily with allyl inversion.

anti-*Selective aldol reactions*.[6] The reaction of the lithium enolate **2** (BuLi, $-78°$) of ethyl trityl ketone (**1**) with aromatic aldehydes results in the *syn*-aldols (**3**) in 96–99:4–1 selectivity. The reaction of **1** with $Al(CH_3)_3$ and an aromatic

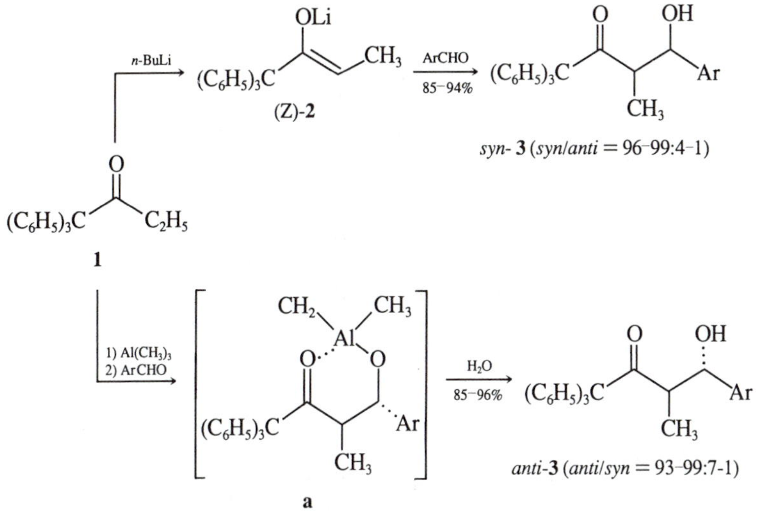

aldehyde for short reaction periods also results in *syn*-aldols (**3**), but if the reaction stands for several days before quenching with water the *anti*-aldol is obtained. The stereoselectivity is ascribed to isomerization to a chelated intermediate (**a**).

The MEM ethers of **3** are cleaved by lithium triethylborohydride to the MEM ethers of 1,3-diols.

Example:

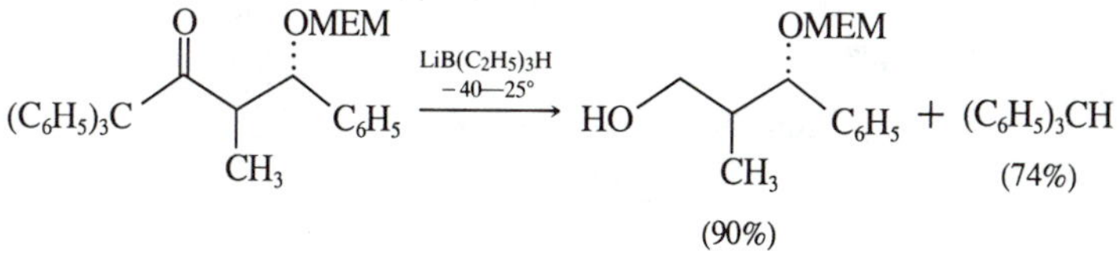

***Asymmetric enolates of meso-ketones*.[7]** The ketal **1**, formed from 4-ethyl-cyclohexanone and (2R,4R)-2,4-pentanediol, when treated with triisobutylalu-minum in CH_2Cl_2 at $-78°$ gives the enol ethers (S)-**2** and (R)-**2** in the ratio 6:1. Somewhat superior selectivity obtains by use of a dialkylaluminum amide such as **3**.[8]

$Al[CH_2CH(CH_3)_2]_3$ 99%
3, toluene, 0°   91%

86:14
92:8

$$c\text{-}C_6H_{11}CH \overset{\overset{\textstyle C(CH_3)_3}{|}}{\underset{\underset{\textstyle C(CH_3)_3}{|}}{}}\!\!-NAl[CH_2CH(CH_3)_2]_2$$

3

[1] S. Matsubara, K. Wakamatsu, Y. Morizawa, N. Tsuboniwa, K. Oshima, and H. Nozaki, *Bull. Chem. Soc. Japan*, **58**, 1196 (1985).
[2] J. J. Eisch and R. Sanchez, *J. Org.*, **51**, 1848 (1986).
[3] K. Maruoka, T. Itoh, and H. Yamamoto, *Am. Soc.*, **107**, 4573 (1985).
[4] G. H. Posner and S. R. Haines, *Tetrahedron Letters*, **26**, 1823 (1985).
[5] B. M. Trost and M. R. Ghadiri, *Am. Soc.*, **108**, 1098 (1986).
[6] M. Ertas and D. Seebach, *Helv.*, **68**, 961 (1985).
[7] Y. Naruse and H. Yamamoto, *Tetrahedron Letters*, **27**, 1363 (1986).
[8] R. R. Fraser and T. S. Mansour, *J. Org.*, **49**, 3442 (1984) have prepared the secondary amine from which **3** is obtained.

Organocerium reagents.

(*2-Trimethylsilylethynyl)cerium(III) chloride*, $(CH_3)_3SiC\equiv CCeCl_2$ (**1**); *tris(2-trimethylsilylethynyl)cerium(III)*, $[(CH_3)_3SiC\equiv C]_3Ce$ (**2**).   The reagents are prepared by reaction of $(CH_3)_3SiC\equiv CLi$ with 1 equiv. or 0.33 equiv. of $CeCl_3$, respectively. The reagents are particularly useful for the conversion of $\rangle\!\!=\!O$ into $\overset{\displaystyle COCH_3}{\underset{\displaystyle OH}{\diagup\!\!\diagdown}}$, a group present in anthracyclinones.[1]

Example:

1   62%
2   55%

$$100\% \downarrow \begin{array}{l} \text{HgO, H}_2\text{SO}_4, \\ \text{H}_2\text{O} \end{array}$$

The simpler reagents $HC\equiv CCeCl_2$ and $(HC\equiv C)_3Ce$ are less useful because of lower yields of 1,2-adducts.

[1] M. Suzuki, Y. Kimura, and S. Terashima, *Chem. Pharm. Bull.*, **34**, 1531 (1986).

Organocopper reagents.

L-*Amino acids*.[1]   A wide variety of optically pure L-amino acids (**2**) can be obtained by addition of lithium dialkylcuprates to the bromo derivative (**1**) of homoserine obtained from L-α-amino-γ-butyrolactone.

A similar reaction with halo or tosyl derivatives of L-serine is less useful because of competing β-elimination.

Coupling of vinylcopper reagents with acyl halides.[2]   (Z)-Vinylcuprates can couple directly with acetyl bromide to give (Z)-α,β-enones in moderate yield (equation I). Vinylcopper reagents (complexed with MgX_2) also undergo coupling with acyl halides or mixed anhydrides with retention of geometry of the double bond

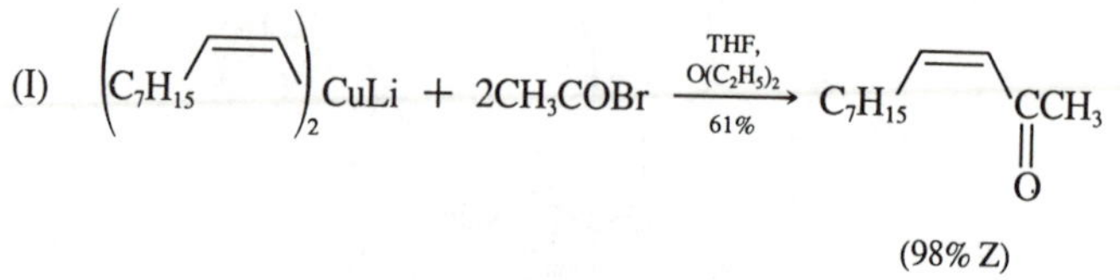

(I) $\left(C_7H_{15} \diagup\!\!\!\diagdown \right)_2 CuLi + 2CH_3COBr \xrightarrow[\substack{61\%}]{\substack{THF, \\ O(C_2H_5)_2}} C_7H_{15} \diagup\!\!\!\diagdown \underset{\underset{O}{\|}}{CCH_3}$

(98% Z)

when catalyzed by Pd(0). This coupling can be extended to synthesis of divinyl ketones (Nazarov reagents) and conjugated dienones.

Examples:

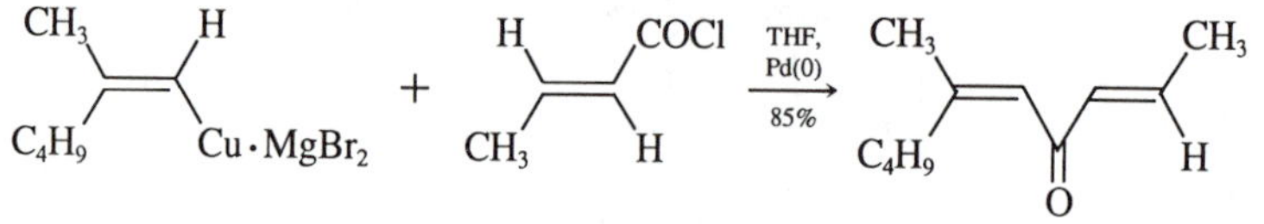

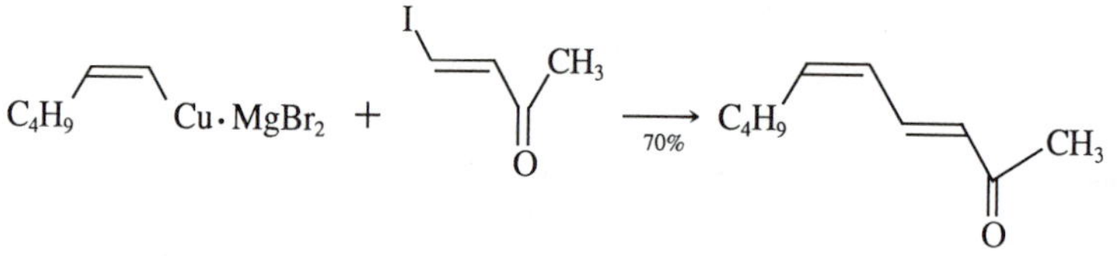

Alkylation of aziridines; amine synthesis.[3]   N-Substituted aziridines (alkyl, benzyl, silyl, Boc) react with lithium dialkylcuprates in the presence of BF$_3$ etherate (excess) with opening of the ring to give primary or secondary amines.

Examples:

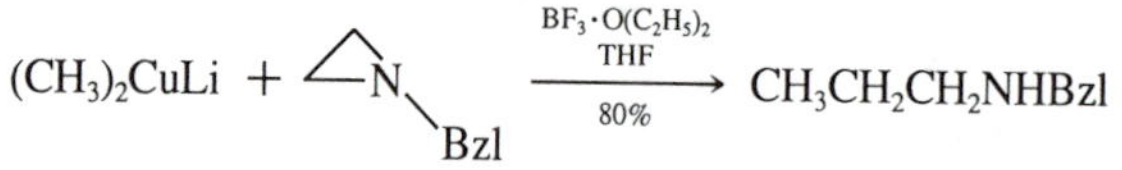

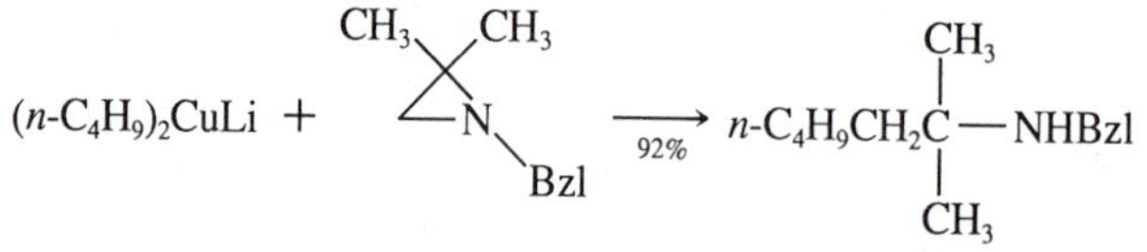

$(CH_3)_2CuLi + \underset{Bzl}{\overset{\triangle}{N}} \xrightarrow[\substack{80\%}]{\substack{BF_3 \cdot O(C_2H_5)_2 \\ THF}} CH_3CH_2CH_2NHBzl$

$(n\text{-}C_4H_9)_2CuLi + \underset{Bzl}{\overset{CH_3 \diagup\!\!\diagdown CH_3}{N}} \xrightarrow[92\%]{} n\text{-}C_4H_9CH_2\underset{\underset{CH_3}{|}}{\overset{\overset{CH_3}{|}}{C}}-NHBzl$

Reaction with allylic sulfoxides (1) or sulfones (2).[4]   The allylic sulfoxides or sulfones undergo γ-substitution on reaction with lithium dialkylcuprates in ether to provide trisubstituted (E)-alkenes selectively.

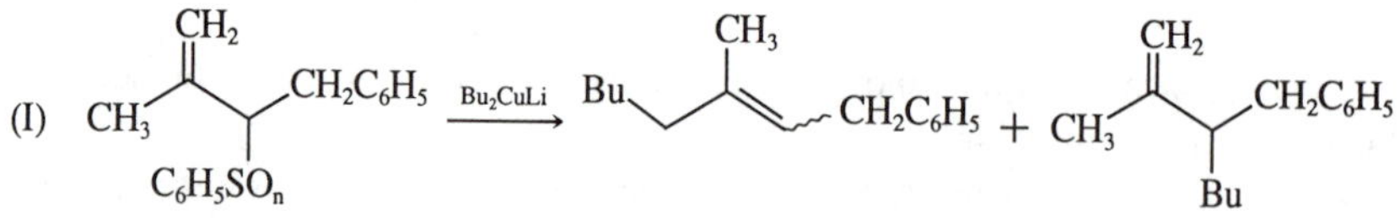

1, n = 1	75%	E/Z = 93/7	95:5
2, n = 2	89%	E/Z = 86/14	92:8

ArSCu(RMgX)$_n$.   These heterocuprates are more useful than lithium dialkyl-cuprates for conjugate addition to enones.[5] They are also useful for conjugate addition to the less reactive cinnamates and crotonates.[6] Yields are markedly improved by use of 2-methoxyphenylthio as the ligand in additions to the crotonates.

Example:

$$CH_3CH=CHCOOCH_3 \xrightarrow[\text{ether}]{ArSCu(CH_3MgBr)_n} \begin{array}{c} CH_3 \\ \diagdown \\ CH_3 \diagup \end{array} CHCH_2COOCH_3$$

Ar = C$_6$H$_5$ 33%
Ar = o-CH$_3$OC$_6$H$_4$ 67%

Acylcuprate reagents.[7]   The reaction of cuprates of the type R$_2$(CN)CuLi$_2$ with carbon monoxide at $-110°$ results in carbonylated reagents, possibly with the composition (RCO)R(CN)CuLi$_2$. In any case, these cuprates effect 1,4-acylation of α,β-enones and -enals to provide 1,4-dicarbonyl compounds.

Example:

$$(n\text{-}C_4H_9)_2CNCuLi_2 \xrightarrow[66\%]{\substack{1)\ CO,\ THF,\ ether \\ 2)\ CH_2=CHCOCH_3}} n\text{-}C_4H_9CCH_2CH_2CCH_3$$

with the two carbonyl groups shown as:

$$\underset{O}{\overset{\|}{}} \qquad \underset{O}{\overset{\|}{}}$$

Review.[8]   Taylor has reviewed the conjugate addition–enolate trapping reactions of organocopper reagents, in particular of lithium dialkylcuprates (131 references).

[1] J. A. Bajgrowicz, A. El Hallaovui, R. Jacquier, C. Pigiere, and P. Viallefont, *Tetrahedron*, **41**, 1833 (1985).
[2] N. Jabri, A. Alexakis, and J. F. Normant, *ibid.*, **42**, 1369 (1986); J. P. Foulon, M. Bourgain-Commercon, and J. F. Normant, *ibid.*, **42**, 1399 (1986).
[3] M. J. Eis and B. Ganem, *Tetrahedron Letters*, **26**, 1153 (1985).
[4] Y. Masaki, K. Sakuma, and K. Kaji, *J.C.S. Perkin I*, 1171 (1985).
[5] G. H. Posner, C. E. Whitten, and J. J. Sterling, *Am. Soc.*, **95**, 7788 (1973).
[6] M. Behforouz, T. T. Curran, and J. L. Bolan, *Tetrahedron Letters*, **27**, 3107 (1986).
[7] D. Seyferth and R. C. Hui, *Am. Soc.*, **107**, 4551 (1985).
[8] R. J. K. Taylor, *Synthesis*, 364 (1985).

Organolithium reagents.

Quinone synthesis.   Two laboratories[1,2] have found that the adducts formed by addition of an aryl-, alkynyl-, or heteroaryllithium to a cyclobutenedione rearrange when heated (138–160°) to hydroquinones, which are usually isolated as the quinone after air or chemical oxidation. The rearrangement involves an intermediate vinylketene. Similar rearrangement of the adducts of benzocyclobutenedione provides anthraquinones.

Examples:

See also Methyllithium (this volume).

[1] L. S. Liebeskind, S. Iyer, and C. F. Jewell, Jr., *J. Org.*, **51**, 3065 (1986).
[2] S. T. Perri, L. D. Foland, O. H. W. Decker, and H. W. Moore, *ibid.*, **51**, 3067 (1986).

Organomanganese(II) iodides.

Acylation of RMnI (**10**, 290).[1] Acylating agents such as acyl halides and anhydrides react with RMnI to form ketones in high yield. This reaction can be used to provide acetoxy ketones, keto esters, keto nitriles, and diketones in generally high yield.

Examples:

$$BuMnI + C_2H_5O\overset{O}{\overset{\|}{C}}\overset{O}{\overset{\|}{C}}(CH_2)_6CO_2C_2H_5 \xrightarrow[97\%]{} Bu\overset{O}{\overset{\|}{C}}(CH_2)_6CO_2C_2H_5$$

$$C_2H_5MnI + ClO\overset{O}{\overset{\|}{C}}(CH_2)_6\overset{}{C}Bu \xrightarrow[87\%]{} C_2H_5\overset{O}{\overset{\|}{C}}(CH_2)_6\overset{O}{\overset{\|}{C}}Bu$$

2,4,6-Cycloheptatrienyl ketones can be prepared by reaction of cycloheptatri-

enecarbonyl chloride (**2**) with RMnI in ether.[2] They can also be obtained in similar yield by reaction of a Grignard reagent catalyzed by tris(acetylacetonate)iron.[3]
Example:

2

[1] G. Friour, G. Cahiez, and J. F. Normant, *Synthesis*, 50 (1985).
[2] K. Ritter and M. Hanack, *Tetrahedron Letters*, **26**, 1285 (1985).
[3] V. Fiandanese, G. Marchese, V. Martina, and L. Ronzini, *ibid.*, **25**, 4805 (1984).

Organotellurium reagents.

Review.[1] This review covers preferred methods of preparation of the more important organotellurium reagents. Most of these are aromatic derivatives of tellurium, which are more stable and less volatile than aliphatic ones. Even though synthetic use of tellurium reagents is relatively new, the review covers 119 references, including several for the year 1985.

[1] N. Petragnani and J. V. Comasseto, *Synthesis*, 1 (1986).

Organotin reagents.

Trialkyl(trialkylsilyl)tin, $R_3^1SnSiR_3^2$. The reagents are readily available by reaction of R_3^1SnLi (**8**, 495) with $ClSiR_3^2$. These silylstannanes undergo *cis*-addition to 1-alkynes in the presence of $Pd[P(C_6H_5)_3]_4$ to give *cis*-1-trialkylsilyl-2-trialkyl-stannylalkenes (65–95% yield).[1] These silyltin alkenes, when R^2 is larger than CH_3, couple with acyl chlorides in the presence of bis(acetonitrile)dichloropalladium to give β-silyl-α,β-unsaturated ketones in good yield.[2]
Example:

A similar coupling with α,β-unsaturated acid chlorides provides β-silyl divinyl ketones (Nazarov reagents). These ketones cyclize in the presence of a Lewis acid, particularly BF_3 etherate, to cyclopentenones, generally with retention of the silyl group.

1,4-Addition to enones. Cyanide ion (KCN, 18-crown-6 or Bu$_4$NCN) catalyzes conjugate addition of these silyltin reagents to α-β-enones.[3] The addition is also catalyzed to a less degree by potassium t-butoxide, but fluoride ion is not effective. The reaction is sensitive to the steric bulk on silicon. Thus (CH$_3$)$_3$Si, C$_6$H$_5$(CH$_3$)$_2$Si, and Bu(CH$_3$)$_2$Si groups permit addition, but (C$_2$H$_5$)$_3$Si and (CH$_3$)$_3$C(CH$_3$)$_2$Si groups prevent addition. However, steric crowding in the enone is not a factor.
    Example:

Compounds of this type can serve as a synthon for an α,β-dianion of cyclohexanone.
    Example:

OSiMe$_2$-t-Bu

1) NaH; CH$_3$I
2) CH$_2$=CHCH$_2$Br (TASF)

56%

Review. Stille[4] has reviewed methods for synthesis of organotin reagents, particularly recent ones. Most involve addition of tin derivatives to alkenes and alkynes. The reagents are particularly useful for palladium-catalyzed coupling with acid chlorides, allylic, vinyl, and aryl halides. Essentially only one group of the organotin reagent undergoes transfer in the coupling. Alkynyl, alkenyl, aryl, and allyl groups are more readily transferred than simple alkyl groups.

[1] B. L. Chenard, E. D. Laganis, F. Davidson, T. V. RajanBabu, *J. Org.*, **50**, 3666 (1985); B. L. Chenard and C. M. Van Zyl, *ibid.*, **51**, 3561 (1986).
[2] B. L. Chenard, C. M. Van Zyl, and D. R. Sanderson, *Tetrahedron Letters*, **27**, 2801 (1986).
[3] B. L. Chenard, *ibid.*, **27**, 2805 (1986).
[4] J. K. Stille, *Angew. Chem. Int. Ed.*, **25**, 508 (1986).

Organotitanium reagents.

n-*Alkyltriisopropoxytitanium reagents*, RTi(O-*i*-Pr)$_3$ (1).   These reagents are readily prepared by reaction of *n*-alkyllithium or -magnesium chloride with chlorotriisopropoxytitanium. When R is methyl (**10**, 422, **12**, 297) the reagent can be distilled, but analogs are prepared and used in solution. Reagents containing *sec*- or *tert*-alkyl groups are not useful because of extensive β-hydride elimination. The great value of RTi(O-*i*-Pr)$_3$ reagents is their ability to undergo Grignard-type and aldol reactions with aldehyde groups selectively in the presence of keto groups (**12**, 356). However, the allyltitanium reagent **1** (R = allyl) is not completely aldehyde-

(I) $C_6H_5\overset{\text{O}}{\overset{\|}{C}}(CH_2)_4CHO + CH_2{=}CHCH_2\bar{T}i(O\text{-}i\text{-}Pr)_4\overset{+}{M}gCl \xrightarrow[90\%]{\text{THF, } -78°} C_6H_5\overset{\text{O}}{\overset{\|}{C}}(CH_2)_4\overset{\text{OH}}{\overset{\|}{C}}HCH_2CH{=}CH_2$

2

selective, but the ate complex (**2**), obtained by reaction of allylmagnesium chloride and titanium(IV) isopropoxide, is highly aldehyde-selective (equation I).[1]

Complete details are available concerning stereoselective addition of these reagents to chiral aldehydes or ketones and of crotyltitanium compounds to carbonyl groups (**12**, 354).[2] The most diastereoselective additions to cyclohexanones known to date are effected with organotitanium reagents.

Examples:

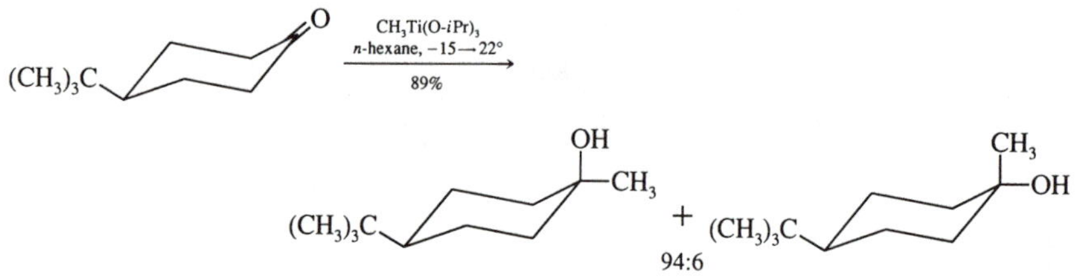

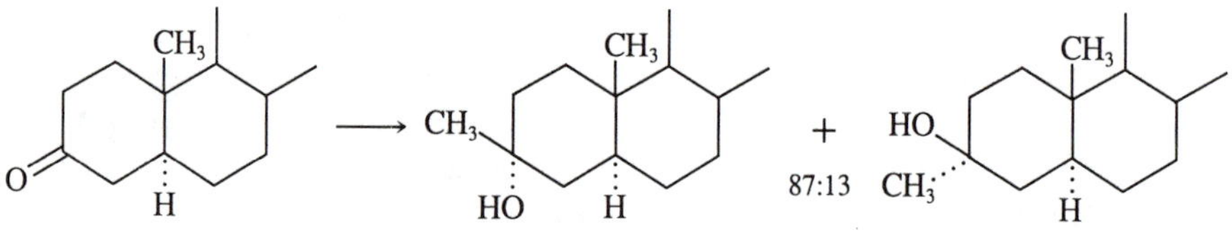

These reagents add to the chiral ketones **3** with high 1,5-asymmetric induction. Grignard and organolithium reagents show slight diastereoselectivity.[3]

Example:

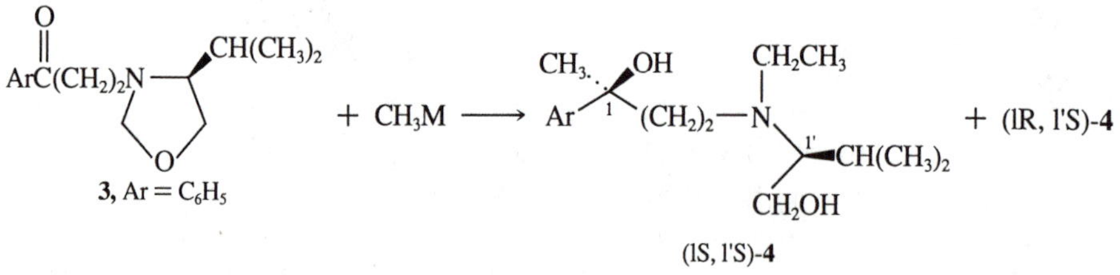

M = Ti(O-i-Pr)$_3$	86%	93:7
= MgX	93%	47:53
= Li	95%	54:46

Methyltriisopropoxytitanium.[4]   Full details are available for *in situ* preparation of CH$_3$Ti[OCH(CH$_3$)$_2$]$_3$ from CH$_3$Li and ClTi[OCH(CH$_3$)$_2$]$_3$ and use for addition to aldehydes. In comparison, use of CH$_3$Li or CH$_3$MgI results in the adduct in only 10–30% yield.

$$3Ti[OCH(CH_3)_2]_4 \; + \; TiCl_4 \; \longrightarrow \; ClTi[OCH(CH_3)_2]_3 \; \xrightarrow{\text{CH}_3\text{Li,} \atop \text{THF}}$$

$$CH_3Ti[OCH(CH_3)_2]_3 \; \xrightarrow[\substack{85\text{-}89\% \\ \text{overall}}]{}$$

A critical step in a synthesis of the benzoquinone antibiotic sarubicin A (**5**) is addition of a methyl nucleophile to the aldehyde group of **2**. Reaction with CH₃MgBr is very slow at low temperatures and the yield and stereoselectivity are low. In contrast, methyltriisopropoxytitanium reacts with **2** to give the desired triol **3** in 80% yield. Oxidative cyclization of **3** to **4** was carried out in two steps: benzylic bromination followed by silver-catalyzed ring closure.[5]

2

$$\xrightarrow[\substack{80\%}]{\text{CH}_3\text{Ti(O-}i\text{-Pr)}_3 \\ \text{THF, 25}^\circ}$$

3

$$\xrightarrow[\substack{53\%}]{\text{1) NBS, AIBN} \\ \text{2) AgClO}_4}$$

4

$$\xrightarrow{\text{several steps}}$$

5

***Crotyltriisopropoxytitanium* (1).**[6]   The reaction of diethyl ethylidenemalonate (**2**) with **1** produces the *anti*-adduct (**3**) as the major product. Comparable selectivity is shown by B-crotyl-9-borabicyclo[3.3.1]nonane, but crotylzirconium or -magnesium reagents show less stereoselectivity.

3 (*anti/syn* = 9:1)

Trichloromethyltitanium, CH_3TiCl_3 (**10**, 270, **12**, 355).   This reagent, previously available from $(CH_3)_2Zn$ and $TiCl_4$ in CH_2Cl_2, is more readily prepared from CH_3Li or CH_3MgCl and $TiCl_4$ in ether. It is aldehyde-selective, but does add to ketones slowly at $-20-0°$, often in high yield.
   Example:

$$CH_3\overset{\overset{O}{\|}}{C}(CH_2)_3COOC_2H_5 \xrightarrow[86\%]{CH_3TiCl_3} \underset{CH_3}{\overset{HO}{\diagdown}}\underset{(CH_2)_3COOC_2H_5}{\diagup}CH_3$$

This reagent also adds to chiral α-substituted aldehydes with Cram diastereo-selectivity.[7]
   Example:

90:10

Dichlorodimethyltitanium, $(CH_3)_2TiCl_2$ (**1**) (**10**, 138, 270).   Preparation (equation I).

$$(I)\quad Zn \xrightarrow[70-75\%]{\begin{array}{l}1)\ CuO\\2)\ CH_3I,\ 70°,\ CH_2Cl_2\end{array}} Zn(CH_3)_2 \xrightarrow[\geqslant 90\%]{TiCl_4} 1$$

The reagent converts ketones into *gem*-dimethyl compounds in 70–90% yield.[8]
Examples:

$$C_6H_5C(CH_3)_3 \qquad CH_3(CH_2)_2\overset{\overset{CH_3}{|}}{\underset{\underset{CH_3}{|}}{C}}(CH_2)_2CH_3$$

(90%)

(72%)

(88%)

It also effects trimethylation of acid chlorides.

$$C_6H_5COCl \xrightarrow[90\%]{1} C_6H_5C(CH_3)_3$$

β,γ-Unsaturated ketones.[9]   Allyltris(diethylamino)titanium (**1**) reacts with N-acylimidazoles (**2**) to give almost exclusively β,γ-unsaturated ketones (equation I). A similar reaction is observed with the crotyltris(diethylamino)titanium **3** (equation II).

(I) $RC\overset{O}{\underset{\|}{C}}-N\overset{\diagdown}{\diagup}\overset{N}{\diagdown}$
2
 $+ \; CH_2{=}CHCH_2Ti[N(C_2H_5)_2]_3$
1
 $\xrightarrow[85-90\%]{}$
 $RCCH_2CH{=}CH_2$
 (with $\overset{O}{\underset{\|}{}}$)

(II) $RC\overset{O}{\underset{\|}{C}}-N\overset{\diagdown}{\diagup}\overset{N}{\diagdown}$
2
 $+$
 $\underset{H}{\overset{CH_3}{\diagdown}}C{=}C\underset{CH_2Ti[N(C_2H_5)_2]_3}{\overset{H}{\diagup}}$
3
 $\xrightarrow[60-85\%]{}$
 $RC-\underset{CH_3}{\overset{O}{\underset{|}{C}}}CHCH{=}CH_2$

β,γ-Unsaturated ketones are selectively rearranged to (E)-α,β-unsaturated ketones by neutral Al_2O_3. This mild method is particularly useful for the rearrangement of trisubstituted alkenes.

(N₃)₂Ti(O-i-Pr)₂ (1);[10] azidohydrins. Azidotrimethylsilane reacts very slowly in the presence of Ti(O-i-Pr)₄ with epoxides to form azidohydrins. The reaction is more rapid when catalyzed by **1**. Actually **1** is a superior reagent for cleavage of 2,3-epoxy alcohols.[11]

Examples:

(100% *trans*)

Homoaldol reaction of Ti enolates.[12] The reaction of the lithium anion of the chiral crotylcarbamate **1** with isobutyraldehyde gives as the major product the *anti*-adduct (3R,4S)-**2**. However if the lithium enolate is converted into a titanium enolate by exchange with chlorotris(diethylamino)titanium before reaction with the aldehyde, the *anti*-adduct has the (3S,4R)-configuration. Since the configuration at C_3 depends upon the C_1-configuration (1,3-chirality transfer), the lithium and titanium enolates must have different configurations at C_1. One explanation may be that the lithium–titanium exchange occurs with inversion.

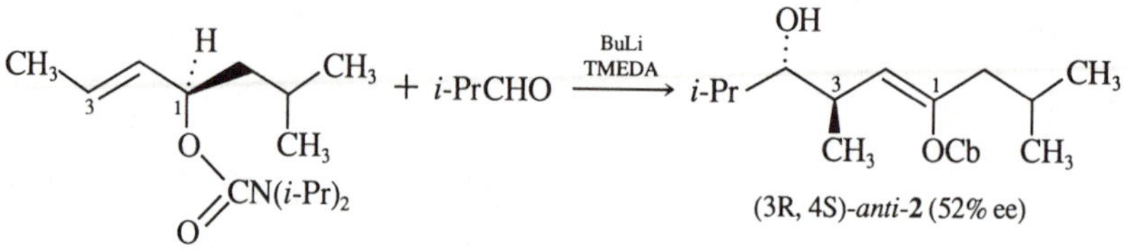

1 (68% ee)

(3R, 4S)-*anti*-2 (52% ee)

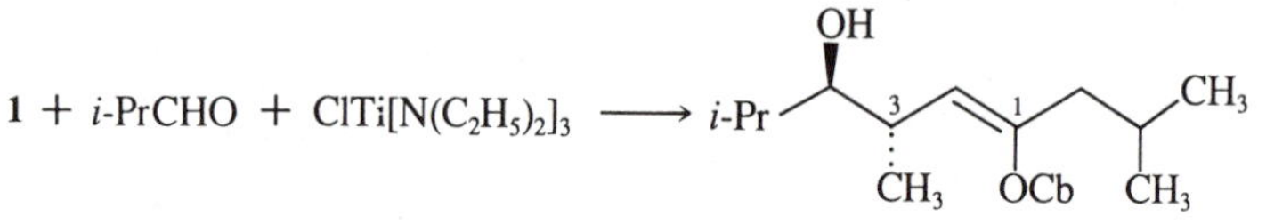

(3S, 4R)-*anti*-2 (35% ee)

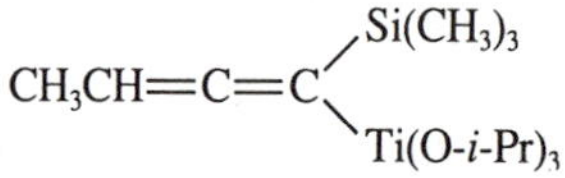

$$CH_3CH=C=C\begin{smallmatrix}Si(CH_3)_3\\Ti(O\text{-}i\text{-}Pr)_3\end{smallmatrix}$$
(**1**). This allenyl titanium reagent is prepared by consecutive reaction of $CH_3CH_2C\equiv CSi(CH_3)_3$ with *t*-BuLi (THF) and Ti(O-*i*-Pr)$_4$. A similar reaction but with B(OCH$_3$)$_3$ provides

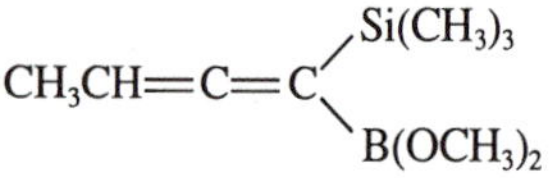

$$CH_3CH=C=C\begin{smallmatrix}Si(CH_3)_3\\B(OCH_3)_2\end{smallmatrix}$$
(**2**). Both reagents react with alkyl pyruvates to form predominantly the *syn*-isomer of alkyl 2,3-dimethyl-2-hydroxy-5-(trimethylsilyl)-4-pentynoates (**3**) (equation I).[13] This reaction is of interest because reaction of alkyl pyruvates with crotyl-9-BBN results mainly in alkyl *anti*-2,3-dimethyl-2-hydroxy-4-pentenoates (**12**, 81).

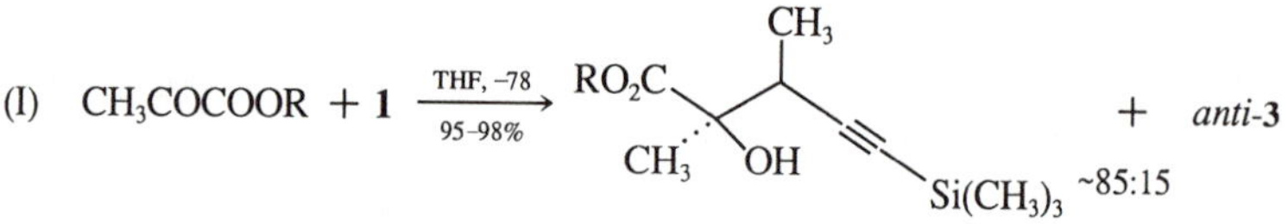

(I) $CH_3COCOOR + 1 \xrightarrow[95-98\%]{THF, -78}$

+   *anti*-3

~85:15

syn-3

Diastereospecific aldol condensations.[14]   The titanium enolate of the chiral ketone **1** reacts with aldehydes to give mainly the *syn*-aldol (~90:10). However, use of excess titanium reagent or addition of 12-crown-4 (which complexes Li$^+$) results in >99:1 diastereoselectivity.

Example:

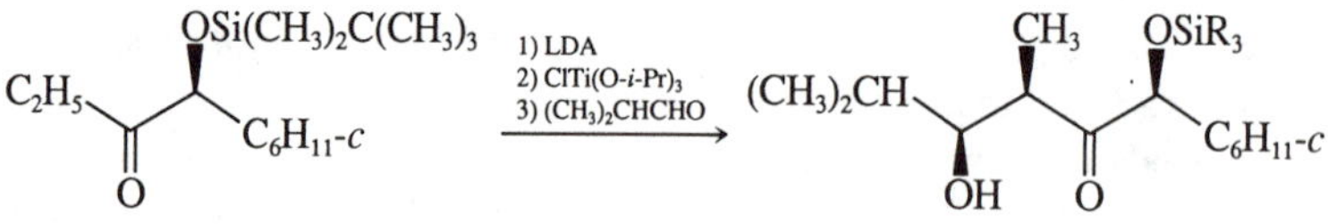

1

2 (*syn/anti* = >99:1)

Dichloro(cyclopentadienyl)methyltitanium, CpTi(CH$_3$)Cl$_2$ (**1**).   The reagent is prepared by reaction of CpTiCl$_3$ with (CH$_3$)$_2$Zn (71% yield).[15] Unlike many organotitanium reagents, **1** is heat-stable. As expected, **1** forms adducts with aldehydes at room temperature but reacts only slowly with ketones even at 50°. The reaction with benzoyl chloride is slow and requires 2 equiv. of **1** to afford 2-phenyl-2-propanol (50% yield).[16]

[1] M. T. Reetz, J. Westermann, R. Steinbach, B. Wenderoth, R. Peter, R. Ostarek, and S. Maus, *Ber.*, **118**, 1421 (1985).
[2] M. T. Reetz, R. Steinbach, J. Westermann, R. Peter, and B. Wenderoth, *ibid.*, **118**, 1441 (1985).
[3] H. Takahashi, K. Tanahashi, K. Higashiyama, and H. Onishi, *Chem. Pharm. Bull.*, **34**, 479 (1986).
[4] R. Imwinkelried and D. Seebach, *Org. Syn.*, submitted (1986).
[5] M. F. Semmelhack, Y. Appapillai, and T. Sato, *Am. Soc.*, **107**, 4577 (1985).
[6] Y. Yamamoto S. Nishii, and K. Maruyama, *J.C.S. Chem. Comm.*, 386 (1985).
[7] M. T. Reetz, *Pure Appl. Chem.*, **57**, 1781 (1985); M. T. Reetz, S. H. Kyung, and M. Hüllmann, *Tetrahedron*, **42**, 2931 (1986).
[8] M. T. Reetz, J. Westermann, and S.-H. Kyrung, *Ber.*, **118**, 1050 (1985).
[9] M. T. Reetz, B. Wenderoth, and R. Urz, *ibid.*, **118**, 348 (1985).
[10] C. Blandy, R. Choukroun, and D. Gervais, *Tetrahedron Letters*, **24**, 4189 (1983).
[11] M. Caron and K. B. Sharpless, *J. Org.*, **50**, 1557, reference 17 (1985).
[12] D. Hoppe and T. Kramer, *Angew. Chem. Int. Ed.*, **25**, 160 (1986).
[13] Y. Yamamoto, K. Maruyama, T. Komatsu, and W. Ito, *J. Org.*, **51**, 886 (1986).
[14] C. Siegel and E. R. Thornton, *Tetrahedron Letters*, **27**, 457 (1986).
[15] G. J. Erskine, G. J. B. Hurst, E. L. Weinberg, B. K. Hunter, and J. D. McCowan, *J. Organomet. Chem.*, **267**, 265 (1984).
[16] G. J. Erskine, B. K. Hunter, and J. D. McCowan, *Tetrahedron Letters*, **26**, 1371 (1985).

Organovanadium compounds.

RVCl$_2$.   Reagents presumably of the structure RVCl$_2$ or ArVCl$_2$ are formed on reaction of Grignard reagents with VCl$_3$ in CH$_2$Cl$_2$ at $-78°$. They react with acid chlorides to form ketones. Similar cross coupling is observed with allyl halides but not with alkyl halides. Allene derivatives are obtained in low yield from reaction with propargyl bromide.[1]

Examples:

$$\text{BuMgBr} \xrightarrow[\underset{78\%}{}]{\substack{\text{1) VCl}_3 \\ \text{2) CH}_3\text{CO(CH}_2)_2\text{COCl}}} \overset{\displaystyle O}{\underset{}{\|}} \overset{\displaystyle O}{\underset{}{\|}} \text{Bu}\overset{O}{\overset{\|}{C}}(\text{CH}_2)_2\overset{O}{\overset{\|}{C}}\text{CH}_3$$

$$\text{C}_6\text{H}_5\text{MgBr} \xrightarrow[\underset{93\%}{}]{\substack{\text{1) VCl}_3 \\ \text{2) BrCH}_2\text{CH}=\text{CH}_2}} \text{C}_6\text{H}_5\text{CH}_2\text{CH}=\text{CH}_2$$

Vanadium acetylides, RC≡CVCl$_2$ (**1**).   These reagents are generated by reaction of lithium acetylides or RC≡CMgBr with VCl$_3$ in CH$_2$Cl$_2$ at $-78°$. They react with aldehydes in refluxing CH$_2$Cl$_2$ to form α,β-alkynyl ketones in ~50–70%

yield. α,β-Enals react with **1** to form 1-ene-4-yne-3-ones. The reaction of ketones results in 1,2-addition.[2]

Example:

$$BuC\equiv CLi \xrightarrow[\substack{2) CH_3CH=CHCHO \\ 50\%}]{1) VCl_3, CH_2Cl_2} BuC\equiv C\overset{\overset{\displaystyle O}{\|}}{C}CH=CHCH_3$$

[1] T. Hirao, D. Misu, K. Yao, and T. Agawa, *Tetrahedron Letters*, **27**, 929 (1986).
[2] T. Hirao, D. Misu, and T. Agawa, *ibid.*, **27**, 933 (1986).

Organozinc compounds.

(*Carboethoxyalkyl)iodozinc*, $[C_2H_5OOC(CH_2)_n]ZnI$ (**1**, n = 2–4). These zinc reagents are obtained by reaction of Zn/Cu with ethyl 3-iodopropionate (n = 2) or ethyl 4-iodobutyrate (n = 3) in C_6H_6/DMF or C_6H_6/DMA (equation I). In the presence of $Pd[P(C_6H_5)_3]_4$, both the homoenolate (**1a**) or bishomoenolate

$$(I) \quad C_2H_5O\overset{\overset{\displaystyle O}{\|}}{C}(CH_2)_nI + Zn/Cu \xrightarrow{\substack{C_6H_6, \\ DMA}} C_2H_5O\overset{\overset{\displaystyle O}{\|}}{C}(CH_2)_nZnI$$

n = 2 or 3

1a, n = 2
1b, n = 3

(**1b**) couple with acid chlorides to provide γ- or δ-keto esters, respectively.[1]

Examples:

$$C_6H_5CH=CHCOCl + 1a \xrightarrow[92\%]{Pd(0)} C_6H_5CH=CH\overset{\overset{\displaystyle O}{\|}}{C}(CH_2)_2COOC_2H_5$$

$$CH_3O_2C(CH_2)_7COCl + 1b \xrightarrow{89\%} CH_3O_2C(CH_2)_7\overset{\overset{\displaystyle O}{\|}}{C}(CH_2)_3COOC_2H_5$$

In the presence of Pd(II) catalysts, **1** couples with aryl or vinyl iodides. For the former coupling, the most satisfactory catalyst is dichlorobis(tri-*o*-tolylphosphine)palladium(II). $Pd[P(C_6H_5)_3]_4$ is satisfactory for coupling with vinyl iodides or triflates.[2]

Examples:

$$C_6H_5I + 1a \xrightarrow[90\%]{\substack{Pd(II), \\ C_6H_6, DMA}} C_6H_5(CH_2)_2COOC_2H_5$$

Stereoselective addition to an α-alkoxy aldehyde.[3]   The addition of organometallic reagents to acrolein dimer **1** can be controlled to a remarkable extent by
the metal, evidently as a result of chelation, with R_2Zn being more stereoselective
than RLi or RMgBr.

		anti/syn =
C_2H_5Li	~80%	0.14–0.39
C_2H_5MgBr	87%	2.3
$(C_2H_5)_2Zn$	76%	5.7

R_3ZnLi (8, 515); *conjugate addition to α,β-enones.*[4]   The use of R_3ZnLi for
this reaction suffers from the fact that only one R group is utilized. This difficulty
is circumvented by use of reagents of the type $R(CH_3)_2ZnLi$. Thus when R is *n*-
or *sec*-Bu, these groups undergo efficient 1,4-addition to α,β-enones. The reagents
are prepared in THF using the complex of $ZnCl_2$ and TMEDA. Transfer is not
observed when the β-position of the enone is disubstituted.
    Example:

(Cyanomethyl)zinc bromide, $BrZnCH_2CN$ (1).   The reagent converts alkyl,
allylic, or propargylic iodides in the presence of HMPT to nitriles. The reaction is
particularly useful for synthesis of γ,δ-unsaturated nitriles.[5]
    Example:

$$(CH_3)_3SiC{\equiv}CCH_2I + 1 \xrightarrow[78\%]{HMPT} (CH_3)_3SiC{\equiv}CCH_2CH_2CN + ZnBrI{\cdot}HMPT$$

Conjugate addition to enals.[6]   In the presence of $Ni(acac)_2$, diarylzinc reagents
undergo 1,4-addition to α,β-enals in satisfactory yield; however, the same reaction
with dialkylzinc reagents gives low yields (15%, one example).
    Example:

$$CH_3(CH_2)_2CH{=}CHCHO + (C_6H_5)_2Zn \xrightarrow[50-65\%]{\substack{Ni(acac)_2, \\ THF, -40°}} CH_3(CH_2)_2\underset{\underset{C_6H_5}{|}}{C}HCH_2CHO$$

[1] Y. Tamaru, H. Ochiai, T. Nakamura, K. Tsubaki, and Z. Yoshida, *Tetrahedron Letters*, **26**, 5559 (1985); Y. Tamaru, H. Ochiai, F. Sanda, and Z. Yoshida, *ibid.*, **26**, 5529 (1985); Y. Tamaru, H. Ochiai, T. Nakamura, and Z. Yoshida, *Org. Syn.*, submitted (1986).
[2] Y. Tamaru, H. Ochiai, T. Nakamura, and Z. Yoshida, *Tetrahedron Letters*, **27**, 955 (1986).
[3] M. Bhupathy and T. Cohen, *Tetrahedron Letters*, **26**, 2619 (1985).
[4] R. A. Watson and R. A. Kjonaas, *ibid.*, **27**, 1437 (1986).
[5] F. Orsini, *Synthesis*, 500 (1985).
[6] J. C. deSouza Barboza, C. Pétrier, and J-L. Luche, *Tetrahedron Letters*, **26**, 829 (1985).

Organozirconium compounds.

Review. Negishi and Takahashi[1] have reviewed the preparation and synthetic uses of organozirconium compounds, mainly of the type $RZrCp_2X$. Because of the high cost of Zr compounds, most stoichiometric applications allow for recovery of the metallic compounds. The review covers more than 100 research papers, all published since 1970.

[1] E. Negishi and T. Takahashi, *Aldrichimica Acta*, **18**, 31 (1985).

Osmium tetroxide.

Oxidation of alcohols.[1] OsO_4 is comparable to RuO_4 for oxidation of secondary alcohols to ketones. It oxidizes primary alcohols only to the corresponding aldehydes; in fact primary alcohols can be oxidized with fair selectivity in the presence of secondary alcohols.

[1] A. M. Maione and A. Romeo, *Synthesis*, 955 (1984).

Osmium tetroxide–*t*-Butyl hydroperoxide.

α-Keto esters.[1]   α-Keto esters are obtained in about 60% yield by oxidation of 1-trimethylsilylalkynes with *t*-butyl hydroperoxide (5 equiv.) and catalytic amounts of OsO_4 in an alcoholic solvent. Presumably the reaction involves an α-keto acylsilane, which undergoes a Brook type rearrangement.

Example:

$$CH_3(CH_2)_4C\equiv CSi(CH_3)_3 \xrightarrow[\text{CH}_3\text{OH}]{\text{OsO}_4,\ t\text{-BuOOH},} \left[CH_3(CH_2)_4\overset{O}{\overset{\|}{C}}-\overset{O}{\overset{\|}{C}}Si(CH_3)_3 \right] \xrightarrow{55\%}$$

$$CH_3(CH_2)_4\overset{O}{\overset{\|}{C}}-COOCH_3$$

[1] P. C. Bulman Page and S. Rosenthal, *Tetrahedron Letters*, **27**, 1947 (1986).

Osmium tetroxide–N-Methylmorpholine N-oxide.

Octoses.[1]   A new route to octoses involves catalytic osmylation (**7**, 256–257) of allylic alcohols derived from a hexose by Wittig olefination to introduce two additional carbon atoms at the nonreducing end. Thus the protected aldehyde **1**, derived from D-galactose, can be converted by known procedures into **2**. Catalytic osmylation of **2** provides the triol **3** as the only product (79% yield). This product on deprotection and reduction is converted into D-erythro-D-galactooctose (**4**), a constituent of the avocado. Note that the *anti*-selectivity of osmylation conforms to Kishi's rule (**12**, 358). Some of the octose derivatives obtained in this way have been converted into decitols with mixed success.

[1] J. S. Brimacombe, R. Hanna, A. K. M. S. Kabir, F. Bennett, and I. D. Taylor, *J.C.S. Perkin I*, 815 (1986); J. S. Brimacombe, R. Hanna, and A. K. M. S. Kabir, *ibid.*, 823 (1986).

Osmium tetroxide–Trimethylamine N-oxide–Pyridine.

Stereoselective dihydroxylation.[1] The OsO_4-catalyzed dihydroxylation of 5-vinyl-4,5-dihydroisoxazoles (**1**) (obtained by reaction of the nitrile oxide derived from nitroethane with 1,3-dienes) is *anti*-selective, and the *anti*-selectivity is markedly enhanced by a *cis*-substituent on the double bond.

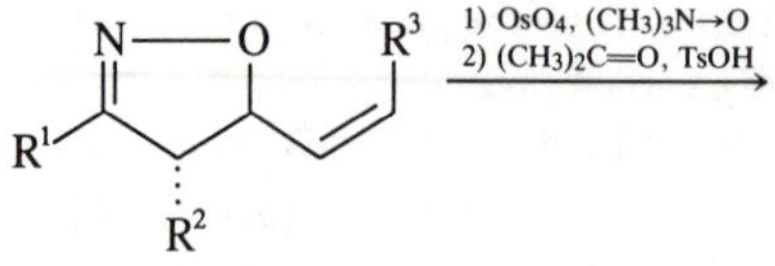

1 (R^2, R^3 = H, CH$_3$)

R^1 = CH$_3$, R^2, R^3 = H	80%	
R^1, R^2, R^3 = CH$_3$	70%	

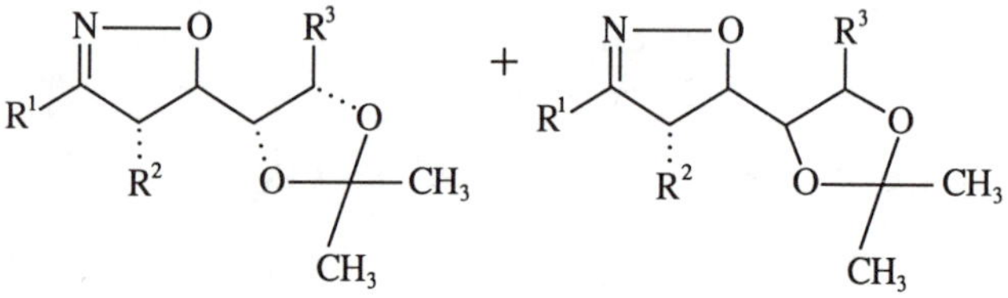

anti-**2**   +   syn-**2**

78:22
92:8

Stereospecific conversion of vinylsilanes to silyl enol ethers.[2] This conversion can be effected by dihydroxylation with OsO$_4$ in combination with (CH$_3$)$_3$NO and pyridine followed by *anti* β-elimination with NaH via an α-oxidosilane. The overall process converts vinylsilanes into silyl enol ethers with preservation of the geometry of the double bond.

Examples:

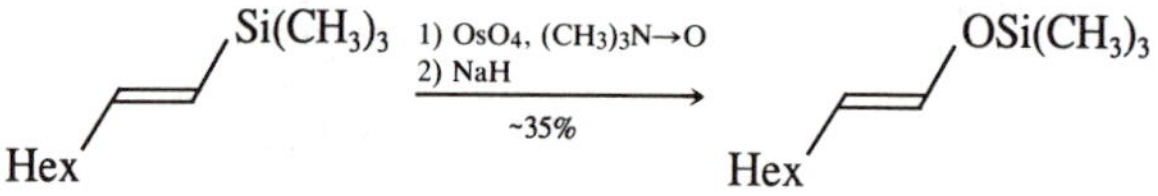

(99% E)

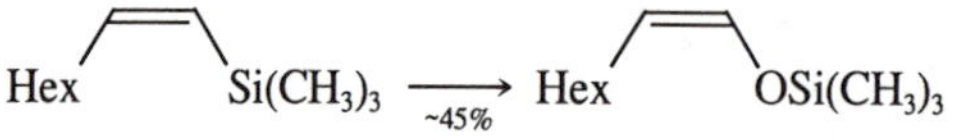

(>99% Z)

Dihydroxylation of allylic β-hydroxy sulfoxides.[3] Osmylation of chiral allylic β-hydroxy sulfoxides results mainly in *anti, syn*-trihydroxy sulfones.

Examples:

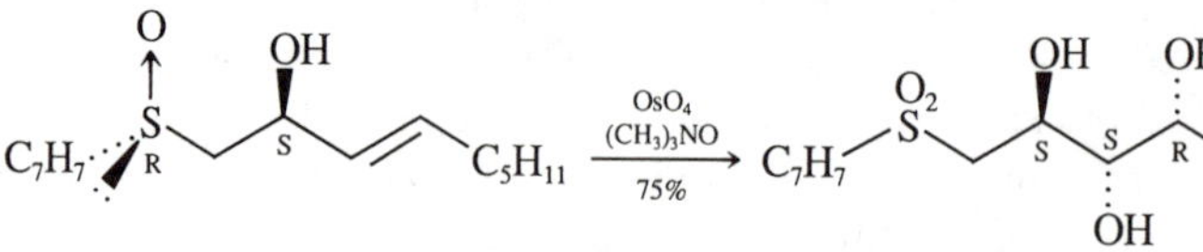

(64% de)

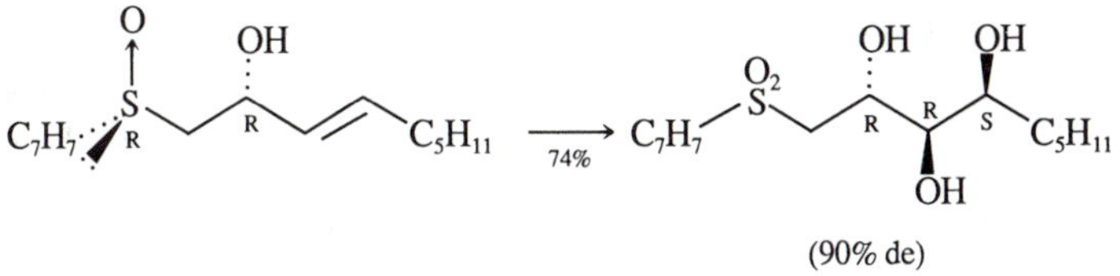

(90% de)

α-Ketoacylsilanes.[4]   Reaction of (Z)-vinylsilanes (**1**), with OsO_4 and tri-methylamine oxide gives the *cis*-1,2-diols (**2**) in moderate yield. These can be oxidized by DMSO/oxalyl chloride to α-keto acylsilanes (**3**, 40–55% yield).

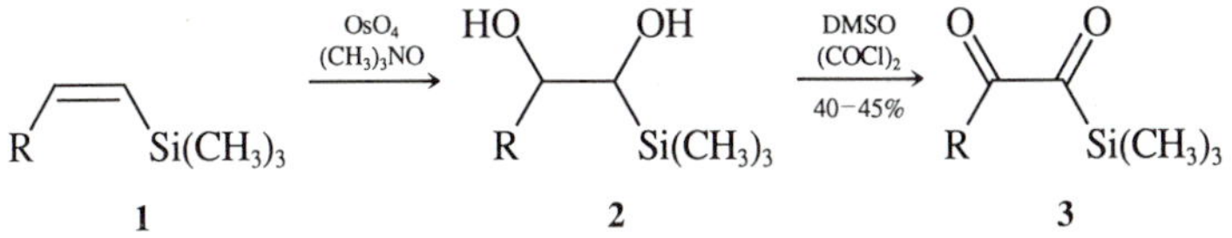

[1] R. Annunziata, M. Cinquini, F. Cozzi, and L. Raimondi, *J.C.S. Chem. Comm.*, 403 (1985).
[2] P. F. Hudrlik, A. M. Hudrlik, and A. K. Kulkarni, *Am. Soc.*, **107**, 4260 (1985).
[3] G. Solladié, C. Fréchou, and G. Demailly, *Tetrahedron Letters*, **27**, 2867 (1986).
[4] P. C. B. Page and S. Rosenthal, *ibid.*, **27**, 2527 (1986).

2-Oxazolidones, chiral.

β-Lactams.[1]   The ketene derived from (S)-phenyloxazolidylacetyl chloride (**1**), prepared from (S)-phenylglycine, undergoes cycloaddition with N-benzyl al-dimines to give two *cis*-azetidinones with high stereochemical control (equation I). The chiral auxiliary and the benzyl group are cleaved by Birch reduction to provide enantiomerically pure azetidinones (**3**).

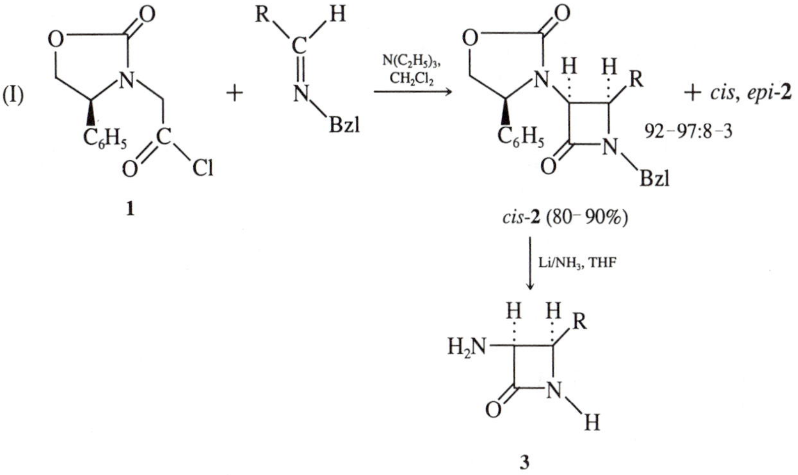

Carbacephalosporins.[2]   The ketene-imine cyclization described above has been extended to a synthesis of a chiral carbacepham (**4**). This synthesis uses a dihydroanisole group as the equivalent of a β-keto ester. Thus the azetidinone **1**, obtained in 80% yield by the above route, was reduced and acylated *in situ* to provide **2**. Ozonization followed by a rhodium-catalyzed cyclization of an α-diazo-β-keto ester provides **3**, which is a useful intermediate to various substituted carbacephams such as **4**.

syn-Aldol condensation.[3]   The titanium enolate of the chiral amide **4**, derived from L-valine, reacts with benzaldehyde to give mainly the *syn*-aldol (**5**), the diastereoisomer of *syn*-aldol (**6**) formed from the boron enolate (**11**, 379–380).

[1] D. A. Evans and E. B. Sjogren, *Tetrahedron Letters*, **26**, 3783 (1985).
[2] *Idem, ibid.*, **26**, 3787 (1985).
[3] M. Nerz-Stormes and E. R. Thornton, *ibid.*, **27**, 897 (1986).

Oxodiperoxymolybdenum(pyridine)(hexamethylphosphoric triamide), $MoO_5 \cdot Py \cdot$ HMPT.

Diastereoselective hydroxylation of ester enolates.[1]   The (E)-enolate of the chiral ester **1**, derived from (+)-camphor, generated with lithium N-isopropyl-cyclohexylamide (LICA) and HMPT (**7**, 209–210), undergoes hydroxylation with this reagent (**11**, 382) from the less hindered side with excellent diastereoselection. The (Z)-enolate as expected shows opposite, but modest, diastereoselectivity.

$$1 = R^*OCO(CH_2)_2C_6H_5 \quad \xrightarrow[\substack{2) \text{ MoOPh} \\ 50\%}]{1) \text{ LICA, HMPT}}$$

(R)-2   7:93   (S)-2

Stereoselective α-hydroxylation of β-trifluoromethyl esters.[2]   Hydroxylation of the ester **1** with the Vedejs reagent results in predominate formation of the (2S,3S)-isomer. $NaBH_4$ reduction of the ketone **3** shows opposite stereoselectivity. Since the steric bulk of CF_3 is similar to that of CH_3, electronic factors may be important.

1

(2S, 3S)-2   97:3   (2R, 3S)-2

(2S, 3R)-2 + (2S, 3S)-2
90:10

3

Epoxidation of **cis-2-butene-1,4-diones**.[3]   Enediones that can assume the all s-cis-conformation (A) are oxidized by $MoO_5 \cdot H_2O \cdot HMPT$ (2 equiv.) stereospecif-ically to cis-2,3-epoxybutane-1,4-diones (**1**).

Example:

A, R = C_6H_5, CH_3,
         OCH_3

1

The *trans*-isomer of the enediones as well as naphthoquinones do not react with the reagent. Reaction of the *cis*-enediones (A) with *m*-chloroperbenzoic acid results in isomerization to the *trans*-isomer.

[1] R. Gamboni, P. Mohr, N. Waespe-Šarčevič, and C. Tamm, *Tetrahedron Letters*, **26**, 203 (1985).
[2] Y. Morizawa, A. Yasuda, and K. Uchida, *ibid.*, **27**, 1833 (1986).
[3] C.-S. Chien, T. Kawasaki, M. Sakamoto, Y. Tamura, and Y. Kita, *Chem. Pharm. Bull.*, **33**, 2743 (1985).

Oxygen, singlet.

Generation from $(C_2H_5)_3SiH$.[1]   Ozonation of a trialkylsilane at low temperatures generates a hydrotrioxide, which decomposes to a trialkylsilanol with release of singlet oxygen. The singlet oxygen can be identified spectroscopically as well as chemically (equation I).

(I) $(C_2H_5)_3SiH \xrightarrow{O_3, -78°} [(C_2H_5)_3SiOOOH] \xrightarrow{-60°} (C_2H_5)_3SiOH + {}^1O_2$

Stereospecific ene reaction.   Singlet oxygen (*meso*-tetraphenylporphine) reacts with anhydrotetracycline (**1**) to give the hydroperoxide **2** in 97% yield. Hydrogenation of **2** provides the antibiotic tetracycline (**3**).[2]

Photosensitized oxidation of furans.[3]   The endoperoxide (**2**) formed in photosensitized (methylene blue) oxidation of 3-substituted or 3,4-disubstituted furans in acetone at −40° rearranges at 20° to 4-hydroxy-2-butenolides in 30–97% yield. Example:

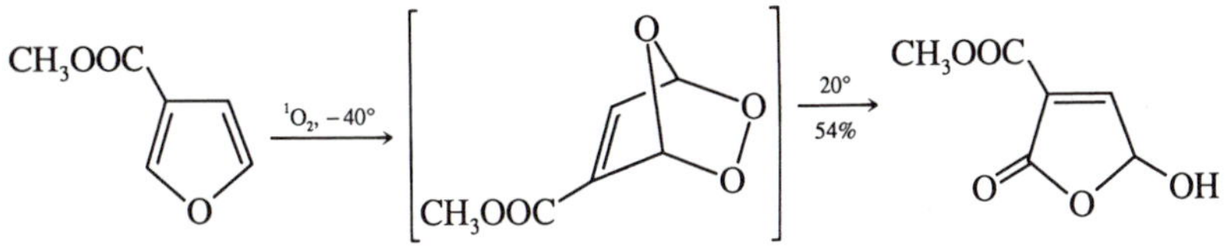

[1] E. J. Corey, M. M. Mehrota, and A. U. Khan, *Am. Soc.*, **108**, 2472 (1986).
[2] H. H. Wasserman, T. J. Lu, and A. I. Scott, *ibid.*, **108**, 4237 (1986).
[3] M. L. Graziano and M. R. Iesce, *Synthesis*, 1151 (1985).

Ozone.

Polyketides.[1]   Polyketides are useful as precursors to acetate-derived natural products, but they are unstable owing to a marked tendency to undergo internal condensation. A new synthesis involves ozonolysis of products of Birch reduction. An example is the synthesis of the polyketide **3** from the aromatic system **1**.

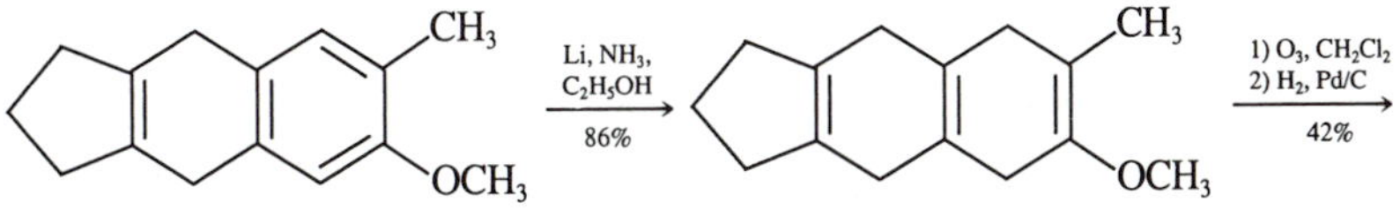

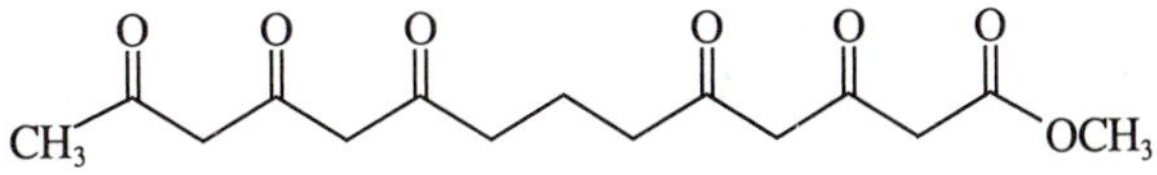

[1] C. L. Kirkemo and J. D. White, *J. Org.*, **50**, 1316 (1985).

P

Palladium catalysts.

Hydrogen-transfer reductions. NaH_2PO_2 can be used as the hydrogen source in Pd/C-catalyzed reduction of alkenes, N-oxides, azides in generally high yield.[1] It is also useful for hydrogenolysis of benzylic and aryl chlorides and benzylic ethers.[1,2] Imines and oximes are not reduced but hydrolyzed to carbonyl compounds. Isolated carbonyl groups are not reduced, but benzylic carbonyl groups are reduced.

Dehalogenation of chlorinated arenes.[3] Palladium (10%) on carbon catalyzes the rapid transfer of hydrogen from ammonium formate to aryl chlorides to provide the parent arene. Dehalogenation of 2,4,6-trichlorophenol proceeds through di-chloro- and chlorophenol and is complete within 10 minutes at ambient temperature and pressure.

β-Hydroxy esters.[4] α-Diazo-β-hydroxy esters (1) can be reduced to β-hydroxy esters in 65–77% yield by hydrogenation in CH_3OH in the presence of 5% Pd on charcoal (equation I).

$$\text{(I)} \quad RCHO + Li\underset{\underset{N_2}{\|}}{C}COOC_2H_5 \xrightarrow[75-95\%]{} R\underset{1}{\underset{\underset{N_2}{\|}}{\overset{\overset{OH}{|}}{CH}C}COOC_2H_5} \xrightarrow[65-75\%]{H_2,\ CH_3OH \atop Pd/C} R\underset{2}{\overset{\overset{OH}{|}}{CH}CH_2COOC_2H_5}$$

Polycyclic hydrocarbons.[5] Perhydrogenation of 2,7-dialkyl-9,10-dihydro-phenanthrenes (1) with Pd/C as catalyst at 200° under pressures of 120–170 atm. results in the all-*trans*-perhydrophenanthrenes (2). A similar hydrogenation (Pd/C, $HClO_4$) converts 2,6-dialkylnaphthalenes (3) into *trans*-dialkyldecalins (4).

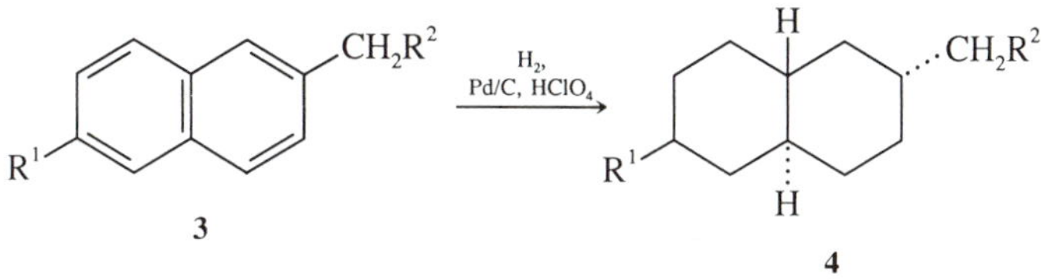

Photochemical cyclization of heterocyclic stilbenes. The photochemical oxidative cyclization of stilbenes to phenanthrenes[6] proceeds in low yield when applied to heterocyclic analogs. Cava et al.[7] have examined a number of oxidants other than O_2 for *in situ* oxidation of the dihydroheterocycle formed initially and find that 5% Pd on carbon is the reagent of choice. In addition, a hydrogen acceptor, *p*-nitrobenzoic acid, can improve yields.

Example:

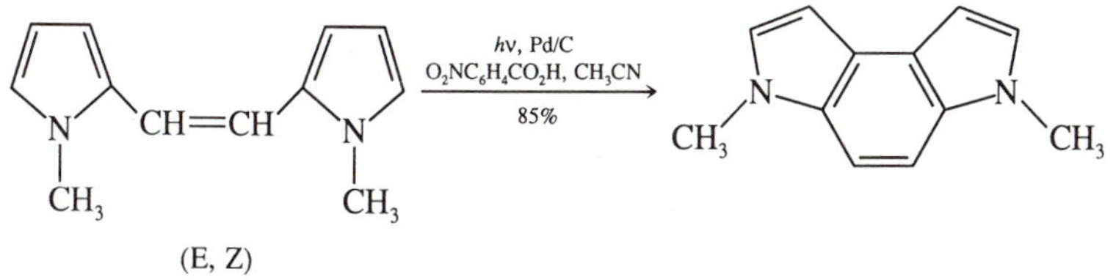

This photocyclization has been used for a synthesis of two of the subunits (B and C) of an antitumor antibiotic CC-1065 (**1**), which is composed of three dipyrrole units. Thus photocyclization and oxidation of **2** provides **3** in 82% yield. The product was converted to the subunits **4** and **5** in several steps.[8]

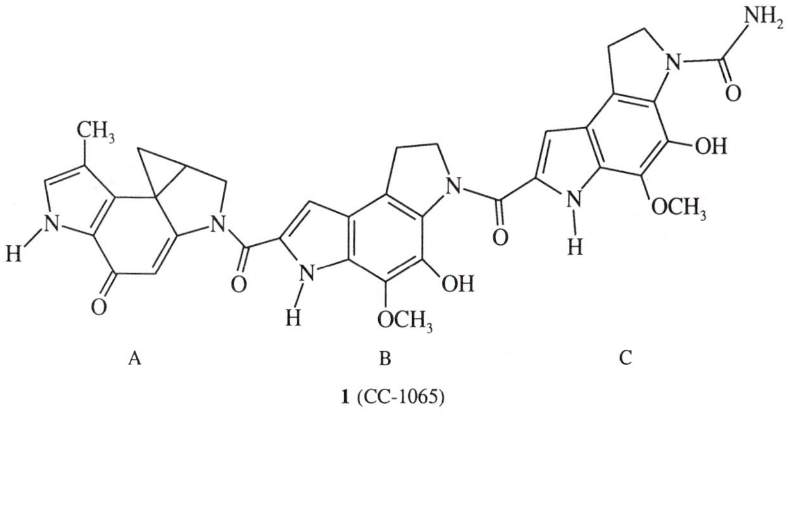

1 (CC-1065)

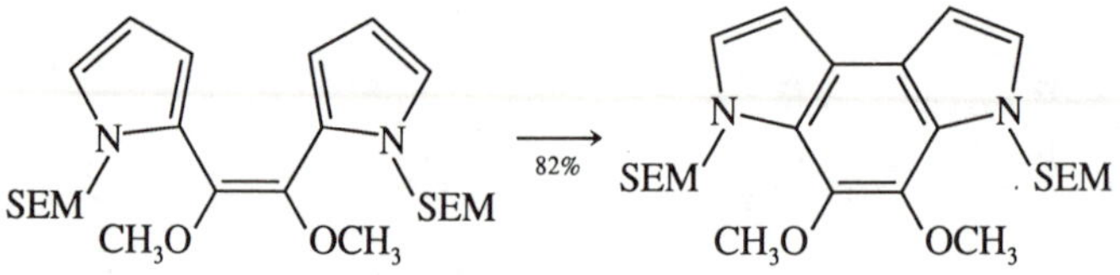

2

3

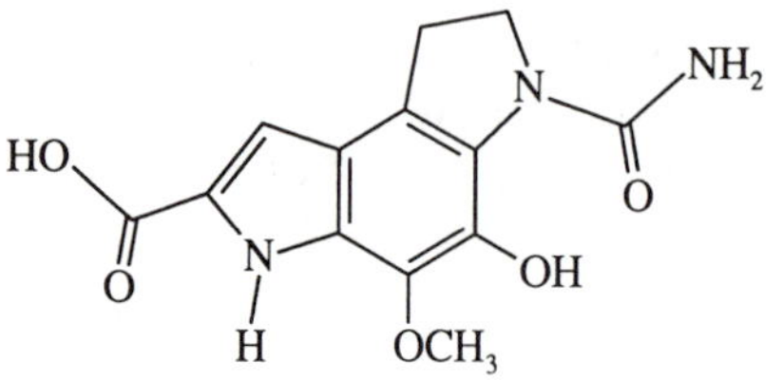

4

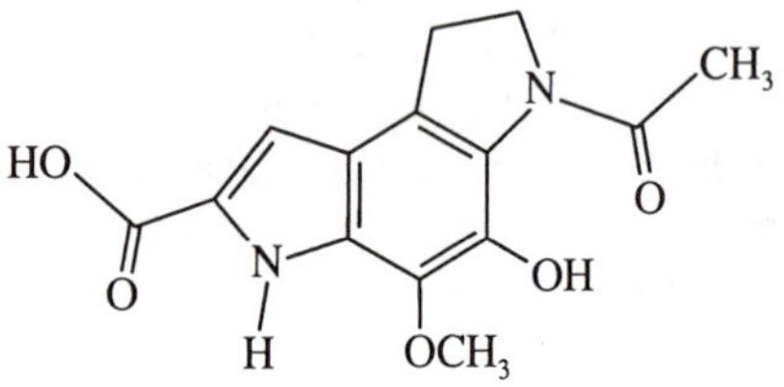

5

[1] S. K. Boyer, J. Bach, J. McKenna, and E. Jagdmann, Jr., *J. Org.*, **50**, 3408 (1985).
[2] R. Sala, G. Doria, and C. Passarotti, *Tetrahedron Letters*, **25**, 4565 (1984).
[3] M. K. Anwer and A. F. Spatola, *ibid.*, **26**, 1381 (1985).
[4] R. Pellicciari, B. Natalini, S. Cecchetti, and R. Fringuelli, *J.C.S. Perkin I*, 493 (1985).
[5] D. Varech, M. J. Brienne, and J. Jacques, *Tetrahedron Letters*, **26**, 61 (1985).
[6] F. B. Mallory and C. W. Mallory, *Org. Reactions*, **30**, 1 (1984).
[7] V. H. Rawal, R. J. Jones, and M. P. Cava, *Tetrahedron Letters*, **26**, 2423 (1985).
[8] V. H. Rawal and M. P. Cava, *Am. Soc.*, **108**, 2110 (1986).

Palladium(II) acetate.

Decarboxylation-dehydrogenation of allyl β-ketocarboxylates (**11**, 391–392). Pd(OAc)$_2$ is now the preferred catalyst for this reaction.[1] An example is a new preparation of 2-methylcyclopentenone (**2**).

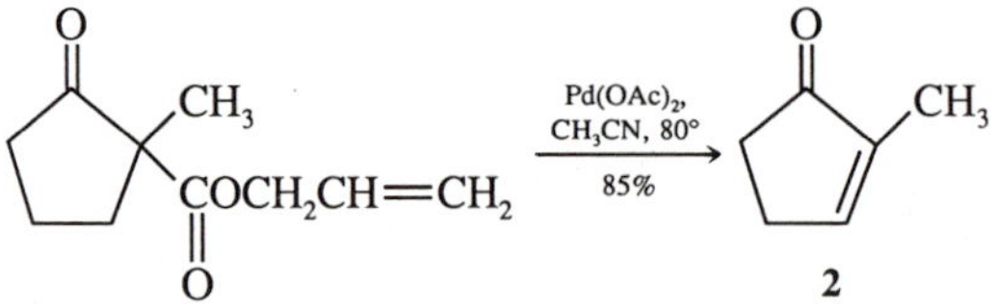

1

2

Vinylation of vinyl halides (cf., **12**, 368).   The original phase transfer catalyzed version of this Heck reaction suffers from partial loss of the stereochemistry of the (Z)-vinyl halide. However, use of K_2CO_3 as the base rather than $NaHCO_3$ accelerates the rate of coupling and improves the stereoselectivity.[2]

Example:

$$(Z)\text{-}C_4H_9CH{=}CHI \; + \; CH_2{=}CHCOOCH_3 \xrightarrow[90\%]{\substack{Pd(OAc)_2, \; K_2CO_3 \\ Bu_4NCl, \; DMF}}$$

$$C_4H_9CH{=}CHCH{=}CHCOOCH_3$$

$$(E, Z/E, E \; = \; 95:5)$$

Oxidative cyclization of **1,5-hexadienes.**[3]   Palladium(II)-catalyzed oxidation of these substrates with *p*-benzoquinone (0.2 equiv.) and MnO_2 (1 equiv.) (*cf.* **12,** 367) results in cyclization to acetoxy-substituted methylenecyclopentanes as the major product.

Example:

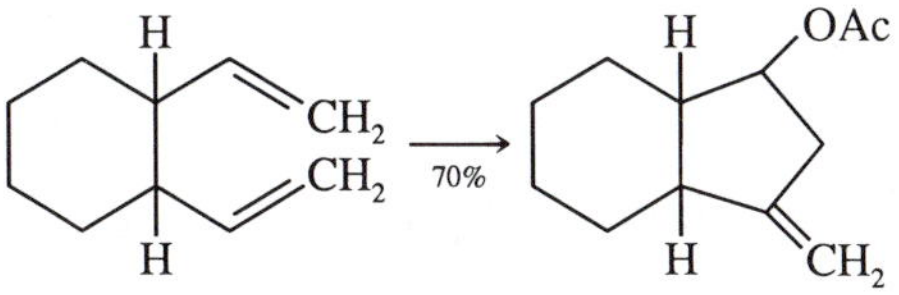

[1] J. Tsuji, I. Minami, I. Shimizu, and H. Kataoka, *Chem. Letters*, 113 (1984); J. Tsuji, M. Nisar, I. Shimizu, I. Minami, *Synthesis*, 1009 (1984).
[2] T. Jeffery, *Tetrahedron Letters*, **26**, 2667 (1985).
[3] T. Antonsson, A. Heumann, and C. Moberg, *J.C.S. Chem. Comm.*, 518 (1986).

Palladium(II) acetate–Triphenylphosphine.

β-*Lactams.*   α-Methylene-β-lactams can be prepared by carbonylation of 2-bromoallylamines catalyzed by $Pd(OAc)_2$ or $Pd(acac)_2$ and $P(C_6H_5)_3$.[1]

Example:

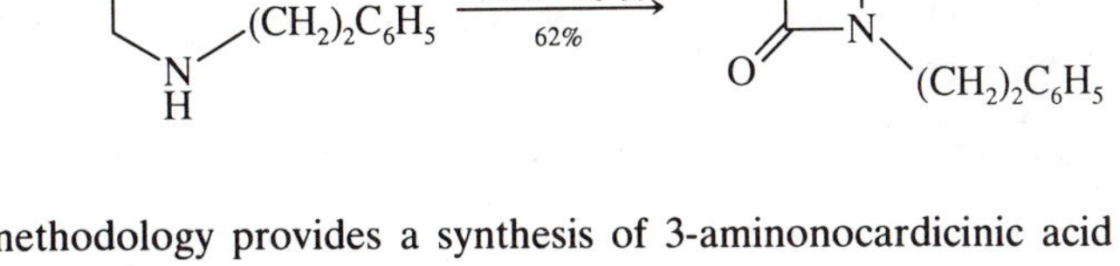

The methodology provides a synthesis of 3-aminonocardicinic acid (**1**), from *p*-hydroxyphenylglycine.[2]

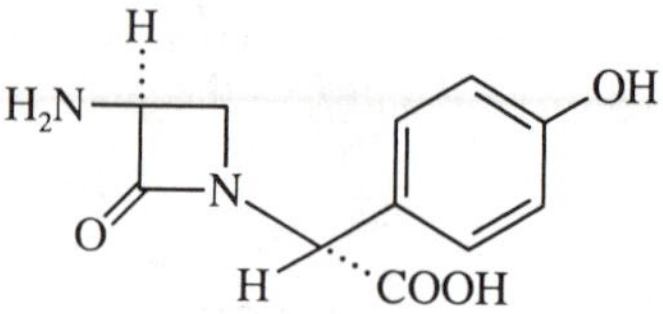

1

Coupling of vinyl bromides; 1,3-dienes.[3] Vinyl bromides can undergo inter- or intramolecular coupling to 1,3-dienes in the presence of Pd(OAc)$_2$, a triaryl-phosphine (1.25 equiv.), and a large excess of K$_2$CO$_3$ (7.5 equiv.).
Examples:

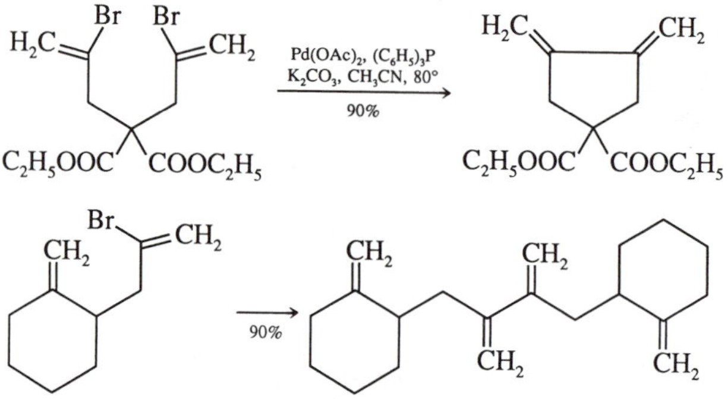

Carbonylation of enol triflates. Carbonylation of enol triflates catalyzed by this Pd(II) complex in the presence of an alcohol or an amine results in one-carbon homologated α,β-unsaturated esters or amides in 70–95% yield.[4]
Example:

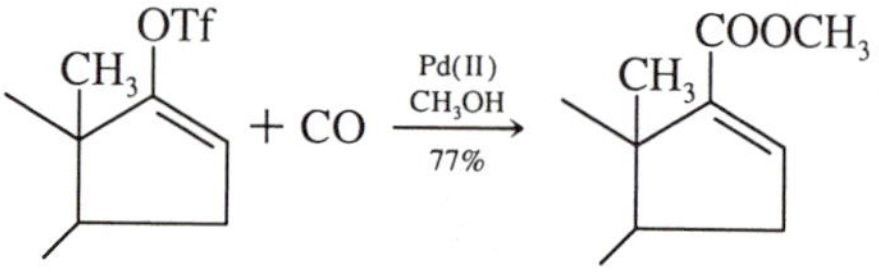

[1] M. Mori, K. Chiba, M. Okita, I. Kayo, and Y. Ban, *Tetrahedron*, **41**, 375 (1985).
[2] K. Chiba, M. Mori, and Y. Ban, *ibid.*, **41**, 387 (1985).
[3] R. Grigg, P. Stevenson, and T. Worakun, *J.C.S. Chem. Comm.*, 971 (1985).
[4] S. Cacchi, E. Morera, and G. Ortar, *Tetrahedron Letters*, **26**, 1109 (1985).

Palladium(II) chloride.
Conjugate addition of 2-bromoarylmercury compounds. A key step in a new

synthesis of 1-indanols involves the Pd-catalyzed conjugate addition of these arylmercury compounds to α,β-enones.[1]

Example:

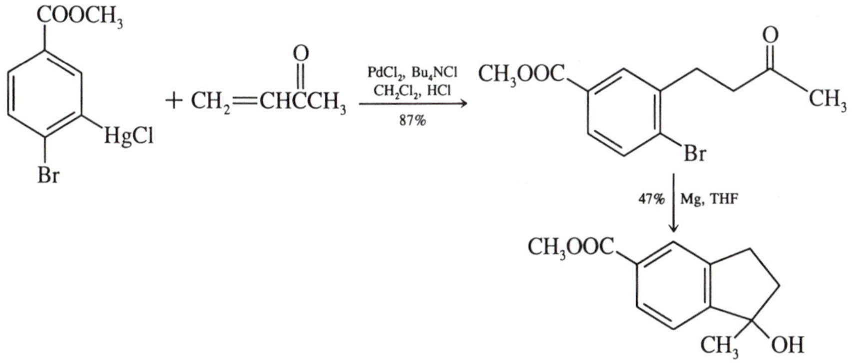

[1] S. Cacchi and G. Palmieri, *J. Organomet. Chem.*, **282**, C3 (1985).

Palladium(II) chloride–Copper(II) chloride.

γ- and δ-Lactones. Reaction of γ,δ- and δ,ϵ-unsaturated alcohols with carbon monoxide and oxygen catalyzed by $PdCl_2$ and $CuCl_2$ results in γ- and δ-lactones, respectively.[1]

Example:

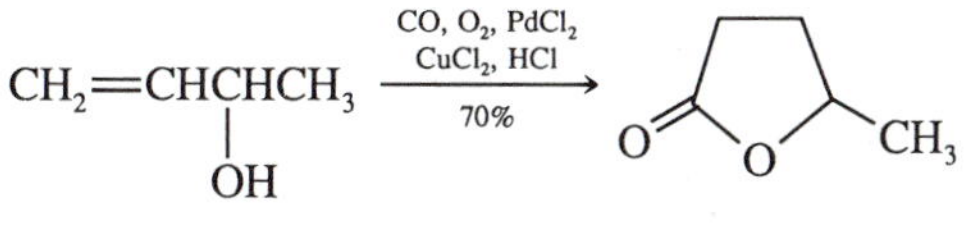

$ArNH_2 \rightarrow ArNHCO_2R$.[2] This transformation can be carried out in one step by reaction of the amine with CO, O_2, ROH, and HCl in the presence of $PdCl_2$ as catalyst and $CuCl_2$ as the reoxidant. Yields range from 16–90%.

Oxidation of alkenes to ketones.[3] Both internal and terminal alkenes are oxidized by $PdCl_2$ (with $CuCl_2$ as reoxidant) in water–polyethylene glycol (PEG), serving as the phase-transfer catalyst as well as the solvent (**9**, 360, 376). This oxidation is more facile than that catalyzed by quaternary ammonium salts, which is applicable only to terminal alkenes.

Examples:

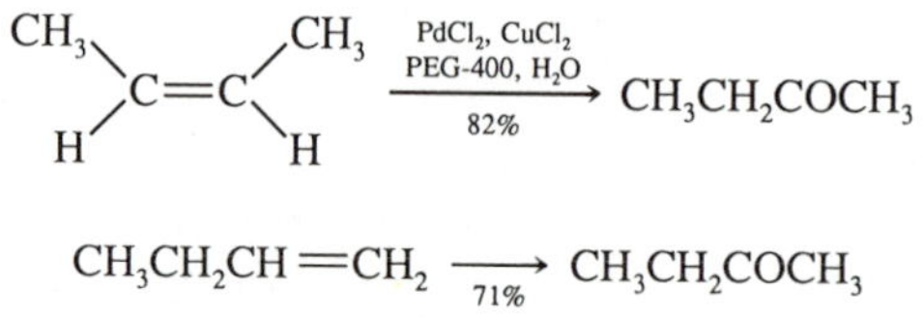

$$\text{CH}_3\text{CH}_2\text{CH}=\text{CH}_2 \xrightarrow[71\%]{} \text{CH}_3\text{CH}_2\text{COCH}_3$$

***Oxycarbonylation of 4-pentene-1,3-diols.*[4]**   Oxycarbonylation of these diols (CO,NaOAc,$CuCl_2$) catalyzed by $PdCl_2$ provides the lactones of *cis*-2-(3-hydroxy-tetrahydrofuranyl)acetic acid.

Examples:

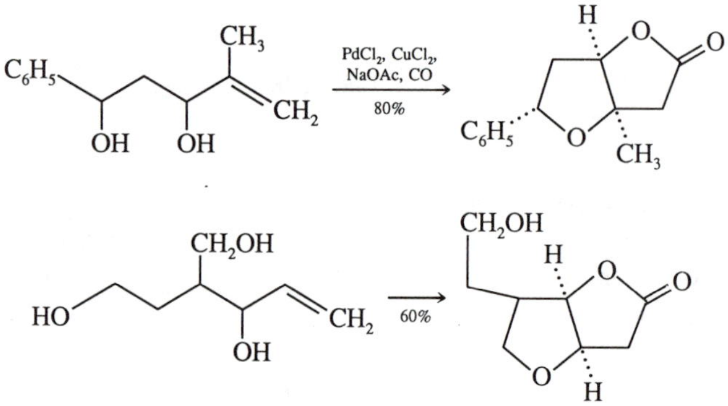

[1] H. Alper and D. Leonard, *J.C.S. Chem. Comm.*, 511 (1985); *Tetrahedron Letters*, **26**, 5639 (1985).
[2] H. Alper and F. W. Hartstock, *J.C.S. Chem. Comm.*, 1141 (1985).
[3] H. Alper, K. Januszkiewicz, and D. J. H. Smith, *Tetrahedron Letters*, **26**, 2263 (1985).
[4] Y. Tamaru, T. Kobayashi, S. Kawamura, H. Ochiai, M. Hojo, and Z. Yoshida, *Tetrahedron Letters*, **26**, 3207 (1985).

Palladium(II) trifluoroacetate.

***Oxygenation of allylsilanes.*[1]**   In the presence of Pd(OCOCF$_3$)$_2$ [or PdCl$_2$(CH$_3$CN)$_2$] photolysis of allylsilanes in oxygenated acetone results in α,β-enals, α,β-enones, or allylic alcohols. However, allylsilanes bearing a tosyl group at the allylic position undergo selective oxidation to α,β-enals or -enones.

Example:

[1] A. Riahi, J. Cossy, J. Muzart, and J. P. Pete, *Tetrahedron Letters*, **26**, 839 (1985).

(2R,4R)-Pentanediol.

Chiral propargylic alcohols.   The ketals (**2**) obtained from propargylic ke-tones and (2R, 4R)-2,4-pentanediol (**1**) are cleaved by DIBAH or Br$_2$AlH to the ethers (**3**) diastereoselectively ($\sim$96:4). The chiral auxiliary is removed by oxidation (PCC) and hydrolysis to furnish optically pure propargylic alcohols with the (R)-configuration.[1]

In contrast, cleavage of **2** (R^1 = C$_4$H$_9$, R^2 = CH$_3$) with triethylsilane and SnCl$_4$ (1:1) gives **3a** and **3b** in the ratio 5:95 and thus provides a stereoselective route to (S)-**4**.[2]

Chiral β-hydroxy carboxylic acids.[3]   The chiral acetals **2**, available from (2R, 4R)-pentanediol (**12**, 375–378), couple with ketene *t*-butyl *t*-butyldimethylsilyl acetal (**3**) in the presence of TiCl$_4$ (0.5 equiv.) to give adducts that are hydrolyzed by TFA–H$_2$O to **4** and **5**. The mixture is converted on oxidation and β-elimination into essentially pure (3R)-β-hydroxy carboxylic acids (**6**).

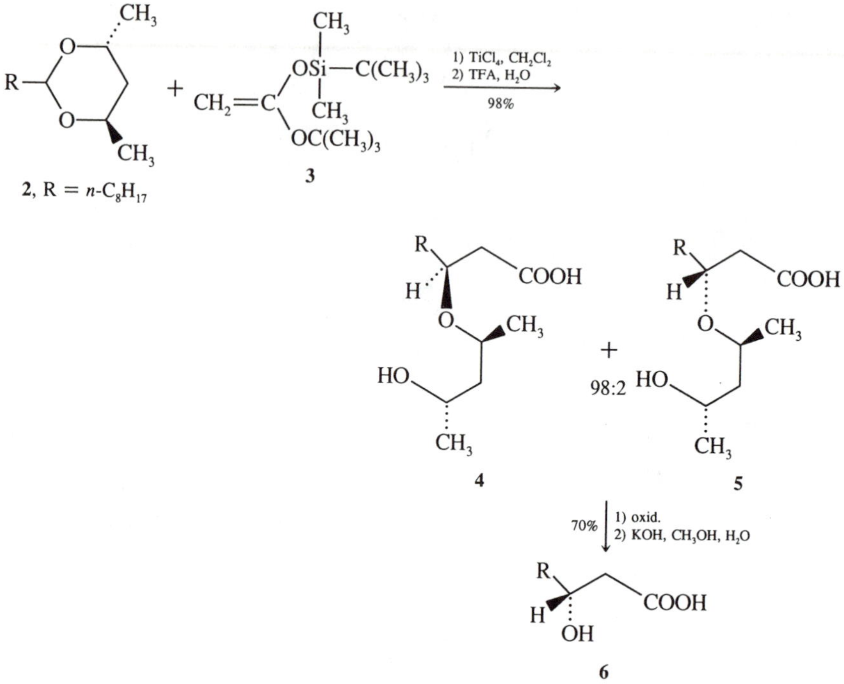

[1] K. Ishihara, A. Mori, I. Arai, and H. Yamamoto, *Tetrahedron Letters*, **27**, 983 (1986).
[2] A. Mori, K. Ishihara, and H. Yamamoto, *ibid.*, **27**, 987 (1986).
[3] J. D. Elliott, J. Steele, and W. S. Johnson, *ibid.*, **26**, 2535 (1985).

Perbenzoic acid.

Hydroxylation.[1]   Perbenzoic acid (C_6H_6, 100°) can effect hydroxylation of hydrocarbons with a regioselectivity of 60–90% in favor of formation of tertiary alcohols. This reaction involves radicals since it is inhibited by hydroquinone. The hydroxylation of *cis*-and *trans*-decalin is stereoselective; thus the former is converted mainly into *cis*-9-decalol (97:3) and the latter is converted mainly *trans*-9-decalol (95:5).

[1] J. Fossey, D. Lefort, M. Massoudi, J.-Y. Nedelec, and J. Sorba, *Can. J. Chem.*, **63**, 678 (1985).

Periodic acid.

Oxidation of naphthols.   1,5-Dihydroxynaphthoquinone (**1**) is oxidized by HIO_4 in aqueous DMF to juglone (**2**) in 55% yield. Under the same conditions β-naphthol is oxidized to 1,2-naphthoquinone in 47% yield.[1]

$$1 \qquad\qquad\qquad 2$$

[1] A. V. Pinto, V. F. Ferreira, and M. C. F. R. Pinto, *Syn. Comm.*, **15**, 1177 (1985).

Phase-transfer catalysts.

Benzoin condensation.[1]   The benzoates of cyanohydrins of aromatic aldehydes undergo benzoin condensation with an aromatic aldehyde in 50% NaOH/C_6H_6 in the presence of a phase-transfer catalyst, benzyltriethylammonium chloride. Theoretically two symmetrical and two unsymmetrical benzoins are possible, but in practice only one unsymmetrical benzoin is formed, that in which the carbonyl group is adjacent to the benzene ring substituted by the more electron-donating group.

Example:

Optically active quaternary ammonium salt (1).[2]   The salt **1** is obtained by alkylation of triethylamine with (S)-(+)-1-bromo-2-methylbutane (α_D + 3.80°) in CH_3CN. Methylation of *sec*-phenethyl alcohol (2) with dimethyl sulfate in pentane/ aqueous NaOH with **1** as catalyst gives the methyl ether **3** in 84% yield in 48%

ee. The enantioselectivity depends upon the solvent; it drops to only 8% when CH_2Cl_2 is used as the organic solvent.

For another example of the use of chiral phase-transfer catalysts in synthesis, see N-(*p*-Trifluoromethylbenzyl)cinchoninium bromide (this volume).

Alkyl azides.   Sodium azide as such is of little use for preparation of alkyl azides by nucleophilic substitution reactions because of solubility problems. The reaction can be carried out under phase-transfer conditions with methyltrioctylammonium chloride/NaN_3.[3] An even more effective polymeric reagent can be obtained by reaction of NaN_3 with Amberlite IR-400.[4] This reagent converts alkyl bromides, iodides, or tosylates into azides at 20° in essentially quantitative yield. The solvents of choice are CH_3CN, $CHCl_3$, ether, or DMF.

[1] M. D. Rozwadowska, *Tetrahedron*, **41**, 3135 (1985).
[2] J. W. Verbicky, Jr., and E. A. O'Neil, *J. Org.*, **50**, 1786 (1985).
[3] W. Reeves and M. L. Bahr, *Synthesis*, 823 (1976).
[4] A. Hassner and M. Stern, *Angew. Chem., Int. Ed.*, **25**, 478 (1986).

Phenyl azide–Aluminum chloride.

Reaction with alkenes.[1] $C_6H_5N_3$ and $AlCl_3$ (1:1.1) react with cyclohexene in CH_2Cl_2 to give after neutralization **1** and **2** in about equal amounts. A similar reaction is observed with cyclopentene. However, in the reaction with *cis*-cyclooctene, the aziridine **3** is isolated in 47% yield. These reactions probably involve an aziridinium–$AlCl_3$ complex.

[1] H. Takeuchi, M. Maeda, M. Mitani, and K. Koyama, *J.C.S. Chem. Comm.*, 287 (1985); H. Takeuchi, Y. Shiobara, M. Mitani, and K. Koyama, *ibid.*, 1251 (1985).

Phenyl dichlorophosphate, $C_6H_5OPOCl_2$ (1).

β-Chloro-α,β-enones.[1]   β-Diketones, particularly cyclic ones, can be converted into β-chloro-α,β-enones by reactions with **1** and LiH as base. Yields are increased substantially by addition of LiCl (2 equiv.).

Example:

¹ H.-J. Liu, G. V. Lamoureux, and M. Llinas-Brunet, *Can. J. Chem.*, **64**, 520 (1986).

β-Phenylethyl chloroformate, $C_6H_5CH_2CH_2OCOCl$. Preparation.[1]

Protection of amines; homobenzyloxycarbonyl (hcbo) group.[2] In contrast to the Cbo group, the homoCbo group is stable to HBr in HOAc at 25° and to HCl in ether. It is also less susceptible to hydrogenolysis, but deblocking can be effected by catalytic transfer hydrogenolysis with a Pd/C catalyst in the presence of ammonium formate. Best results are obtained with a catalyst freshly prepared from $Pd(OAc)_2$.

¹ H. Najer, P. Chabrier, and R. Guidicelli, *Bull Soc.*, 1189 (1955).
² L. A. Carpino and A. Tunga, *J. Org.*, **51**, 1930 (1986).

Phenyl ethylenesulfonate, $CH_2{=}CHSO_2OC_6H_5$ **(1)**. Preparation.[1]

Diels-Alder reactions with furans.[2] This sulfonate ester undergoes cycloaddition to furans at room temperature.

Example:

20°   87% (*endo/exo* = 5.4:1)
70°   48% (*endo/exo* = 2:1)

¹ H. Distler, *Angew. Chem. Intl. Ed.*, **4**, 300 (1965).
² L. L. Klein and T. M. Deeb, *Tetrahedron Letters*, **26**, 3935 (1985).

Phenyliodine(III) bis(trifluoroacetate) (1).

α-Hydroxy ketones. This hypervalent iodine reagent oxidizes terminal alkyl- or arylalkynes to α-hydroxy ketones (equation I). Of greater interest, it oxidizes ethynylcarbinols to α,α'-dihydroxy ketones.[1]

$$(I) \quad HC\equiv CR(Ar) \xrightarrow[\sim 80\%]{\begin{array}{c}\text{1) 1, CHCl}_3\text{, CH}_3\text{CN, H}_2\text{O} \\ \text{2) SiO}_2 \text{ (H}_3\text{O}^+)\end{array}} RCCH_2OH$$

Example:

[1] Y. Tamura, T. Yakura, J. Haruta, and Y. Kita, *Tetrahedron Letters*, **26**, 3837 (1985).

Phenyliodine(III) diacetate.

Cleavage of terminal tyrosyl and tryptophanyl peptides.   L-Tyrosyl-L-alanine (**1**) and related dipeptides are cleaved by $C_6H_5I(OAc)_2$ in CH_3OH containing KOH to 4-(methoxymethyl)phenol (**2**) and an amino acid. Under the same conditions, L-tyrosine yields (4-hydroxyphenyl)acetonitrile, $HOC_6H_4CH_2CN$. The phenolic hydroxyl group and the free amino group are essential for this cleavage.[1]

L-Tryptophan (**3**) and L-tryptophanyl peptides are oxidatively cleaved by C_6H_5I-$(OAc)_2$ to 3-(methoxymethyl)-1*H*-indole (**4**).[2]

Diphenylmethyl (benzhydryl) esters.[3]   These esters can be cleaved either by acid hydrolysis or hydrogenolysis. They have been prepared by reaction of carboxylic acids with diphenyldiazomethane, generated *in situ* by oxidation of benzophenone hydrazone with HgO or CH_3CO_3H. These esters of N-protected amino acids are prepared more conveniently by use of phenyliodine(III) diacetate as the oxidant (equation I).

$$\text{(I) XNHCH(R)COOH } + (C_6H_5)_2C{=}NNH_2 \xrightarrow[\quad CH_2Cl_2, \ I_2 \quad]{C_6H_5I(OAc)_2}$$

$$\text{XNHCH(R)COOCH}(C_6H_5)_2 + C_6H_5I + 2HOAc$$
$$(73\text{–}93\%)$$

1,4-*Epimines*.[4] Irradiation of nitroamines and cyanoamines in the presence of phenyliodine(III) diacetate and iodine generates an aminyl radical which undergoes intramolecular hydrogen abstraction to produce epimines.
   Example:

1) $R^1 = H$, $R^2 = NO_2$ 77%
2) $R^1 = OMOM$, $R^2 = CN$ 64%   31%

   Lead tetraacetate and iodine can also be used, but yields are lower and an excess of lead tetraacetate is required.

[1] R. M. Moriarty, M. Sultana, and Y.-Y. Ku, *J.C.S. Chem. Comm.*, 974 (1985).
[2] R. M. Moriarty and M. Sultana, *Am. Soc.*, **107**, 4559 (1985).
[3] L. Lapatsanis, G. Milias, and S. Paraskewas, *Synthesis*, 513 (1985).
[4] P. de Armas, R. Carrau, J. I. Concepción, C. G. Francisco, R. Hernández, and E. Suárez, *Tetrahedron Letters*, **26**, 2493 (1985).

Phenylmenthol. The preparation of 8-phenylmenthol (**1**) according to Corey and Ensley (**11**, 412) results in formation also of the diastereomer (**2**) that is epimeric at both C_1 and C_2 and hence named 2-*epi*,*ent*-8-phenylmenthol. These products are

1   **2**

usually obtained in the ratio 85:15 and can be separated by preparative HPLC. Although the value of **1** as a chiral auxiliary has been well documented, the use of **2** had not been examined because of its lack of ready availability. In fact, **2** is comparable to **1** as a chiral auxiliary in Diels-Alder and ene reactions. However, only moderate levels of asymmetric induction were observed in Grignard additions to the glyoxylate ester derived from **2**.[1]

Because of the lack of ready access to **2**, Whitesell *et al.*[2] recommend *trans*-2-phenylcyclohexanol (**3**) as an auxiliary. Hydrolysis of the acetate of (+)-**3** with hog liver esterase provides optically pure (−)-**3**,α_D − 56° and (+)-**3**,α_D + 54°. The glyoxylate ester of **3** undergoes ene reactions with high diastereoselectivity.

3

[1] J. K. Whitesell, C.-L. Liu, C. M. Buchanan, H.-H. Chen, and M. A. Minton, *J. Org.*, **51**, 551 (1986).
[2] J. K. Whitesell, H.-H. Chen, and R. M. Lawrence, *ibid.*, **50**, 4663 (1985).

2-(Phenylseleno)acrylonitrile, $CH_2{=}C{<}^{SeC_6H_5}_{CN}$ (**1**). Preparation.[1]

Cyclopentanones.   Enamines (**2**) react with **1** at 25° to give in high yield the keto nitriles **3**, which undergo the expected addition of lithium phenylacetylide to provide **4**. On treatment of **4** with $(C_6H_5)_3SnH$ (AIBN) a δ-acetylenic radical is formed, which cyclizes to a benzylidenecyclopentane (**5**). Ozonolysis of **5** provides cyclopentanone **6**.

Example:

[1] Z. Janousek, S. Piettre, F. Gorissen-Hervens, and H. G. Viehe, *J. Organomet. Chem.*, **250**, 197 (1983).
[2] A. G. Angoh and D. L. J. Clive, *J.C.S. Chem. Comm.*, 941 (1985).

9-(Phenylseleno)-9-borabicyclo[3.3.1]nonane, $\text{B—SeC}_6\text{H}_5$ (1).

The reagent is obtained by reaction of 9-borabicyclo[3.3.1]nonane with C_6H_5SeH in toluene at 60°.

Conjugate addition–aldol reactions. A novel synthesis of α-substituted α,β-enones involves conjugate addition of **1** to an α,β-enone; the resulting β-phenylselenoboron enolate undergoes aldol condensation with aldehydes. The adduct on oxidative elimination furnishes unsaturated β-hydroxy ketones.[1]

Example:

[1] W. R. Leonard and T. Livinghouse, *J. Org.*, **50**, 730 (1985).

N-Phenylselenophthalimide (1).

Bicyclization of an unsaturated amine.[1] Reaction of the unsaturated amine **2** with **1** in CH_2Cl_2 results in a single tricyclic selenide (**3**). The product was used for a synthesis of an aziridinomitosan (**4**), related to some naturally occuring mitomycins.

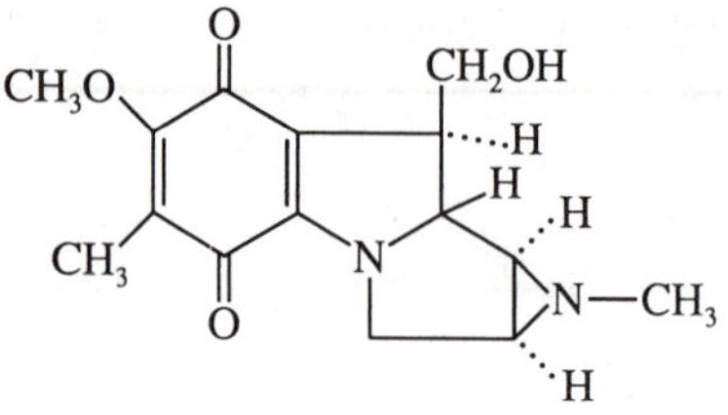

4

γ-Phenylselenoalkylidene-γ-butyrolactones.[2] γ,δ-Acetylenic acids (**2**) are cyclized to these butyrolactones (**3**) on reaction with **1** in CH₂Cl₂ at 25° (4–24 hours). The reaction is (E)-selective in the case of 4-hexynoic acid and 4-heptynoic acid.
Example:

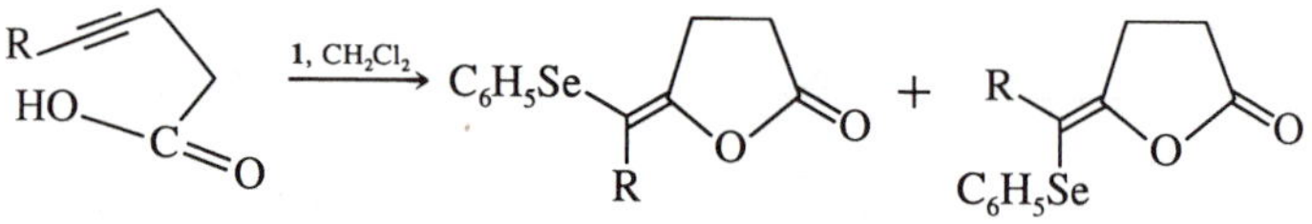

	(E)-**3**	(Z)-**3**
R = H	50%	21%
= CH₃	89%	—
= C₂H₅	76%	—

Selenothiolactonization.[3] Cycloalkenethiolcarboxylic acids react with **1** to form S-acyl phenylselenosulfides (**2**), which undergo cyclization to phenylselenothiolactones (**3**) when refluxed in benzene in the presence of AIBN. The C₆H₅Se group of **3** is selectively oxidized by m-ClC₆H₄CO₃H, and subsequent selenoxide elimination results in thiolactones (**4**).
Example:

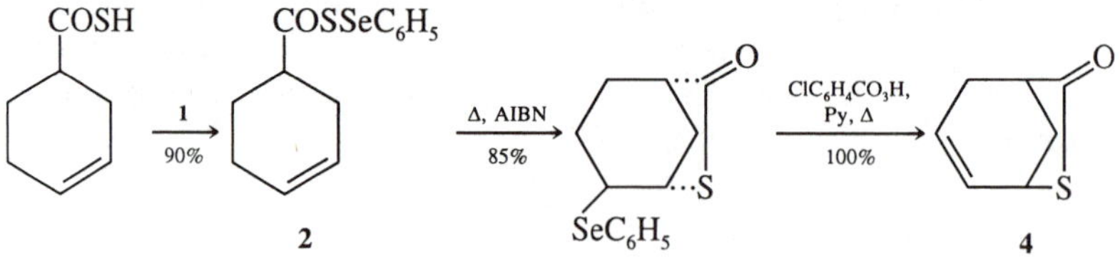

¹ S. Danishefsky, E. M. Berman, M. Ciufolini, S. J. Etheredge, and B. E. Segmuller, *Am. Soc.*, **107**, 3891 (1985).
² T. Toru, S. Fujita, and E. Maekawa, *J.C.S. Chem. Comm.*, 1082 (1985).
³ T. Toru, T. Kanefusa, and E. Maekawa, *Tetrahedron Letters*, **27**, 1583 (1986).

Phenylselenenyl benzenesulfonate (1).

New preparation:[1]

$$C_6H_5SO_2NHNH_2 + C_6H_5SeO_2H \xrightarrow[85\%]{\underset{25°}{CH_2Cl_2}} C_6H_5SeSO_2C_6H_5 + N_2 + 2H_2O$$

1

Vinyl sulfones.[1] The photoadducts of **1** to alkenes on oxidation (H_2O_2) undergo selenoxide elimination to provide vinyl sulfones in 60–95% yield.

Example:

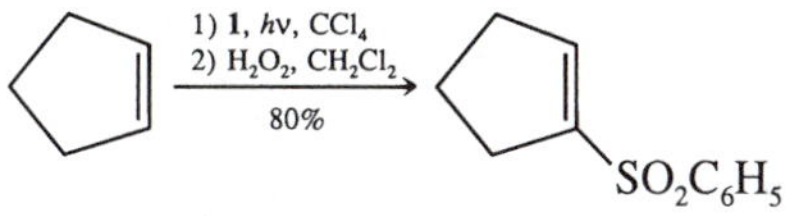

[1] H.-S. Lin, M. J. Coghlan, and L. A. Paquette, *Org. Syn.*, submitted (1986).

(Phenylsulfonyl)allene, $C_6H_5SO_2CH$=C=CH_2 **(1).** The allene, m.p. 44–45°, is obtained by treatment of 3-phenylsulfonylpropyne in CH_2Cl_2 with alumina.[1]

[4 + 2] *and* **[8 + 2]***Cycloadditions.*[2] This allene derivative is considerably more reactive in Diels-Alder reactions than allene itself. Adducts are formed exclusively with the C_1—C_2 double bond, but the stereoselectivity is governed by steric factors.

Example:

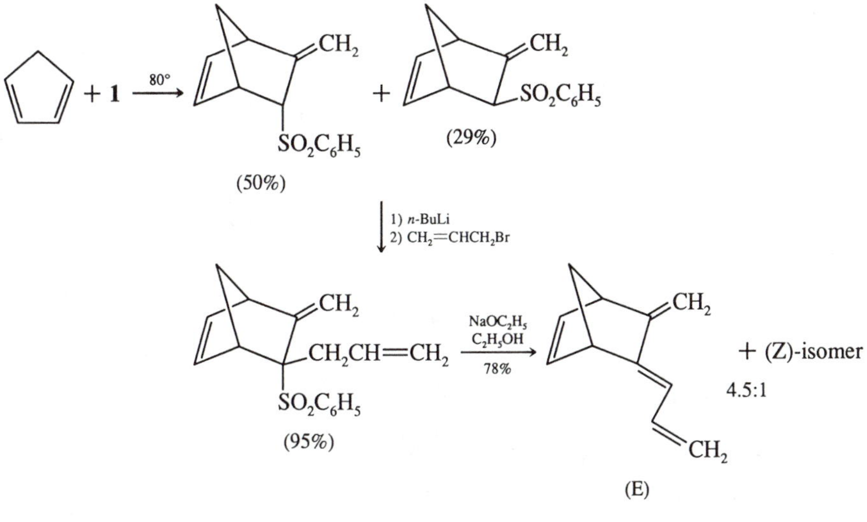

Reaction of **1** with tropone results in a single [8 + 2] cycloadduct (equation I).

[1] C. J. M. Stirling, *J. Chem. Soc.*, 5856, 5863, 5875 (1964).
[2] K. Hayakawa, H. Nishiyama, and K. Kanematsu, *J. Org.*, **50**, 512 (1985).

N-Phenyltrifluoromethanesulfonimide.

Enol triflates from α,β-enones.[1]   The reaction can be carried out by conjugate reduction of the enone with lithium tri-*sec*-butylborohydride followed by reaction of the resulting enolate with N-phenyltriflimide. The method is generally useful, but 1,2-reduction can predominate in the case of β,β- or even γ,γ-disubstituted α,β-enones.

[1] G. T. Crisp and W. J. Scott, *Synthesis*, 335 (1985).

Phosphomolybdic acid–Potassium dichromate–Copper(II) sulfate.

2-Cyclohexene-1,4-diones.   The method of choice for oxidation of the cyclohexenones **1** to the enediones (**2**) is air oxidation catalyzed by phosphomolybdic acid, $K_2Cr_2O_7$, and $CuSO_4 \cdot 5H_2O$.[1,2]

1a, R = H
1b, R = CH₃

2a
2b

[1] M. Seuret and E. Widmer, Belg. Patent **830**, 723 (1975) [C.A., **84**, 164269e (1976)].
[2] V. J. Freer and P. Yates, *Chem. Letters*, 2031 (1984).

Phenyl trimethylsilyl telluride. $C_6H_5TeSi(CH_3)_3$ (**1**). The tellurosilane is accessible by cleavage of diphenyl ditelluride with sodium followed by reaction with $ClSi(CH_3)_3$ (74% yield). It is sensitive to O_2 and H_2O.

Cleavage of esters and ethers.[1]   This silane is more reactive than thio- or selenosilanes. It cleaves lactones in the presence of ZnI_2 at 25°. Ethers are cleaved at 0–25° with some regioselectivity.

Examples:

$$CH_3(CH_2)_{11}OCH_3 \longrightarrow CH_3TeC_6H_5 + CH_3(CH_2)_{11}OSi(CH_3)_3$$
$$\phantom{CH_3(CH_2)_{11}OCH_3 \longrightarrow} 30\% 46\%$$

[1] K. Sasaki, Y. Aso, T. Otsubo, and F. Ogura, *Tetrahedron Letters*, **26**, 453 (1985).

Phosphoryl chloride.

Isocyanides.[1] Phosgene in combination with triethylamine has generally been preferred to phosphoryl chloride and triethylamine for dehydration of formamides to isocyanides (**1,** 857). However, use of phosphoryl chloride in combination with diisopropylamine can give isocyanides in yields comparable to those obtained with phosgene. Even so this new method can fail with some simple alkyl formamides.

[1] R. Obrecht, R. Herrmann, and I. Ugi, *Synthesis*, 400 (1985).

B-3-Pinanyl-9-borabicyclo[3.3.1]nonane.

Enantioselective reductions. The neat reagent (**1**), prepared from (+)-α-pinene, reduces aryl α-halomethyl ketones slowly but in high chemical yield to (R)-halohydrins in 90–96% ee, but optical induction is mediocre in the case of aliphatic α-halo ketones (35–66% ee). The chiral halohydrins are useful precursors to chiral epoxides.

The reagent is also useful for asymmetric reduction of α-keto esters, particularly α-keto *t*-butyl esters. Thus *t*-butyl pyruvate is reduced to (S)-*t*-butyl lactate in 100% ee.[1]

This borane reduces an acyl cyanide to a cyanohydrin-9-BBN adduct, but this intermediate undergoes ready elimination. The desired β-amino alcohols can be obtained in 75–85% yield by reduction of the cyanohydrin intermediate with $NaBH_4$ and $CoCl_2$. In the case of chiral aromatic acyl cyanides the optical yields are 92–98%. One aliphatic acyl cyanide is reduced in 84% ee.[2]

[1] H. C. Brown and G. G. Pai, *J. Org.*, **50**, 1384 (1985).
[2] M. M. Midland and P. E. Lee, *ibid.*, **50**, 3237 (1985).

Pivaldehyde, $(CH_3)_3CCHO$ (**1**), Mol. wt. 86.13, m.p. 6°. Supplier: Aldrich, Fluka.

α-Alkylation of amino acids.[1,2] The N-methyl amide (**2**) of an α-amino acid such as (S)-alanine condenses with (**1**) to form an imine (**3**), which can be cyclized to either the *trans*- or *cis*-imidazolidinone (**4**).

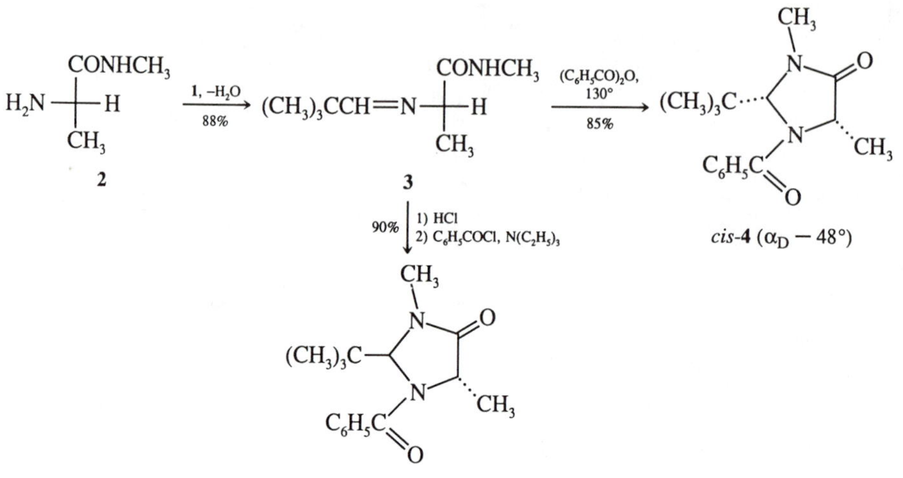

cis-4 ($\alpha_D - 48°$)

trans-4 ($\alpha_D + 45°$)

These products (**4**) can be used to effect α-alkylation of amino acids with retention or inversion of configuration. Thus either (S)- or (R)-α-methyldopa (**6**) can be prepared from (S)-alanine, as shown in equations (I) and (II).

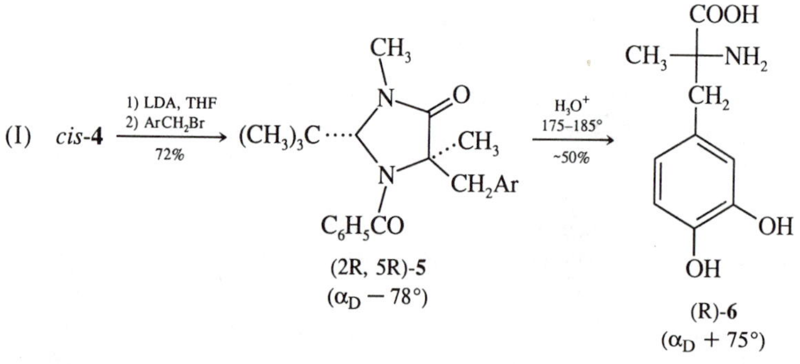

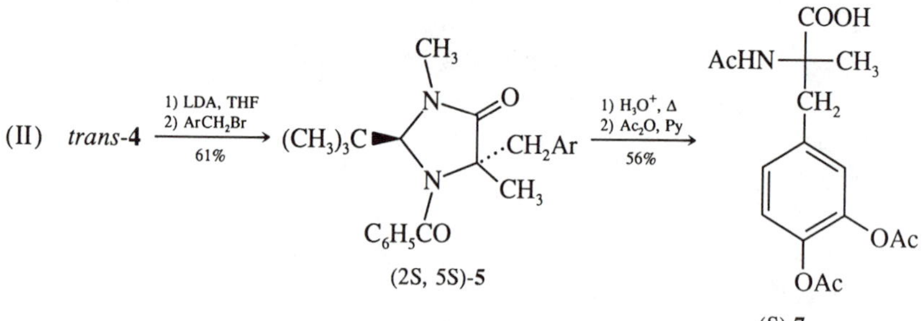

(I)   cis-4 $\xrightarrow[\text{72%}]{\text{1) LDA, THF} \atop \text{2) ArCH}_2\text{Br}}$

(2R, 5R)-5
($\alpha_D - 78°$)

(R)-6
($\alpha_D + 75°$)

(II)   trans-4 $\xrightarrow[\text{61%}]{\text{1) LDA, THF} \atop \text{2) ArCH}_2\text{Br}}$

(2S, 5S)-5

(S)-7

Reaction of the proline derivative **8** with pivaldehyde (**1**) in CH_2Cl_2 catalyzed by TFA results in a single product (**9**). Treatment of **9** with MeLi and then with LDA furnishes the enolate **a**. Alkylation or reaction with aldehydes or acetone followed by acid hydrolysis furnishes enantiomerically pure 2-substituted 4-hydroxyprolines (**10**), with retention of configuration.[3]

---

[1] R. Naef and D. Seebach, *Helv.*, **68**, 135 (1985).
[2] D. Seebach, J. D. Aebi, R. Naef, and T. Weber, *Helv.*, **68**, 144 (1985).
[3] T. Weber and D. Seebach, *Helv.*, **68**, 155 (1985).

Platinum–Titanium.

Dehydrocyclization.[1]   Pt(0) can be used for dehydrogenation of alkanes to alkenes. Ti(0) is known to absorb H_2 to form a dihydride. Paquette *et al.*[1] reasoned that a combination of the two metals could in principal effect dehydrocyclization. Indeed, cyclooctane when heated with Pt(0) and Ti(0) (1:1) adsorbed in Al_2O_3 is converted into bicyclo[3.3.0]octane (equation I).

This dehydrogenation was developed in connection with the synthesis of dodecahedrane (**2**) from secododecahedrane (**1**). Originally this last step was carried out in 40–50% yield by heating **1** with hydrogen-presaturated Pd(0) on carbon at 250°. This yield is increased to about 80% by use of Pt and Ti (2:1).

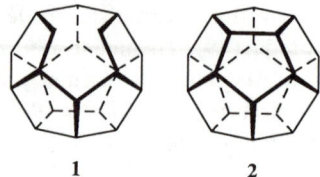

1   2

¹ L. A. Paquette, Y. Miyahara, and C. W. Doecke, *Am. Soc.*, **108**, 1716 (1986).

Potassium–Graphite.

Review.¹ Use of C_8K as a heterogeneous reagent in organic synthesis has been reviewed, particularly as a reducing agent for C=C and C=N bonds, and for reductive cleavage of the C—S bond of α,β-unsaturated sulfones. C_8K has been used to obtain active metals dispersed on graphite by reduction of metal halides.

¹ D. Savoia, C. Trombini, and A. Umani-Ronchi, *Pure Applied Chem.*, **57**, 1887 (1985).

Potassium *t*-butoxide.

Michael intramolecular alkylation.¹ The reaction of *t*-butyl lithioacetate (**1**) with ethyl 6-iodo-2-hexenoate **2** in the presence of potassium *t*-butoxide in THF at −78° results in a single *trans*-1,2-disubstituted cyclopentane (**3**). Reaction of **2** with *t*-butyl lithiopropionate (**4**) under the same conditions results mainly in *syn*-**5**, but when HMPT is also present *anti*-**5** is mainly formed. Similar stereoselectivity is observed in reaction of **1** and **4** with ethyl 6-iodo-2-heptenoate in formation of cyclohexanes. Similar reactions resulting in seven-membered rings require higher temperatures or high dilution and proceed with lower stereoselectivities.

β-*Lactams*.² The mesylate (**1**) of N-(phenylacetyl)serine O-benzylhydroxa-mate can be cyclized directly to **2** in 75% yield by potassium *t*-butoxide (1 equiv.) in *t*-butyl alcohol at −23°. Use of KHCO₃ results in the oxazoline **3** in 94% yield.

Intramolecular conjugate addition of carbamates to α,β-unsaturated esters.³ In the presence of potassium *t*-butoxide or NaH, allylic carbamate esters **1** cyclize stereoselectively to *trans*-oxazolidones (**2**).

Example:

Homoallylic carbamate esters **3** cyclize to 6-membered carbamates **4** with high 1,3-*syn* induction.

Example:

$$CH_3 \diagdown CH_2CH{=}CHCOOCH_3 \xrightarrow{\text{NaH}}$$

OCONH$_2$

(E)-3	53%
(Z)-3	70%

syn-4   anti-4

10:1
>20:1

Wittig methylenation.[4] Potassium *t*-butoxide has seen limited use for generation of stabilized ylides. Actually, it is a superior base for generation of methylenetriphenylphosphorane. For methylenation of moderately hindered ketones, the phosphorane is generated and used in ether. For reactions of strongly hindered ketones, the ylide is generated in benzene and the reaction is conducted at 90–120° in benzene.

Catalysis of Michael reactions.[5] Michael reactions proceed at 25° in THF when catalyzed by potassium *t*-butoxide absorbed on xonotlite, $Ca_6Si_6O_{17}(OH)_2$,[6] or by KF adsorbed on alumina.

[1] M. Yamaguchi, M. Tsukamoto, and I. Hirao, *Tetrahedron Letters*, **26**, 1723 (1985).
[2] M. A. Krook and M. J. Miller, *J. Org.*, **50**, 1126 (1985).
[3] M. Hirama, T. Shigemoto, Y. Yamazaki, and S. Ito, *Am. Soc.*, **107**, 1797 (1985).
[4] L. Fitjer and U. Quabeck, *Syn. Comm.*, **15**, 855 (1985).
[5] P. Laszlo and P. Pennetreau, *Tetrahedron Letters*, **26**, 2645 (1985).
[6] E. Lippmaa, M. Mägis, A. Samoson, G. Engelhardt, and A.-R. Grimmer, *Am. Soc.*, **102**, 4889 (1980).

Potassium carbonate.

Wittig-Horner reaction with 1,4- and 1,5-dialdehydes.[1] Reaction of these aldehydes with 1 equiv. of a Wittig-Horner phosphonate in an aqueous medium with K_2CO_3 as base is accompanied by an intramolecular aldolization to provide five- or six-membered cycloalkenols.

Example:

W = CN, CO$_2$C$_2$H$_5$,
COCH$_3$

W = CN, 68% yield

[1] M. Graff, A. Al Dilaimi, P. Seguineau, M. Rambaud, and J. Villieras, *Tetrahedron Letters*, **27**, 1577 (1986).

Potassium diisopropylamide.

Metallation of aryl selenides and selenoacetals.[1]   This base (KDA) is the most efficient for metallation of these substrates.
Example:

[1] M. Clarembeau and A. Krief, *Tetrahedron Letters*, **27**, 1723 (1986).

Potassium ferricyanide.

Oxidative coupling of phenolic β-diketones.[1]   The phenolic β-diketone **1** does not undergo oxidative coupling with several of the standard oxidants (*e.g.*, FeCl$_3$, VOF$_3$, VOCl$_3$, CuCl$_2$), but on oxidation with K$_3$Fe(CN)$_6$ in Na$_2$CO$_3$ solution, the *para*-coupling product **2** is obtained in 67% yield together with the *ortho*-coupling product **3** (8% yield).

1 $\xrightarrow{\text{K}_3\text{Fe(CN)}_6,\ \text{Na}_2\text{CO}_3}$ **2** (67%)   +   **3** (8%)

[1] A. S. Kende, F. H. Ebetino, and T. Ohta, *Tetrahedron Letters*, **26**, 3063 (1985).

Potassium fluoride.

Nucleophilic fluorination of alkyl bromides.[1,2]   KF is ineffective for fluorine transfer reactions probably because of low solubility in organic solvents. Addition of CaF_2 markedly enhances the rate of reaction of KF with $C_6H_5CH_2Br$ in sulfolane. An even more reactive supported reagent can be obtained by evaporation of a mixture of KF and CaF_2 (1 : 5) in CH_3OH. CaF_2 also enhances the rate of fluorination with CsF. The KF–CaF_2 reagent converts a primary alkyl bromide or benzyl bromide into the fluorides in sulfolane at 120–160° in 2 hours in 85–90% yield. It converts CH_3COCl into CH_3COF in 75% yield at 30° in 12 minutes.

Stereoselective benzylation of a methyl dithionate (**1**).[3]   Benzylation of **1** is effected with the highest (Z)-stereoselectivity with KF absorbed on alumina as base, either under dry conditions or in CH_3CN (equation I).

LDA, THF	70%		70:30
KOC(CH$_3$)$_3$, THF	65%		57:43
KF, DMF, 153°	95%		42:58
KF-Al$_2$O$_3$	95%		85:15

[1] J. H. Clark, A. J. Hyde, and D. K. Smith, *J.C.S. Chem. Comm.*, 791 (1986).
[2] J. Ichihara, T. Matsuo, T. Hanafusa, and T. Ando, *ibid.*, 793 (1986).
[3] D. Villemin, *ibid.*, 870 (1985).

Potassium hexamethyldisilazide, KN[Si(CH₃)₃]₂.

Regioselective deprotonation of α,β-*enones.*[1] This base selectivily deproton-ates cyclic α,β-enones at the γ-position rather than the α'-position.

Example:

KN[Si(CH₃)₃]₂	34%	>99%	—
LiN[Si(CH₃)₃]₂	68%	75%	25%
LDA	44%	—	99%

[1] M. Kawanisi, Y. Itoh, T. Hieda, S. Kozima, T. Hitomi, and K. Kobayashi, *Chem. Letters,* 647 (1985).

Potassium hydride.

Purification.[1] Commercial KH is prepared by hydrogenation of K(0) at 230–400°; several laboratories report variation in the activities of different lots. Mac-donald *et al.*[1] reasoned that the impurities could be unreduced metal or KO_2. In any case, treatment of commercial KH (suspended in mineral oil) with I_2 until the iodine color persists can markedly improve some reactions involving KH, in par-ticular, the oxy-Cope rearrangement. Iodine-modified KH is also useful for *in situ* generation of $(Bu)_3SnK$ from KH and Bu_3SnH.

Anionic oxy-Claisen rearrangement.[2] The [3,3] Claisen rearrangement of en-olates of α-allyloxy ketones is markedly dependent on the nature of the metal hydride used, and to a less extent, the solvent. An example is the rearrangement of α-(allyloxy)propiophenone (**1**) to the α-hydroxy ketone **2**. The rearrangement

occurs more readily in THF than in toluene, but the effect of the counterion is more striking. The rearrangement with KH occurs at − 42°, the lowest temperature known to date for such a rearrangement.

This rearrangement has been used for a tandem 1,4-addition followed by Claisen rearrangement (equation I).

[1] T. L. Macdonald, K. J. Natalie, Jr., G. Prasad, and J. S. Sawyer, *J. Org.*, **51**, 1124 (1986).
[2] M. Koreeda and J. I. Luengo, *Am. Soc.*, **107**, 5572 (1985).

Potassium hydroxide.

Preparation and saponification of hindered esters.[1]   Mesitoic acid can be esterified in virtually quantitative yield by reaction with an alkyl iodide, chloride, or sulfate and powdered KOH (2.5 equiv.) containing ~15% water and 2% Aliquat 336 in the absence of a solvent. The same system but with 5.0 equiv. of KOH and temperatures of 85° effects saponification of alkyl mesitoates in high yield.

[1] A. Loupy, M. Pedoussaut, and J. Sansoulet, *J. Org.*, **51**, 740 (1986).

Potassium hydroxide–18-Crown-6.

Alkylation of aryl 1-alkynes.[1]   These alkynes undergo rapid alkylation with alkyl iodides in the presence of powdered KOH and a catalytic amount of 18-crown-6 (C_6H_6, 80°); yields 60–75%.

[1] M. Lissel, *Tetrahedron Letters*, **26**, 1843 (1985).

Potassium nitrosodisulfonate (Fremy's salt).

Selective oxidation of benzylic alcohols.[1]   These alcohols are oxidized by Fremy's salt in benzene–water with Adogen 464 as the phase-transfer catalyst to aldehydes or ketones in generally high yields. Allylic alcohols are completely stable under these conditions. See also sodium nitrosodisulfonate (this volume).

[1] J. Morey, A. Dzielenziak, and J. M. Saa, *Chem. Letters*, 263 (1985).

Potassium permanganate.

Selenones.[1]   Oxidation of selenides to selenones is hampered by the ready decomposition of the intermediate selenoxides. The most useful oxidants are peracids (m-$ClC_6H_4CO_3H$, CF_3CO_3H) and $KMnO_4$ (CH_2Cl_2–H_2O, 20°).

Alkyl phenyl selenones undergo substitution reactions with a number of nucleophiles with concomitant formation of benzeneselenenic acid.

Examples:

$$CH_3(CH_2)_9I \xleftarrow[97\%]{\substack{NaI, \\ CH_3COCH_3}} CH_3(CH_2)_9\underset{\underset{O}{\|}}{\overset{\overset{O}{\|}}{Se}}C_6H_5 \xrightarrow[81\%]{\substack{MgBr_2, \\ ether}} CH_3(CH_2)_9Br$$

1

$$94\% \downarrow C_6H_5SNa$$

$$CH_3(CH_2)_9SC_6H_5$$

The reaction of **1** with benzaldehyde in the presence of base results in an epoxide (equation I).

(I) $1 + C_6H_5CHO \xrightarrow[74\%]{KOC(CH_3)_3, THF}$

[1] A. Krief, W. Dumont, J.-N. Denis, G. Evrard, and B. Norberg, *J.C.S. Chem. Comm.*, 569 (1985); A. Krief, W. Dumont, and J.-N. Denis, *ibid.*, 571 (1985).

Potassium peroxomonosulfate (Oxone), $KHSO_5$.

Epoxidation.[1]   Oxone alone (1.5-2 equiv.) can effect epoxidation of alkenes. Yields are generally higher if the pH is maintained at 6 by addition of KOH. Water-soluble substrates (sorbic acid) are epoxidized in water; aqueous methanol is used for water-insoluble substrates.

Examples:

[1] R. Bloch, J. Abecassis, and D. Hassan, *J. Org.*, **50**, 1544 (1985).

Potassium ruthenate.

Oxidation of a δ-lactone to a 1,5-dicarbonyl compound; Nef oxidation.[1]   A key step in a synthesis of Antheridium-inducing factor (A_{An}, **3**) is the conversion of the nitro lactone **1** into **2** by saponification, potassium ruthenate oxidation ($K_2S_2O_8$ and $RuCl_3$ in *t*-BuOH–H_2O), and finally esterification. Rearrangement of the double bond to the more stable position probably occurs during the first step. The oxidation of the nitronate group to a ketone group is a novel Ru(VI) oxidation.

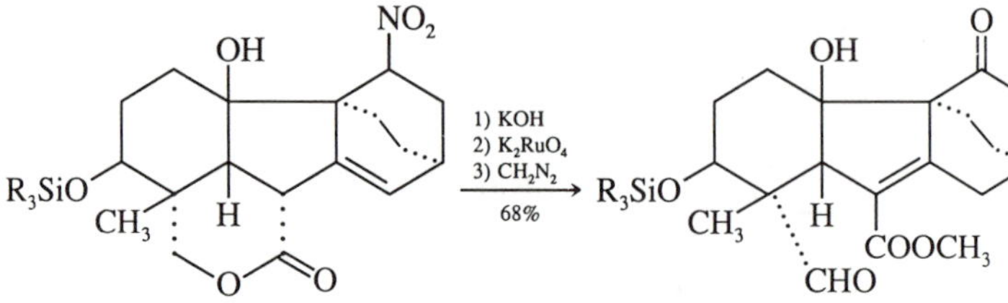

1 **2**

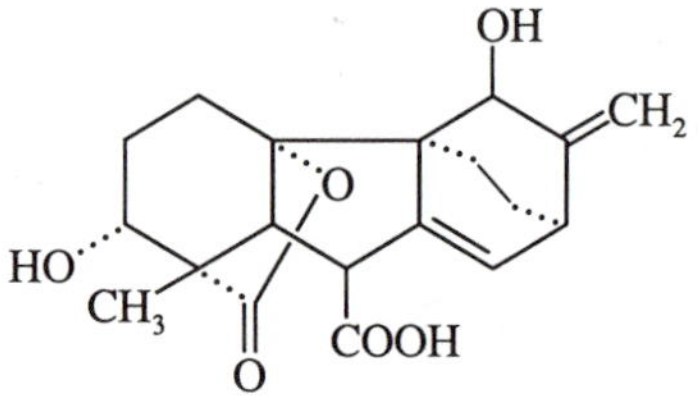

3

[1] E. J. Corey and A. G. Myers, *Am. Soc.*, **107**, 5574 (1985).

Potassium superoxide.

Desulfuration.[1] The reaction of KO_2 with *o*-nitrobenzenesulfonyl chloride results in a peroxysulfur reagent that converts thioureas, thioamides, and thiocarbamates in CH_3CN at $-30°$ into the corresponding carbonyl compounds.

Example:

$$o\text{-}O_2NC_6H_4SO_2Cl + KO_2 \xrightarrow[86\%]{C_6H_5C(=S)NHCH_3,\ CH_3CN,\ -35°} C_6H_5C(=O)NHCH_3$$

[1] Y. H. Kim, B. C. Chung, and H. S. Chang, *Tetrahedron Letters*, **26**, 1079 (1985).

Potassium triethylborohydride.

anti-*Selective reduction of* 2-*alkyl*-3-*keto amides.*[1] Optically active amides of this type, derived from optically active *trans*-2,5-bis(methoxymethoxymethyl)-pyrrolidine, are reduced by $KB(C_2H_5)_3H$ with high *anti*-diastereoselectivity.

Example:

(97% ee, *anti/syn* =
>99:1)

[1] Y. Ito, T. Katsuki, and M. Yamaguchi, *Tetrahedron Letters,* **26**, 4643 (1985).

(S)-Prolinol.

Chiral cyclohexadienes.[1]   The optically active amide **1**, obtained by acylation of L-prolinol with *o*-anisoyl chloride, on reductive methylation gives **2** with high diastereoselectivity. The high stereoselectivity is attributed to chelation between

the OCH_3 on the benzene ring and the prolinol oxygen, since a similar reaction with **3** results in the opposite, but reduced stereoselectivity. This sequence provides chiral cyclohexadienes in both enantiomeric forms.

[1] A. G. Schultz, P. Sundararaman, M. Macielag, F. P. Lavieri, and M. Welch, *Tetrahedron Letters*, **26**, 4575 (1985).

1,3-Propanedithiol.

Demercuration.[1]   The product (**1**) of oxymercuration of an α,β-unsaturated ester is demetallated by 1,3-propanedithiol (1 equiv.) in ethanol in the presence of NaHCO$_3$ with net retention of the stereochemistry (80–95%). Demetallation with the usual reagent, NaBH$_4$, proceeds with predominant inversion of stereochemistry (70–86%).

Example:

	99%	95:5
HS(CH$_2$)$_3$SH, NaHCO$_3$		
NaBH$_4$	85%	20:80

[1] F. H. Gouzoules and R. A. Whitney, *Tetrahedron Letters*, **26**, 3441 (1985).

Pyridinium chlorochromate–Benzotriazole, PyH$^+$CrO$_3$Cl$^-$

Oxidation of allylic alcohols.[1]   Allylic alcohols are oxidized to the corresponding enones in 80–95% yield by PCC complexed with benzotriazole (CH$_2$Cl$_2$, 23°).

[1] E. J. Parish and S. Chitrakorn, *Syn. Comm.*, **15**, 393 (1985).

4-Pyrrolidinopyridine.

Cyclopeptides.[1]   Cyclization of polypeptides under conditions of high dilution usually proceeds in low yield. A new method involves cyclization of ω-peptide-pentafluorophenyl esters in dioxane in the presence of 4-pyrrolidinopyridine and an alcohol (ethanol or *t*-butyl alcohol). *p*-Nitrophenyl esters can also be employed, but separation of *p*-nitrophenol from product is more difficult than that of pentafluorophenol.

Example:

[1] U. Schmidt, A. Lieberknecht, H. Griesser, R. Utz, T. Beuttler, and F. Bartkowiak, *Synthesis*, 361 (1986).

Pyrylium perchlorate, ClO$_4^-$ (**1**), potentially explosive.

This reagent can be prepared in ~85% yield by reaction of HClO$_4$ with the sodium enolate of gluconic dialdehyde, NaOCH=CHCH=CHCHO.[1]

Dienes and trienes.[2]   Alkyl- or vinyllithium reagents react with **1** to form a 1,2-adduct (**a**) that rearranges to a (2Z,4E)-5-substituted pentadienal (**2**) in 50–65% yield. The aldehyde can be used for a Wittig reaction or allowed to react with

a second alkyllithium reagent. By use of vinyl or alkynyl reagents, a variety of conjugated dienes and trienes can be prepared stereoselectively.

[1] F. Klages and H. Trager, *Ber.*, **86**, 1327 (1953).
[2] M. Furber and R. J. K. Taylor, *J.C.S. Chem. Comm.*, 782 (1985).

Q

Quina alkaloids.

Asymmetric Michael addition.[1]   The highest optical yield for addition of thiophenol to maleates is obtained with diisopropyl maleate catalyzed by cinchonine (equation I). The succinate obtained in this way was used for synthesis of (R)-(+)-3,4-epoxy-l-butanol.

[1] H. Yamashita and T. Mukaiyama, *Chem. Letters*, 363 (1985).

R

Raney nickel.

   Reduction of ketimines.[1]   Reduction of N-cyclohexylideneaniline (**1**) with aluminum isopropoxide and isopropyl alcohol (Meerwein-Ponndorf reduction, **1**, 35–36) results in N-isopropylaniline as the major product (equation I). However, if

(I) 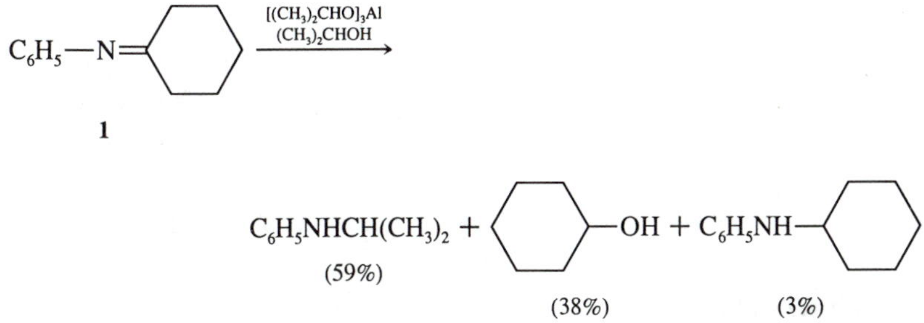 C_6H_5—N=⟨cyclohexylidene⟩ $\xrightarrow{\substack{[(CH_3)_2CHO]_3Al \\ (CH_3)_2CHOH}}$

   1

$$C_6H_5NHCH(CH_3)_2 \;+\; ⟨cyclohexyl⟩{-}OH \;+\; C_6H_5NH{-}⟨cyclohexyl⟩$$
   (59%)   (38%)   (3%)

W-2 Raney nickel is added to this reaction, the major product is the secondary amine, formed by reduction of the double bond.

   Examples:

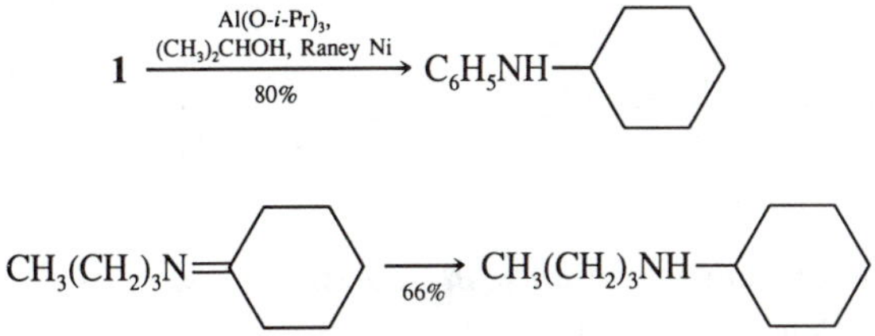

1 $\xrightarrow[80\%]{\substack{Al(O\text{-}i\text{-}Pr)_3, \\ (CH_3)_2CHOH, \ Raney\ Ni}}$ $C_6H_5NH{-}⟨cyclohexyl⟩$

$CH_3(CH_2)_3N$=⟨cyclohexylidene⟩ $\xrightarrow{66\%}$ $CH_3(CH_2)_3NH{-}⟨cyclohexyl⟩$

   Isomerization of* cis- *to* trans,vic-*diols.[2]   Taylor has reported two examples of this isomerization, effected with W2 Raney nickel at 60–90° (equations I and II).

(I) 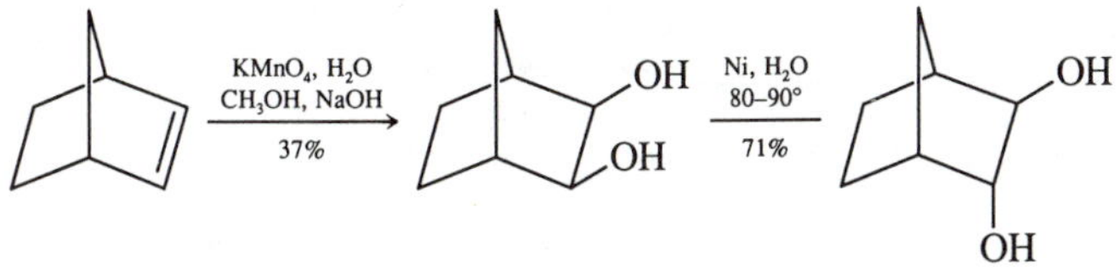 ⟨norbornene⟩ $\xrightarrow[37\%]{\substack{KMnO_4, \ H_2O \\ CH_3OH, \ NaOH}}$ ⟨norbornane-cis-diol⟩OH, OH $\xrightarrow[80-90°]{Ni, \ H_2O}$ ⟨norbornane-trans-diol⟩OH, OH
                                                                                                                            71%

(II)

[1] M. Botta, F. De Angelis, A. Gambacorta, L. Labbiento, and R. Nicoletti, *J. Org.*, **50**, 1916 (1985).
[2] J. E. Taylor, *Synthesis*, 1142 (1985).

Raney nickel–2-Propanol.

Reduction of arenes.[1]   Raney nickel (Mozingo type) in combination with 2-propanol (reflux) effects reduction of aromatic rings in 2–18 hours. Naphthalene is reduced in 18 hours to tetralin (90% yield) and *cis*- and *trans*-decalin (10% yield). Anisole is reduced in 110 hours to cyclohexyl methyl ether (90% yield). Nitrobenzene is reduced quickly to aniline and then further to cyclohexylamine and cyclohexylisopropylamine.

[1] S. Srivastava, J. Minore, C. K. Cheung, and W. J. le Noble, *J. Org.*, **50**, 394 (1985).

Rhodium(II) carboxylates.

Benzyl ketones.   Benzyl ketones can be prepared by alkylation of benzene with an α-diazo ketone catalyzed by $Rh(O_2CCF_3)_2$ at 25° followed by workup with TFA.[1]

Example:

Cyclization-cycloaddition of an α-diazoacetophenone.   Reaction of **1** with $Rh_2(OAc)_4$ in C_6H_6 at 25° results in **2** in 87% yield. This reaction is considered to involve a carbonyl ylide (**a**), which is trapped internally by the neighboring double bond.[2]

Chiral 3,3-disubstituted cyclopentanones.   Taber *et al.*[3] have extended a synthesis of cyclopentanones by a rhodium catalyzed intramolecular C—H insertion (**11**, 459) to a synthesis of (+)-α-cuparenone (**3**), which contains a chiral quaternary center. Thus the chiral α-diazo-β-keto ester **1**, prepared by alkylation of a chiral oxazolidone (**11**, 379–381) on treatment with $Rh_2(OAc)_4$ is converted into **2** in 67% yield. This product is converted in several steps into (+)-**3**.

Intramolecular C—H insertion.   The key step in a recent synthesis of pentalenolactone E methyl ester (**3**) is the reaction of the α-diazo-β-keto ester (**1**), prepared in several steps from 4,4-dimethylcyclohexanone, with $Rh_2(OAc)_4$. A single product (**2**) is obtained in high yield even though insertion involves bond formation with a nonfunctionalized carbon atom.[4]

Oxepanes and oxocanes.[5]   These seven- and eight-membered ring ethers can be obtained by cyclization of α-diazo carbonyl compounds induced with $Rh_2(OAc)_4$. Examples:

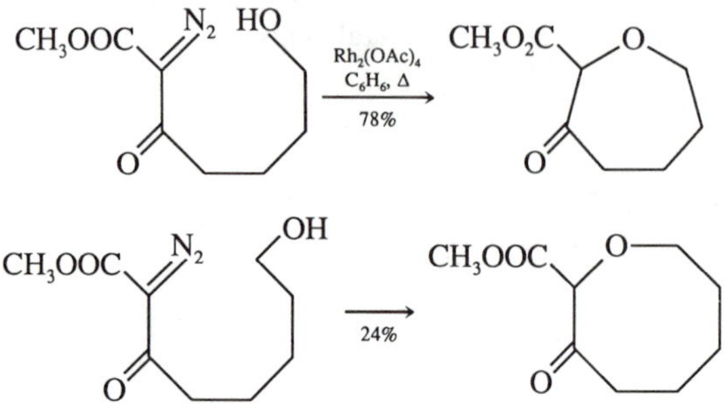

[1] M. A. McKervey, D. N. Russell, and M. F. Twohig, *J.C.S. Chem. Comm.*, 491 (1985).
[2] A. Padwa, S. P. Carter, and H. Nimmesgern, *J. Org.*, **51**, 1157 (1986).
[3] D. F. Taber, E. H. Petty, and K. Raman, *Am. Soc.*, **107**, 196 (1985).
[4] D. F. Taber and J. L. Schuchardt, *ibid.*, **107**, 5289 (1985).
[5] J. C. Heslin, C. J. Moody, A. M. Z. Slawin, and D. J. Williams, *Tetrahedron Letters*, **27**, 1403 (1986).

Ruthenium(III) chloride.

Quinoline synthesis.   In the presence of $RuCl_3 \cdot 3H_2O$ and Bu_3P, 1,3-propanediol and an aniline condense to form a quinoline. Yields are low in the absence of a hydrogen acceptor. Yields are improved by use of the nitroarene which is reduced to the aniline undergoing reaction.[1]
    Example:

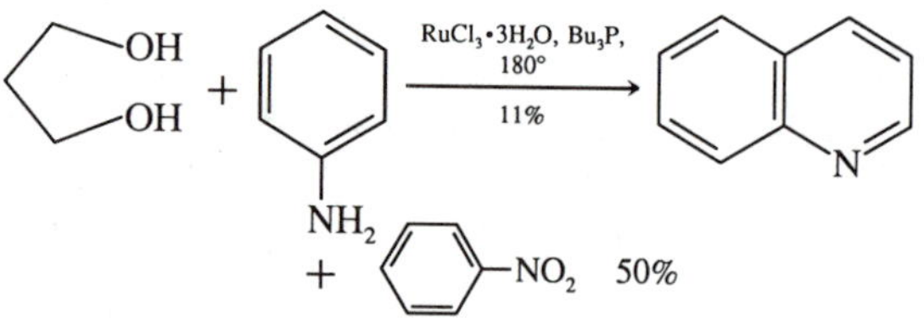

[1] Y. Tsuji, H. Nishimura, K. Huh, and Y. Watanabe, *J. Organomet. Chem.*, **286**, C44 (1985).

Ruthenium tetroxide.

ω-Amido-α-amino acids.[1]   Protected derivatives (**1**) of α,ω-diamino acids are oxidized by RuO_2 in combination with excess $NaIO_4$ in a two-phase system (ethyl acetate–H$_2$O) to the protected ω-carbamoyl-α-amino acids **2**. Thus asparagine (**3**, n = 1) is obtained from L-2,4-diaminobutyric acid.

Oxidation of cyclic α-amino acid esters.[2] The oxidation of N-acyl cyclic amines to lactams by RuO$_2$–NaIO$_4$ (**6**, 504–505) has been extended to oxidation of N-acyl cyclic amino acid esters. Thus N-acyl-L-proline esters (**1**) are oxidized to the corresponding lactams (**2**) by RuO$_2$–NaIO$_4$ in a two-phase system, preferably aqueous ethyl acetate. The rate is dependent on the bulk of the acyl group. Analogous oxidation of 2-azetidinecarboxylic acid esters fails, but homoproline derivatives (**4**) are oxidized successfully to lactams (**5**). The lactams are hydrolyzed by aqueous hydrochloric acid to α-amino dicarboxylic acids. Thus L-glutamic acid (**3**) is obtained from **2** in ~85% yield.

[1] S. Yoshifuji, K. Tanaka, and Y. Nitta, *Chem. Pharm. Bull.*, **33**, 1749 (1985); N. Sakura, K. Hirose, and T. Hashimoto, *ibid.*, **33**, 1752 (1985).
[2] S. Yoshifuji, K. Tanaka, T. Kawai, and Y. Nitta, *ibid.*, **33**, 5515 (1985).

S

Samarium(II) iodide.

Intramolecular Barbier-type cyclization (**12**, 429–430).[1] Cyclizations of 2-(ω-iodoalkyl)cyclopentanones induced with SmI_2 result almost entirely in *cis*-fused products (equation I). *cis*-Fused bicyclic alcohols are also formed selectively from 2-substituted 2-(ω-iodoalkyl)cycloalkanones (equation II).

(I)

$$CH_2(CH_2)_nCH_2I \xrightarrow[\text{THF}]{SmI_2}$$

(CH$_2$)$_n$

OH

H

cis/trans

n = 1	90%	99.5:0.5
n = 2	67%	18:1

(II)

R

$(CH_2)_3I \longrightarrow$

OH

R

R = CH$_3$	75%
R = COOC$_2$H$_5$	81%

Reduction of α-alkoxy ketones.[2] α-Hydroxy ketones are reduced by SmI_2 in THF/CH$_3$OH in rather low yield (~30%). In contrast α-acetoxy ketones are reduced in >75% yield. A simple expedient is to reduce the α-hydroxy ketone in the presence of Ac_2O with SmI_2 (equation I). α-Silyloxy, benzoyloxy, and tosyloxy

(I) $\quad C_5H_{11}\overset{O}{\overset{\|}{C}}CHC_5H_{11} \xrightarrow[29\%]{2SmI_2} C_5H_{11}\overset{O}{\overset{\|}{C}}CH_2C_5H_{11}$

$\quad\quad\quad\quad\quad\quad\underset{OH}{|}$

$\quad\quad +Ac_2O \quad\quad 59\%$

ketones are reduced by SmI_2 in essentially quantitative yield. The reagent also effects reduction of α-heterosubstituted ketones (equation II).

(II)

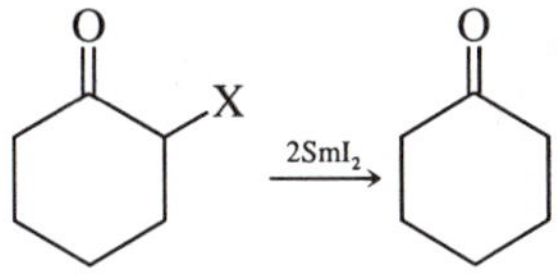

$$2SmI_2 \longrightarrow$$

X = Cl	100%
SC$_6$H$_5$	75%
SO$_2$C$_6$H$_5$	88%
HgCl	94%

Reduction of α,β-epoxy ketones; chiral aldols.[3] α,β-Epoxy ketones are reduced by SmI$_2$ in THF/CH$_3$OH at $-90°$ to β-hydroxy ketones with retention of
stereochemistry at the β-, but not at the α-position.

Example:

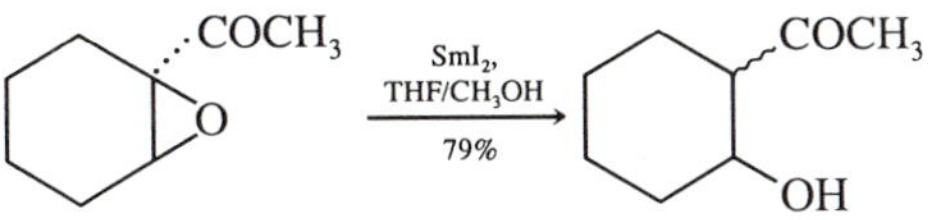

(2.3:1)

Reduction of an optically active epoxy ketone gives a chiral β-hydroxy ketone
(equation I).

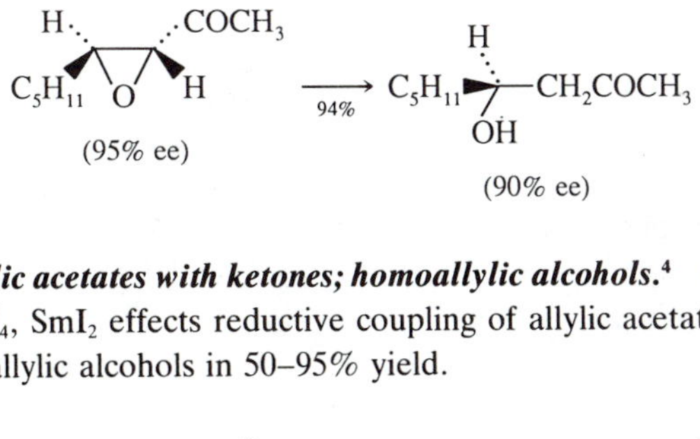

Coupling of allylic acetates with ketones; homoallylic alcohols.[4] In the presence of Pd[P(C$_6$H$_5$)$_3$]$_4$, SmI$_2$ effects reductive coupling of allylic acetates with ketones to form homoallylic alcohols in 50–95% yield.

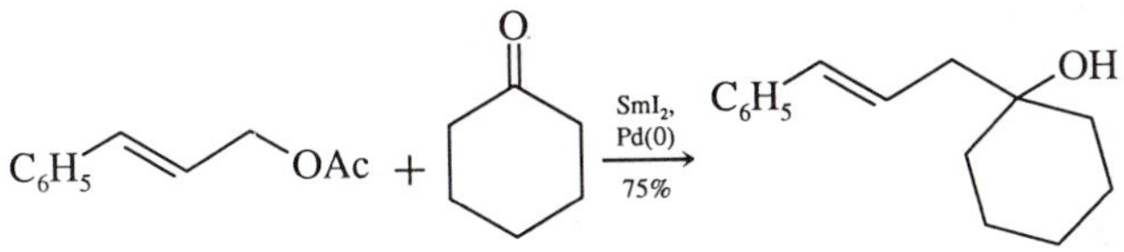

Oppenauer oxidations.[5] Oppenauer oxidations and Meerwein-Ponndorf-Verley reductions are usually carried out in the presence of aluminum alkoxides in at
least stoichiometric amounts. Kagan *et al.* report that both reactions can be carried

out with only catalytic amounts of *t*-butoxydiiodosamarium, obtained by reaction of SmI_2 with di-*t*-butyl peroxide. Later the same laboratory found that the combination of SmI_2 and *t*-BuOSmI$_2$ can be more effective than the alkoxide alone in Oppenauer oxidations and that SmI_2 is a useful catalyst, presumably as a precursor to a samarium alkoxide formed *in situ* from the secondary alcohol.

Coupling of carbonyl compounds with alkenes.[6] In the presence of SmI_2, aldehydes and ketones couple with electron-deficient alkenes to form γ-lactones. An alcohol functions as the essential proton donor.

Example:

$$C_6H_5\overset{\overset{\displaystyle O}{\|}}{C}CH_3 + CH_2{=}CHCOOC_2H_5 \xrightarrow[70\%]{\underset{(CH_3)_3COH}{SmI_2}}$$

[1] G. A. Molander and J. B. Etter, *J. Org.*, **51**, 1778 (1986).
[2] G. A. Molander and G. Hahn, *ibid.*, **51**, 1135 (1986).
[3] *Idem, ibid.*, **51**, 2596 (1986).
[4] T. Tabuchi, J. Inanaga, and M. Yamaguchi, *Tetrahedron Letters*, **27**, 1195 (1986).
[5] J. L. Namy, J. Souppe, J. Collin, and H. B. Kagan, *J. Org.*, **49**, 2045 (1984); J. Collin, J.-L. Namy, and H. B. Kagan, *Nouv. J. Chem.*, **10**, 229 (1986).
[6] S. Fukuzawa, A. Nakanishi, T. Fujinami, and S. Sakai, *J.C.S. Chem. Comm.*, 624 (1986).

Selenium dioxide.

Regioselective oxidation of isoprenyl groups.[1] A short synthesis of several ergot alkaloids involves the common intermediate **1**. Thus isochanoclavine (**3**) is obtained by oxidation of **2** with selenium dioxide in dioxane. The regioselectivity is ascribed to coordination of SeO_2 with the amino group. Similar oxidation of the corresponding nitro compound **1** results in exclusive oxidation of the (E)-methyl group of the isoprenyl system.

1α-*Hydroxylated vitamin* D *derivatives*.[2]   The (5E, 7E)-triene derivative **2** of (5E)-cholecalciferol (**1**) is oxidized by SeO_2 (0.7 equiv.) and N-methylmorpholine N-oxide (NMO, 4 equiv.) in CH_3OH/CH_2Cl_2 to the 1α-hydroxy derivative (**3**). The corresponding (5Z)-isomer of **1** affords both (5E)- and (5Z)-products on similar oxidation. However, **3** is converted into the desired (5Z)-isomer on irradiation in the presence of acridine. Desilylation of the product affords 1α-hydroxycholecalciferol (**4**).

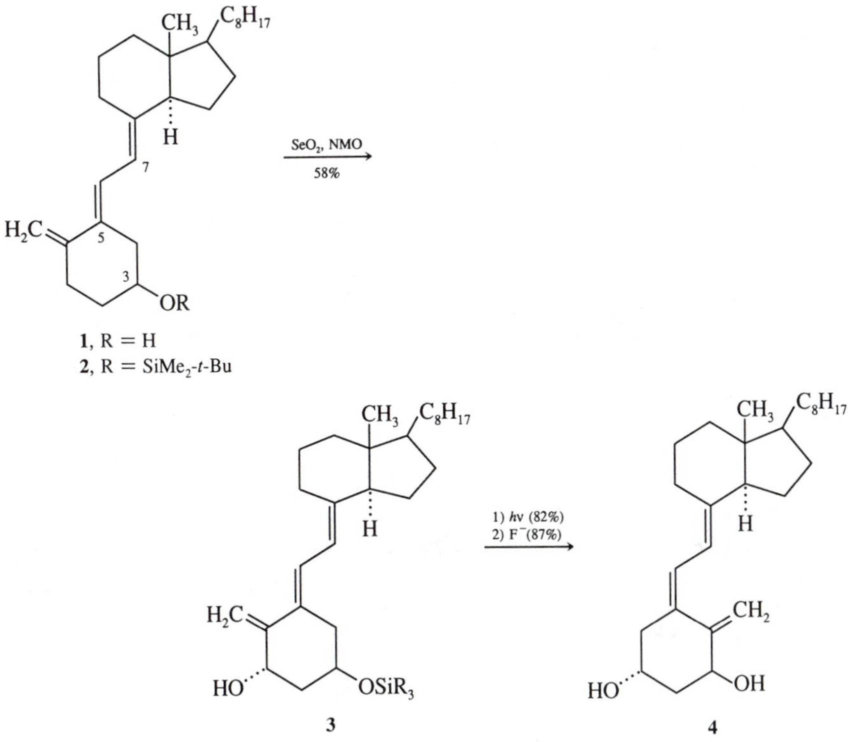

1, R = H
2, R = SiMe$_2$-*t*-Bu

3

4

[1] M. Somi, Y. Makita, and F. Yamada, *Chem. Pharm. Bull.*, **34**, 948 (1986).
[2] D. R. Andrews, D. H. R. Barton, K. P. Cheng, J.-P. Finet, R. H. Hesse, G. Johnson, and M. M. Pechet, *J. Org.*, **51**, 1635 (1986).

Silver tetrafluoroborate.

Lactone N,N-*dialkylhydrazones*.[1]   These hydrazones (**2**) have been obtained for the first time by reaction of halohydrazides (**1**) with 2 equiv. of AgBF$_4$ in THF. Cyclization of **1** with NaH results in N-(dialkylamino)lactams (**3**).

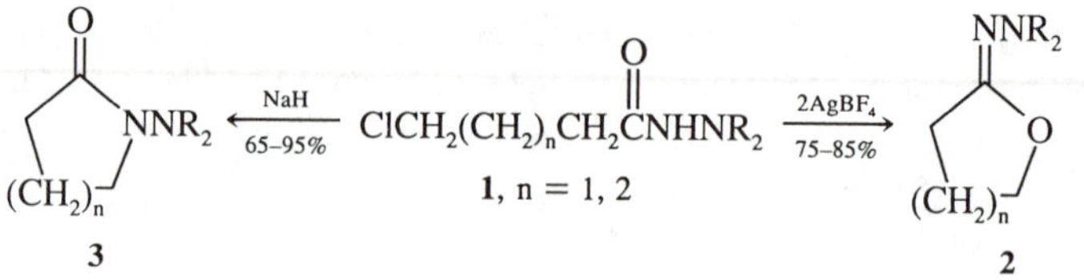

[1] D. Enders, S. Brauer-Scheib, and P. Fey, *Synthesis*, 393 (1985).

Silver(I) trifluoromethanesulfonate.

C-Glycosides.[1]   The (1-thiopyridyl)-β-D-glucopyranose **1** on activation with silver(I) triflate reacts with various carbon nucleophiles to yield C-glycosides in which the α-isomer usually predominates, regardless of the nucleophile, the solvent, or the temperature.

Example:

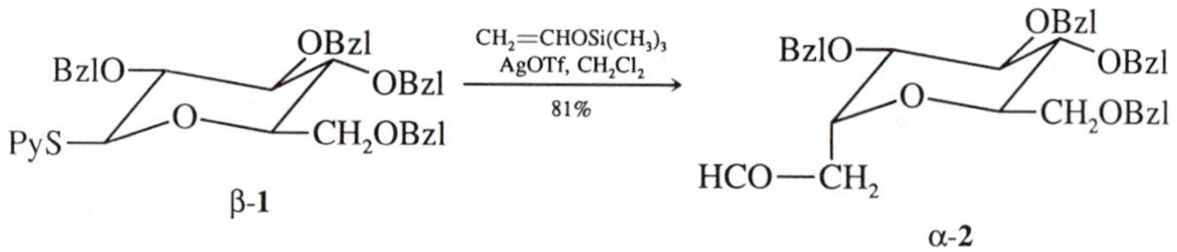

Acetylenic oxy-Cope rearrangement (**12**, 51). The oxy-Cope rearrangement of 5-hexen-l-yn-3-ols is accelerated in refluxing N-methyl-2-pyrrolidone (**9**, 316), but the required temperature (165°) can result in rearranged products. The rearrangement can proceed at 20–60° in the presence of silver triflate (1 equiv.), which is known to complex with triple bonds.[2]

Example:

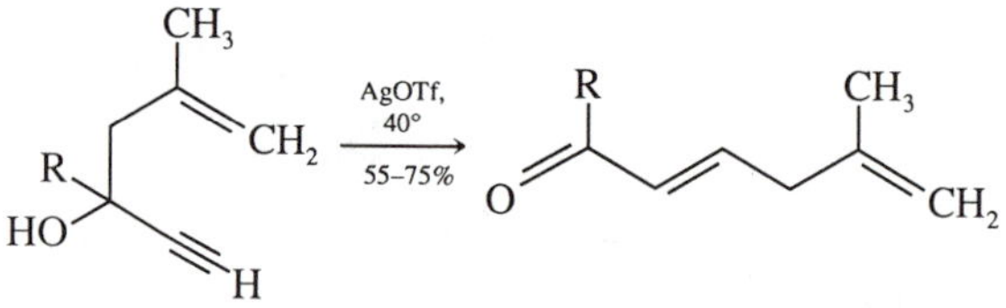

This rearrangement can be effected also with $AgNO_3$ in a catalytic amount (0.1 equiv.) and KNO_3 (1 equiv.) in only slightly lower yield.

Glycosidation.[3]   Glycosidation of the protected actinamine **1** with the pyranosyl chloride **2** with AgOTf in CH_2Cl_2 proceeds in 60–65% yield. The usual catalyst, silver carbonate, affords less than a 10% yield.

[1] A. O. Stewart and R. M. Williams, *Am. Soc.*, **107**, 4289 (1985).
[2] N. Bluthe, J. Gore, and M. Malacria, *Tetrahedron*, **42**, 1333 (1986).
[3] S. Hanessian and R. Roy, *Can. J. Chem.*, **63**, 163 (1985).

Simmons-Smith reagent.

Diastereoselective cyclopropanation of enones.[1] The homochiral ketals of cyclic α,β-enones obtained by reaction with 1,4-di-O-benzyl-L- or D-threitol[2] undergo diastereoselective cyclopropanation on reaction with the Simmons-Smith reagent (8-20:1). The diastereoselectivity is less in the case of ketals of cyclic β, γ-enones (2:1).

Example:

This reaction has been used for a synthesis of (R)-muscone (**1**) from cyclopentadecanone in 60% overall yield (equation I).[3]

The above process proceeds with low enantioselectivity when applied to acyclic enones. However the acetals of α,β-enals formed from L-diethyl or L-diisopropyl tartrate react enantioselectively with $CH_2I_2/Zn(C_2H_5)_2$ (**2**, 134).[4]

Example:

(88–94% ee)

[1] E. A. Mash and K. A. Nelson, *Am. Soc.*, **107**, 8256 (1985).
[2] N. Ando, Y. Yamamoto, J. Oda, and Y. Inouye, *Synthesis*, 688 (1978).
[3] K. A. Nelson and E. A. Mash, *J. Org.*, **51**, 2721 (1986).
[4] I. Arai, A. Mori, and H. Yamamoto, *Am. Soc.*, **107**, 8254 (1985).

Sodium–Alcohol.

Reduction of ketones.   Reduction of ketones with metals in an alcohol is one of the earliest methods for effecting reduction of ketones, and is still useful since it can proceed with stereoselectivity opposite to that obtained with metal hydrides.[1] An example is the reduction of the 3α-hydroxy-7-ketocholanic acid **1** to the diols **2** and **3**. The former, ursodesoxycholic acid, a rare bile acid found in bear bile, is used in medicine for dissolution of gallstones. The stereochemistry is strongly dependent on the nature of the reducing agent (equation I).[2] Sodium dithionite and sodium borohydride reductions result mainly in the 7α-alcohol, whereas reductions with sodium or potassium in an alcohol favor reduction to the 7β-alcohol. More recently[3] reduction of **1** to **2** and **3** in the ratio 96:4 has been achieved with K, Rb, and Cs in *t*-amyl alcohol. Almost the same stereoselectivity can be obtained by addition of potassium, rubidium, or cesium salts to reductions of sodium in *t*-amyl alcohol. This cation effect has not been observed previously.

(I)

Na, *n*-PrOH	100%	85:15
K, *t*-BuOH	100%	94:6
$Na_2S_2O_4$, H_2O NaHCO$_3$	100%	7:92
NaBH$_4$, H_2O NaHCO$_3$	100%	6:94

[1] H. O. House, "*Modern Synthetic Reactions,*" 150–172 (1972).
[2] G. Castaldi, G. Perdoncin, and C. Giordano, *Tetrahedron Letters*, **24**, 2487 (1983).
[3] *Idem, Angew. Chem. Int. Ed.*, **24**, 499 (1985).

Sodium acyloxyborohydrides.

Review.   These reducing agents are commonly prepared *in situ* from NaBH$_4$ and carboxylic acids. The order of decreasing hydride-denoting ability is NaBH$_3$(OCOR) ⟩ NaBH$_2$(OCOR)$_2$ ⟩ NaBH(OCOR)$_3$. The review also covers reducing agents obtained from carboxylic acids and NaBH$_3$CN, LiBH$_4$, and Bu$_4$NBH$_4$. These reagents were first used for reduction of enamines. Their most useful property is the ability to alkylate amines. Unsymmetrical tertiary amines can be obtained from primary amines by a two-step reaction with two different carboxylic acids in combination with NaBH$_4$.[1]

See also Sodium triacetoxyborohydride (this volume), for a specific use of these reagents in synthesis.

[1] G. W. Gribble and C. F. Nutaitis, *Org. Prep. Proc. Int.*, **17**, 319 (1985).

Sodium amide.

cis-*Selective Wittig reaction*.[1]   Ylides generated from phosphonium salts with solid sodium amide in THF or liquid ammonia undergo Wittig reactions with a *cis*-selectivity of 97–98%.[2] Addition of traces of KNH_2 generally accelerates ylide formation. Chloromethyl, methoxymethyl, and methylmercaptomethyl groups are tolerated. The salt can also contain an ω-hydroxyl group if the sodium amide is coated with paraffin (equation I).[3]

$$(I)\quad RCHO + (C_6H_5)_3P{=}CH(CH_2)_nCH_2ONa \xrightarrow[65-85\%]{} \overset{H}{R}C\underset{(Z)}{=}\overset{H}{C}(CH_2)_nCH_2OH$$

***Glutarimides*.**[4]   Sodium amide (3 equiv.) in liquid ammonia is the base of choice for conversion of glutaric acid diesters into glutarimides (45% yield in the case of glutarimide itself).

[1] M. Schlosser and B. Schaub, *Chimia*, **36**, 396 (1982).
[2] A. M. Moiseenkov, B. Schaub, C. Margot, and M. Schlosser, *Tetrahedron Letters*, **26**, 305 (1985).
[3] B. Schaub, G. Blaser, and M. Schlosser, *ibid.*, **26**, 307 (1985).
[4] T. Kinoshita, K. Okamoto, and J. Clardy, *Synthesis*, 402 (1985).

Sodium borohydride.

$R^1CH{=}CHCOR^2{\rightarrow}R^1CH{=}CHCH(OH)R^2$.   This reduction of enones to al-lylic alcohols can be accomplished by $NaBH_4$ in the mixed solvent CH_3OH–THF (1:10) in 70–90% yield.[1]

***Reduction of steroid tosylhydrazones*.**[2]   Steroid tosylhydrazones can be re-duced to the corresponding methylene compounds in 55–75% yield by $NaBH_4$ in HOAc without reduction of alkoxycarbonyl substituents.

***Reduction of —CN and —COOR*.**[3] $NaBH_4$ does not ordinarily reduce nitriles, but it reduces selectively one cyano group of 1,1-dicyano epoxides, that which is *trans* to an aryl or alkyl substituent.

Example:

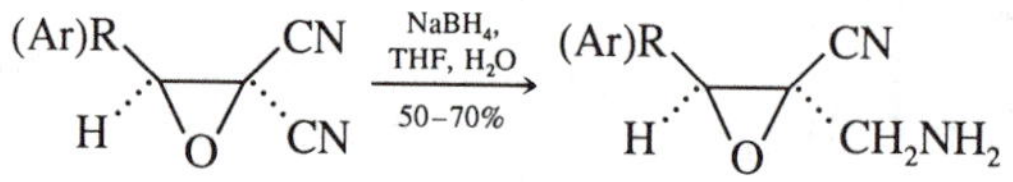

An ester group *trans* to an aryl or alkyl group in epoxides of type **1** is reduced by $NaBH_4$ in ethanol to an alcohol. These reductions do not occur in ethyl acetate, and therefore involve a hydroxy- or alkoxyborohydride.

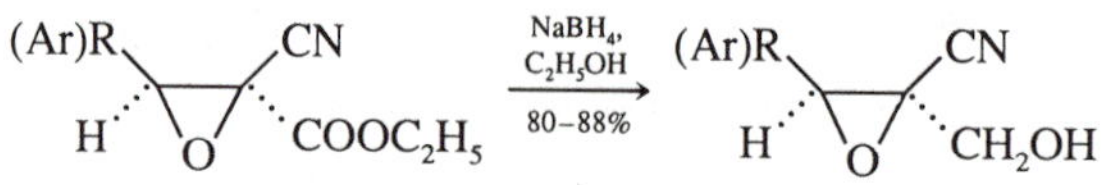

[1] R. S. Varma and G. W. Kabalka, *Syn. Comm.*, **15**, 985 (1985).
[2] T. Iida, T. Tamura, T. Matsumoto, and F. C. Chang, *Synthesis*, 957 (1984).
[3] J. Mauger and A. Robert, *J.C.S. Chem. Comm.*, 395 (1986).

Sodium borohydride–Copper(II) acetate.

Reduction of t-*nitro groups.*[1] $NaBH_4$ in combination with $Cu(OAc)_2$ reduces aryl and aliphatic tertiary nitro groups to amino groups in 75–80% yield.

[1] J. A. Cowan, *Tetrahedron Letters*, **27**, 1205 (1986).

Sodium borohydride–1,3-Dicyanobenzene.

Photo-Birch reduction.   Dihydroxylenes are formed on irradiation of a mixture of xylene, 1,3-dicyanobenzene, and sodium borohydride (large excess) in CH_3CN/H_2O (9:1).[1] The major products of reduction are isomeric with those, the more stable isomers, formed on Birch reduction.

Example:

Photo-Birch   (81.5%)   (14%)

[1] G. A. Epling and E. Florio, *Tetrahedron Letters*, **27**, 1469 (1986).

Sodium borohydride–Molybdenum(VI) oxide.

$\geqslant C{=}NOH \rightarrow \geqslant CHNH_2$.   Oximes are reduced selectively to amines in the presence of carbon–carbon double bonds by $NaBH_4$ and MoO_3. $NaBH_4$ and $NiCl_2 \cdot 6H_2O$ effect reduction of both groups in 90–95% yield.[1] This reduction was used in the final steps in a synthesis[2] of solenopsin A (**1**), a constituent of fire ant venom (equation I).

[1] J. Ipaktschi, *Ber.*, **117**, 856 (1984).
[2] B. P. Mundy and M. Bjorklund, *Tetrahedron Letters*, **26**, 3899 (1985).

Sodium borohydride–Nickel boride.

RNO₂→RNH₂.[1] Nickel boride catalyzes reductions of aliphatic nitro compounds by $NaBH_4$. Ni_2B alone is ineffective. Reduction with $NaBH_4$–$NiCl_2$ in aqueous 2-propanol is much slower and results in more by-products.

Example:

$$C_8H_{17}NO_2 \xrightarrow[\substack{61\%}]{\substack{NaBH_4, \\ Ni_2B, \ CH_3OH}} C_8H_{17}NH_2$$

[1] J. O. Osby and B. Ganem, *Tetrahedron Letters*, **26**, 6413 (1985).

Sodium chlorite, NaClO₂.

—CHO→—COOH.[1] This oxidation can be effected with $NaClO_2$ if a scavenger is also present to reduce the HOCl formed to HCl and O_2. For this purpose either 35% H_2O_2 or DMSO is more useful than sulfamic acid, H_2NSO_3H, 2-methyl-2-butene, or resorcinol. Reactions with $NaClO_2$–H_2O_2 are carried out usually in aqueous acetonitrile, those with DMSO, in aqueous DMSO. Yields of >90% can be obtained with aliphatic and aromatic aldehydes as well as the α,β-enals. Only chlorinated products are obtained from aldehydes containing isolated double bonds.

[1] E. Dalcanale and F. Montanari, *J. Org.*, **51**, 567 (1986).

Sodium cyanoborohydride–Tin(II) chloride.

Reduction of halides.[1] The reagent prepared from $NaBH_3CN$ and $SnCl_2$ in a 2:1 ratio does not reduce primary or secondary alkyl halides or aryl halides in ether at 25°, but does reduce tertiary, allyl, and benzyl halides. It is thus comparable to $NaBH_3CN$–$ZnCl_2$ (**12**, 446). Aldehydes, ketones, and acid chlorides are reduced to alcohols, but esters and amides are inert.

[1] S. Kim and J. S. Ko, *Syn. Comm.*, **15**, 603 (1985).

Sodium cyanoborohydride–Zinc iodide.

Deoxygenation.[1] Aryl aldehydes and ketones are reduced to arylalkanes by reaction with $NaBH_3CN/ZnI_2$ (5:1) in dichloroethane at 22–83°. The rate of reduction is facilitated by electron-releasing groups. Use of $NaBH_3CN/ZnCl_2$ (2:1) effects reduction only to the corresponding alcohol, but tosylhydrazones are completely reduced.[2] The more potent $NaBH_3CN$–ZnI_2 reagent also reduces benzylic, allylic, and tertiary alcohols to hydrocarbons, often in high yield.

[1] C. K. Lau, C. Dufresne, P. C. Bélanger, S. Piétré, and J. Scheigetz, *J. Org.*, **51**, 3038 (1986).
[2] S. Kim, C. H. Oh, J. S. Ko, K. H. Ahn, and Y. J. Kim, *ibid.*, **50**, 1927 (1985).

Sodium dithionite.

Conjugate reduction of enones.[1]   Under phase-transfer conditions (Aliquat, NaHCO$_3$), α,β-enones undergo exclusive 1,4-reduction with Na$_2$S$_2$O$_4$ (yields 70–85%).

Hydrogenolysis of vinyl sulfones.[2]   Vinyl sulfones are reduced by Na$_2$S$_2$O$_4$ and NaHCO$_3$ in aqueous ethanol with retention of configuration. The reaction is facilitated by use of a phase-transfer catalyst such as Adogen. The reduction involves addition of HSO$_2^-$ to form a sulfone sulfinate salt which can be trapped by CH$_3$I. The overall reaction thus involves a β-*syn*-addition followed by an *anti*-elimination. This hydrolysis was used in a synthesis of several insect pheromones, such as (Z)-8-dodecenyl-1-acetate (**1**).

$$CH_3(CH_2)_2 \diagdown \diagup SO_2C_6H_5 \underset{(CH_2)_7OAc}{} \xrightarrow[57\%]{\substack{Na_2S_2O_4, \\ NaHCO_3}} CH_3(CH_2)_2 \diagdown \diagup \diagdown (CH_2)_7OAc$$

1 (98% Z)

[1] O. Louis-Andre and G. Gelbard, *Tetrahedron Letters*, **26**, 831 (1985).
[2] M. Julia, H. Lauron, J.-P. Stacino, and J.-N. Verpeaux, *Tetrahedron*, **42**, 2475 (1986).

Sodium dodecyl sulfate, CH$_3$(CH$_2$)$_{11}$OSO$_3$Na (**1**).

Cycloaddition reactions.[1]   This surfactant increases the rate of Diels-Alder reactions. In addition, the *endo/exo* ratio is improved in the addition of acrylic acid derivatives to cyclopentadiene.

Example:

	Hexane	68:32
	1	90:10

[1] R. Braun, F. Schuster, and J. Sauer, *Tetrahedron Letters*, **27**, 1285 (1986).

Sodium hydride–Sodium *t*-amyl oxide–Zinc chloride (ZnCRA).

Reduction of epoxides.[1]   ZnCRA and the related NiCRA, prepared from NaH, C$_2$H$_5$C(CH$_3$)$_2$ONa, and NiCl$_2$, both reduce epoxides, but ZnCRA is the more reactive reagent. Surprisingly the two complex reducing agents show different regioselectivity.

Example:

$$
\underset{\substack{\text{ZnCRA}\\ \text{NiCRA}}}{\overset{\displaystyle C_6H_5\diagdown\ \diagup CH_3}{\underset{O}{\triangle}}} \longrightarrow \underset{OH}{C_6H_5CHCH_2CH_3} + \underset{OH}{C_6H_5CH_2CHCH_3}
$$

ZnCRA		95%	—
NiCRA	95%	35:65	

[1] Y. Fort, R. Vanderesse, and P. Caubere, *Tetrahedron Letters*, **26**, 3111 (1985).

Sodium hydrogen telluride.

Reduction of t-nitro groups.[1] This reduction can be accomplished with NaHTe, generated *in situ* in C_2H_5OH by reduction of tellurium with excess $NaBH_4$, in 80–100% yield. However, nitro groups that are in the γ-position to a carbonyl, ester, or nitrile group are inert.

[1] H. Suzuki, K. Takaoka, and A. Osuka, *Bull. Chem. Soc. Japan*, **58**, 1067 (1985).

Sodium iodide–Boron trifluoride etherate.

Allylic iodides.[1] Allylic and benzylic alcohols (primary and secondary) are converted into the corresponding iodides by reaction with NaI and BF_3 etherate in 70–95% yield. The same system converts sulfoxides to sulfides in 90–98% yield.

[1] Y. D. Vankar and C. T. Rao, *Tetrahedron Letters*, **26**, 2717 (1985).

Sodium nitrite.

(S)-2-Chloroalkanoic acids.[1] Diazotization of (S)-2-amino acids ($NaNO_2$) in water containing HCl at 0→25° results in (S)-2-chloroalkanoic acids in 60–65% yield and with 95–98% ee (equation I).

$$
(I) \quad \underset{R}{\overset{COOH}{H_2N\blacktriangleright C\blacktriangleleft H}} \xrightarrow[60-65\%]{\substack{NaNO_2,\\ HCl,\ H_2O}} \underset{R}{\overset{COOH}{Cl\blacktriangleright C\blacktriangleleft H}}
$$

(S), R = CH_3, $CH(CH_3)_2$,   (95–98% ee)
$CH_2CH(CH_3)_2$, $CH(CH_3)C_2H_5$

The products are reduced by $LiAlH_4$ in ether to (S)-2-chloroalkanols, which are converted by base into 2-alkyloxiranes with inversion of configuration (equation II).

$$(II) \quad \underset{(S)}{Cl \blacktriangleright \overset{\overset{\displaystyle COOH}{|}}{\underset{\underset{\displaystyle R}{|}}{C}} \blacktriangleleft H} \quad \xrightarrow[55-70\%]{LiAlH_4} \quad \underset{(S)}{Cl \blacktriangleright \overset{\overset{\displaystyle CH_2OH}{|}}{\underset{\underset{\displaystyle R}{|}}{C}} \blacktriangleleft H} \quad \xrightarrow[73-85\%]{KOH} \quad \underset{(R),\ 93-97\%\ ee}{H\blacktriangleleft\overset{\diagup\diagdown O}{\underset{R}{\diagup}}}$$

[1] B. Koppenhoefer and V. Schurig, *Org. Syn.*, submitted (1985).

Sodium nitrosodisulfonate (Fremy's salt).

This salt can be obtained by air oxidation of an equimolar mixture of $NaNO_2$ and $NaHSO_3$ in an alkaline aqueous solution.[1]

See also Potassium nitrosodisulfonate (this volume).

[1] T. Ozawa and T. Kwan, *J.C.S. Chem. Comm.*, 54 (1985).

Sodium triacetoxyborohydride, $NaBH(OAc)_3$ (1).

Stereoselective reduction of ketones.[1] The germine triester **2** is reduced by **1** in HOAc to the 7β-alcohol (**3**) in 92% yield. In contrast, reduction with $NaBH_4$ gives the 7α-alcohol as the major product. The difference in stereoselectivity is attributed to the α-C_{14}-hydroxy group.

2, R^1, R^2, $R^3 = i$-Bu   **3**

For additional information on this class of reducing agent, see sodium acyloxy-borohydrides (this volume).

[1] A. K. Saksena and P. Mangiaracina, *Tetrahedron Letters*, **24**, 273 (1983).

Squaric acid.

(2-*Alkynylethenyl*)*ketenes.*[1]   These ketenes (**3**) are available by reaction of dimethyl squarate (**1**) with a lithium acetylide followed by quenching with $ClSi(CH_3)_3$. They undergo cyclization to silyl-1,4-quinones (**4**) and/or the cyclo-pentenediones **5**. The mode of ring closure depends on the R substituent; formation of **5** is favored by electron-withdrawing groups (R = $CO_2C_2H_5$) whereas electron-releasing groups promote formation of **4**. The yields of **4** are 75–80% when R = C_4H_9-*n*, $CH_2C_6H_5$, or $CH_2OSi(CH_3)_3$.

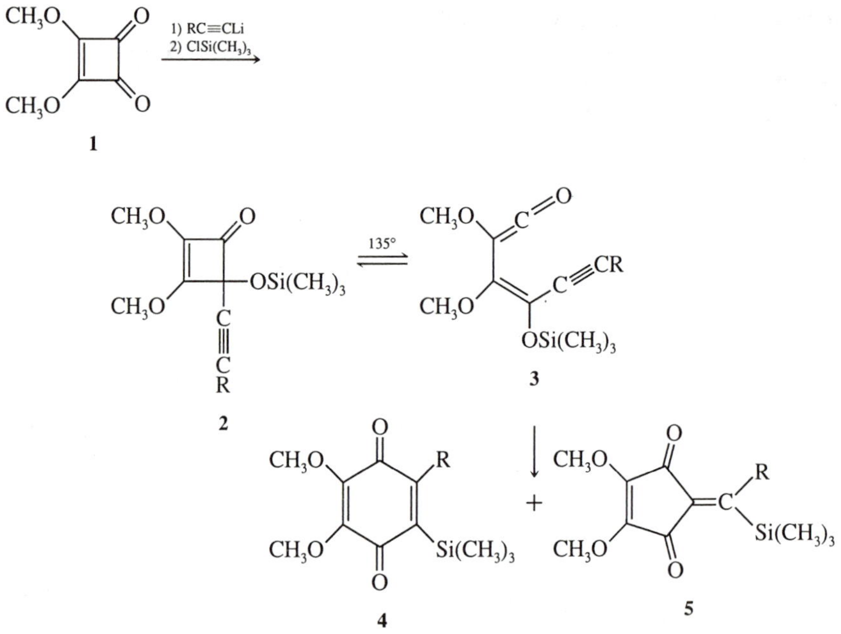

¹ J. O. Karlsson, N. V. Nguyen, L. D. Foland, and H. W. Moore, *Am. Soc.*, **107**, 3392 (1985).

Sulfuryl chloride.

Sulfinyl chlorides.[1]   These reagents are generally prepared by reaction of sulfinic acids with thionyl chloride. They can be prepared more conveniently by reaction of the readily available disulfides in acetic acid with sulfuryl chloride; yields are nearly quantitative.

¹ J.-H. Youn and R. Herrmann, *Tetrahedron Letters*, **27**, 1493 (1986).

T

2,4,4,6-Tetrabromo-2,5-cyclohexadienone (1).
   Regioselective bromination of polyenes.[1] The first steps in a biomimetic synthesis of concinndiol (**4**) from methyl geranylgeranate (**2**) involve regioselective bromination with **1** followed by cyclization with AgOAc in HOAc to **3** in 20% overall yield.

[1] Y. Yamaguchi, T. Uyehara, and T. Kato, *Tetrahedron Letters*, **26**, 343 (1985).

Tetrabutylammonium dihydrogentrifluoride, $Bu_4N^+H_2F_3^-$ (**1**). The reagent is obtained by addition of Bu_4NF in $ClCH_2CH_2Cl$ to an aqueous solution of HF/KF (2:1). The salt is isolated from the $ClCH_2CH_2Cl$ layer after distillation.

   Hydrofluorination of alkynes.[1] The reagent adds to alkynes activated by CN, $COOCH_3$, COR, or CHO groups to form fluoroalkenes.

   Example:

[1] P. Albert and J. Cousseau, *J.C.S. Chem. Comm.*, 961 (1985).

Tetrabutylammonium fluoride.

Selenoaldehydes.[1]   The first step in a synthesis of selenoaldehydes involves addition of dimethylphenylsilyllithium[2] to aldehydes followed by trapping of the adduct with TsCl. The α-silyl tosylates are converted into α-silyl selenocyanates by reaction with KSeCN catalyzed by 18-crown-6. These products are easily purified, and are converted by fluoride ion into selenoaldehydes, which can be trapped by Diels-Alder cycloadditions (equation I).

(*endo/exo* = 3–9:1)

Thioaldehydes.[3]   Thioaldehydes can be generated from α-silyl disulfides (**1**) by treatment with CsF in THF at 25° or with Bu₄NF in THF at 0°. The thioaldehyde is identified by cycloaddition to cyclopentadiene.
    Example:

exo/endo = 1:7

Cleavage of vinylsilanes.[4]   Desilylation of 1-silylalkynes is relatively facile but desilylation of vinylsilanes generally requires strong acid catalysts, HI or TsOH. Cleavage of (trimethyl)vinylsilanes under neutral conditions with Bu₄NF is difficult. However the presence of a phenyl, allyl, or alkoxyl group on Si facilitates the cleavage to a fluorosilane and then cleavage of the C—Si bond. This cleavage is further facilitated by a hydroxyl group in positions β, γ, or δ to Si.
    Example:

Aromatic fluorodenitration.[5]   This substitution of nitroarenes by fluorine can be accomplished with "anhydrous" Bu_4NF obtained by evaporation of the commercial trihydrate at 50–60° and <1 mm. for 18–24 hours. However, only activated or sterically hindered nitroarenes undergo this substitution.

Example:

[1] G. A. Krafft and P. T. Meinke, *Am. Soc.*, **108**, 1314 (1986).
[2] M. V. George, D. J. Peterson, and H. Gilman, *ibid.*, **82**, 403 (1960).
[3] G. A. Krafft and P. T. Meinke, *Tetrahedron Letters*, **26**, 1947 (1985).
[4] H. Oda, M. Sato, Y. Morizawa, K. Oshima, and H. Nozaki, *Tetrahedron*, **41**, 3257 (1985).
[5] J. H. Clark and D. K. Smith, *Tetrahedron Letters*, **26**, 2233 (1985).

Tetrabutylammonium iodide–Boron trifluoride etherate.

Cleavage of dialkyl ethers.[1]   Methyl, allyl, and benzyl alkyl ethers are cleaved by this combination to the alcohols and methyl, allyl, and benzyl iodide (75–90% yield). The reagent also cleaves epoxides to *trans*-2-iodoalkanols (~70% yield).

[1] A. K. Mandal, N. R. Soni, and K. R. Ratnam, *Synthesis*, 274 (1985).

Tetrabutylammonium nitrate, $Bu_4N^+NO_3^-$. Preparation.[1]

Inversion of mesylates or tosylates.[2]   Mesylates or tosylates undergo displacement with inversion by $Bu_4N^+NO_3^-$ or Amberlyst A-26 NO_3^- form. Milder conditions and shorter reaction times are possible using pyridylsulfonates. The nitrate esters are readily converted into alcohols by Zn-HOAc or by catalytic hydrogenation.

Example:

[1] R. E. Buckles and L. Harris, *Am. Soc.*, **79**, 886 (1957).
[2] G. Cainelli, F. Manescalchi, G. Martelli, M. Panunzio, and L. Plessi, *Tetrahedron Letters*, **26**, 3369 (1985).

Tetrabutylammonium tribromide, $Bu_4N^+Br_3^-$ (1), m.p. 84°. Supplier: Janssen Chimica.

Addition to alkenes[1] and alkynes.[2]   The reagent converts (E)-alkenes into *meso*- or *erythro*-dibromides in CHCl₃ in 80–95% yield. It is also useful for selective α-bromination of ketals (75–80% yield).

This reagent adds to alkynes in CHCl₃ at 20° to give exclusively (E)-1,2-dibromoalkenes.

[1] M. Fournier, F. Fournier, and J. Berthelot, *Bull. Soc. Chim. Belg.*, **93**, 157 (1984).
[2] J. Berthelot and M. Fournier, *Can. J. Chem.*, **64**, 603 (1986).

Tetrabutylphosphonium bromide, $Bu_4\overset{+}{P}Br^-$, m.p. 100–103°. Supplier: Aldrich.

Dealkoxycarbonylation.[1]   This reaction can be conducted by heating a malonic or β-keto ester at 200° in the presence of a catalytic amount of this phosphonium salt and a high molecular weight carboxylic acid such as stearic acid as the solvent and proton source. The reaction involves catalyzed transesterification followed by decarboxylation. Yields are typically 70–90%.

[1] E. V. Dehmlow and E. Kunesch, *Synthesis*, 320 (1985).

Tetra-μ³-carbonyldodecacarbonylhexarhodium, $Rh_6(CO)_{16}$. Supplier: Strem.

Transfer hydrogenation of dienes to monoenes.[1]   1,5-Cyclooctadiene is selectively reduced to cyclooctene by transfer hydrogenation with isopropanol catalyzed by this metal carbonyl cluster. The first step is isomerization to conjugated diene isomers. 1,5-Hexadiene is reduced under these conditions to *trans*-3-hexene (19%), *cis*-2-hexene (21%), *trans*-2-, and *cis*-3-hexene (56%). $Ru_3(CO)_{12}$, $Os_3(CO)_{12}$, and $Ir_4(CO)_{12}$ catalyze isomerization of 1,5-cyclooctadiene, but are less active than $Rh_6(CO)_{16}$ for transfer hydrogenation.

[1] J. Kaspar, R. Spogliarich, and M. Graziani, *J. Organomet. Chem.*, **281**, 299 (1985).

Tetrachlorotris[bis(1,4-diphenylphosphine)butane]diruthenium, $Ru_2Cl_4(dppb)_3$ **(1).**

Regioselective hydrogenation of anhydrides to lactones.[1]   The last step in the synthesis of arylnaphthalene lignans is usually selective hydrogenation of an an-

2   3 (>99:1)

hydride to a lactone. Highly selective catalytic hydrogenation of the less hindered carbonyl group of **2** to give **3** can be effected with **1** as the catalyst. Use of $RuCl_2[P(C_6H_5)_3]_3$ provides similar selectivity but the chemical yield is only 12%.

[1] Y. Ishii, T. Ikariya, M. Saburi, and S. Yoshikawa, *Tetrahedron Letters*, **27**, 365 (1986).

Tetracyanoethylene (TCNE).

Acid anhydrides.[1]   Carboxylic acids are converted into anhydrides on reaction with TCNE in benzene containing pyridine. Yields are 50–80%.

[1] D. Voisin and B. Gastambide, *Tetrahedron Letters*, **26**, 1503 (1985).

Tetrakis(trifluoroacetate)ruthenium, $Ru(O_2CCF_3)_4$ **(1).** The reagent is prepared *in situ* by reaction of RuO_2 with trifluoroacetic acid and trifluoroacetic anhydride.

Oxidative biaryl coupling.[1]   The last step in a synthesis of neoisostegane **3** involves coupling of the dibenzylbutanolide **2**. TTFA or $VOCl_3$ has generally been used for this nonphenolic oxidative coupling, but in this case give yields of about 50%. Coupling with **1** is slower than with V or Tl catalysts, but results in a 98% yield of **3**. In this and two related cases this Ru(IV) catalyst is also superior to TTFA.

2   **3**

[1] Y. Landais and J.-P. Robin, *Tetrahedron Letters*, **27**, 1785 (1986).

Tetrakis(triphenylphosphine)palladium(0).

Denitro-sulfonylation; allylic sulfones.[1,2]   Two groups have reported that in the presence of this Pd(0) catalyst, allylic nitro compounds (**1**) undergo denitro-sulfonylation on reaction with $C_6H_5SO_2Na\cdot2H_2O$ to afford allylic sulfones (**2**) and (**3**). The reaction occurs with high regioselectivity to afford the product of direct displacement of the nitro group.

Example:

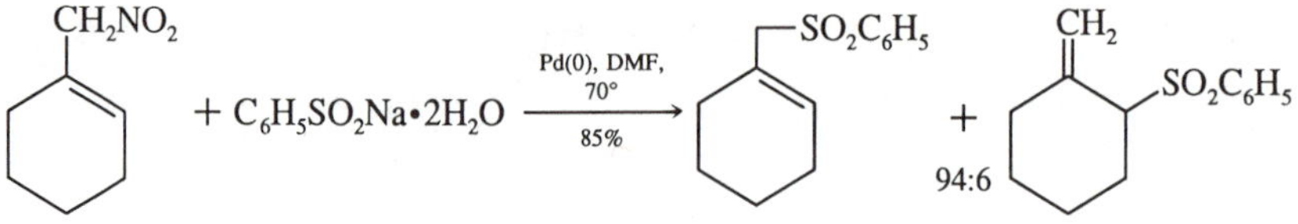

In contrast, reaction of allylic nitro compounds **4** with sodium benzenethiolate in HMPT leads to rearranged allylic sulfides **5**.[2]

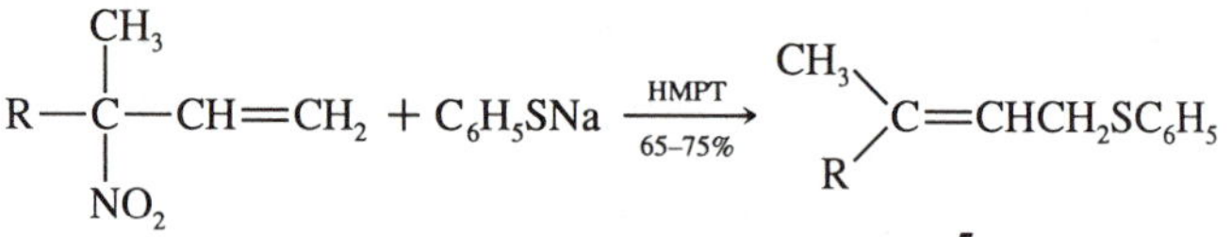

$$R-\underset{\underset{NO_2}{|}}{\overset{\overset{CH_3}{|}}{C}}-CH=CH_2 + C_6H_5SNa \xrightarrow[65-75\%]{HMPT} \underset{R}{\overset{CH_3}{>}}C=CHCH_2SC_6H_5$$

4 (R = H, CH$_3$)

5

α-Nitroalkenes react with sodium benzenesulfinate (2 equiv.) and triethylamine (1 equiv.) in DMF in the presence of Pd(0) and dppe [1,2-bis(diphenylphosphine)-ethane] to form (E)-allylic sulfones. Yields are low in the absence of the ligand and the amine.[3]

Example:

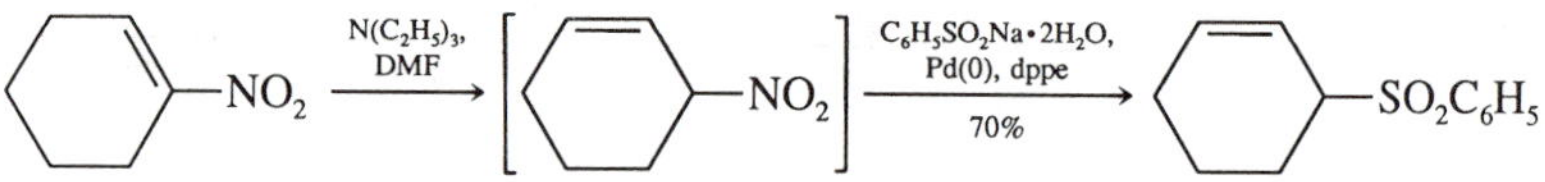

***Coupling of* 1-*alkenylboranes with* 1-*bromoalkenes or -alkynes*.**[4,5] This re-action fails when catalyzed by Pd(0) alone, but proceeds in high yield in the presence of an added sodium alkoxide or sodium hydroxide in refluxing benzene (or THF). This coupling can be used to prepare (E,E)-, (E,Z)-, (Z,E)-, or (Z,Z)-1,3-dienes stereo- and regiospecifically.

Examples:

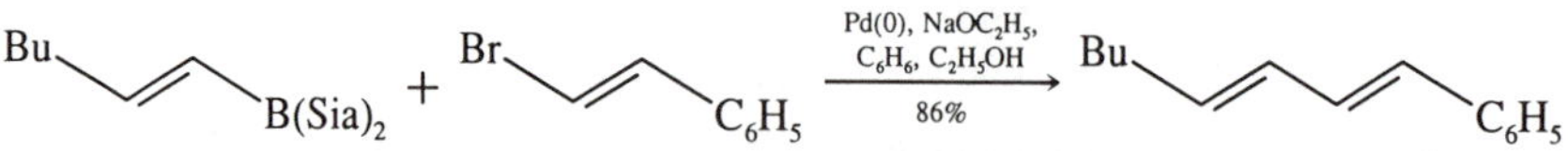

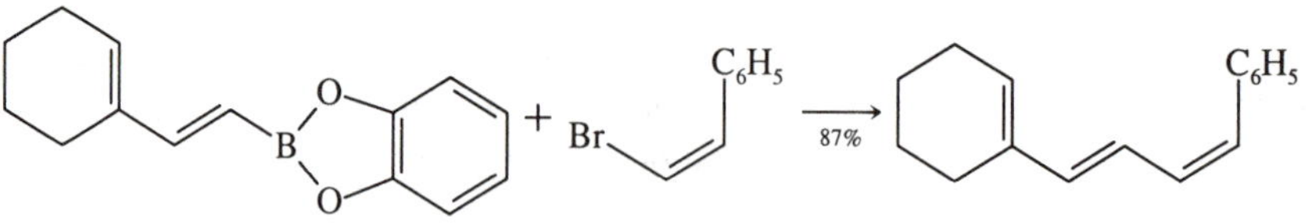

Cross-coupling of organoalanes with allylic acetals or ortho esters (cf., **10**, 388).[6] The Pd(0)-catalyzed coupling of alkenylalanes with **1** proceeds with exclusive attack at the γ-position to provide alkenyl ethers.

Example:

$$n\text{-}C_5H_{11}CH\!\!=\!\!CHAl(i\text{-}Bu)_2 \;+\; CH_2\!\!=\!\!CHCH(OCH_3)_2 \xrightarrow[54\%]{\substack{Pd(0) \\ THF}}$$

$$\text{(E)} \qquad\qquad\qquad\qquad\quad \mathbf{1}$$

$$n\text{-}C_5H_{11}CH\!\!=\!\!CHCH_2CH\!\!=\!\!CHOCH_3$$

$$\text{(E)} \qquad\qquad E/Z \,=\, 2\!:\!1$$

Carbonylation; 1,4-diazepines.[7]   A key step in the synthesis of the antitumor agents neothramycin A and B (**3**) is the Pd(0)-catalyzed carbonylation of **1**, prepared by condensation of 2-bromo-4-methoxy-5-tosyloxyaniline with N,O-bistrifluoroac-etyl-4-hydroxyproline followed by deprotection and protection with methoxymethyl chloride. The reaction provides the 1,4-diazepine **2**, which can be converted into a mixture of **3A** and **3B**.

3A, $R^1 = OH$, $R^2 = H$
3B, $R^1 = H$, $R^2 = OH$

Decarbonylative cross coupling of acyl halides.[8]   This unusual reaction is observed on coupling aroyl chlorides with alkyl(phenyl)acetyl chlorides in the presence of $N(C_2H_5)_3$ and catalyzed by Pd(0). The reaction involves loss of the carbonyl group of the aroyl chloride, and may involve an intermediate ketene.

Example:

$$C_2H_5CHCOCl + p\text{-}CH_3C_6H_4COCl \xrightarrow[77\%]{\substack{Pd(0) \\ N(C_2H_5)_3, \text{ THF}}} CH_3CH=C-C-C_6H_4CH_3\text{-}p$$

(the first reactant bears a C_6H_5 substituent; the product bears a C_6H_5 substituent on the central carbon and a $\parallel O$ carbonyl)

$$(E/Z) = 1{:}1$$

Decarbonylation of acyl cyanides.[9] Aromatic acyl cyanides, which are easily obtained by oxidation of cyanohydrins with t-butyl hydroperoxide catalyzed by $RuCl_2[P(C_6H_5)_3]_3$, undergo decarbonylation to nitriles in high yield when heated in the presence of Pd(0).

Allylation of lithium 1-cyclopentenolates.[10] In the presence of this Pd(0) complex and 2 equiv. of $B(C_2H_5)_3$, lithium cyclopentenolate reacts with the (E)- or (Z)-allylic acetate **1** to provide (E)- or (Z)-**2** in about 75% isolated yield and

$$\text{(structure OLi-cyclopentene)} + n\text{-}C_5H_{11}CH=CHCH_2OAc \xrightarrow[\substack{B(C_2H_5)_3}]{Pd(0),} \text{(cyclopentanone with } CH_2CH=CHC_5H_{11}\text{-}n\text{ substituent)}$$

(E)-**1**	77%	**2** E/Z = ≤98:2
(Z)-**1**	73%	Z/E = ≥97:3

with high retention of stereochemistry. Both the Pd catalyst and $B(C_2H_5)_3$ are essential for this stereospecific allylation. This reaction can be used for stereo- and regioselective allylation of lithium 3-vinyl-1-cyclopentenolate (equation I). Use of

(I)   (3-vinyl-1-cyclopentenolate, OLi, CH=CH₂) $+ $ (Z)-**1** $\xrightarrow{64\%}$ (cyclopentanone product with C_5H_{11}-n allyl and CH=CH₂ substituents)

the corresponding allylic bromide or iodide suffers from low ring-regioselectivity. This allylation was used to obtain methyl (Z)-jasmonate in 70% yield from cyclopentenone.

Asymmetric allylation.[11] The chiral enamine **1**, derived from the allyl ester of (S)-proline, when treated with this Pd(0) complex at 25° in various solvents provides (S)-(−)-2-allylcyclohexanone (**2**) in 80–100% ee, the highest enantioselectivity being observed in $CHCl_3$.

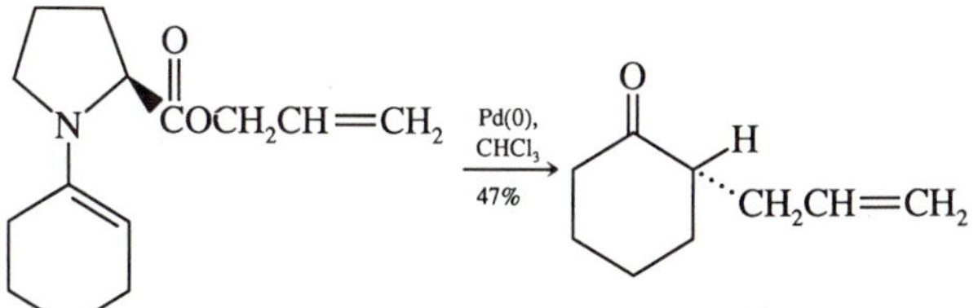

2 (100% ee)

1

Bisethynylation of vic-dichloroethylenes.[12]   Enediynes can be prepared by the palladium-catalyzed coupling of terminal alkynes with vic-dichloroethylenes, particularly with the cis-isomer (equation I).

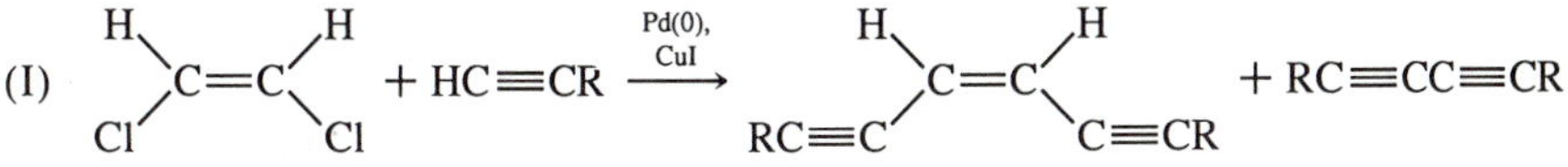

R = Si(CH₃)₃	78.8%	2.5%
R = C₆H₅	80.6%	7.4%

Pyrrolidines; piperidines.[13]   Suitably unsaturated α-iodo (or α-bromo) esters cyclize to pyrrolidines or piperidines in the presence of this Pd(0) catalyst and 1,8-bis(dimethylamino)naphthalene (proton sponge).

Examples:

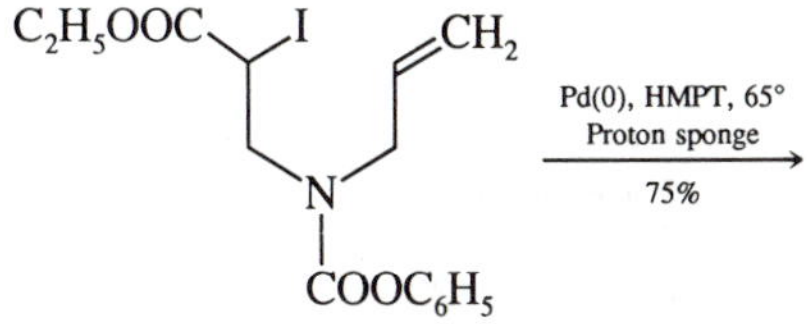

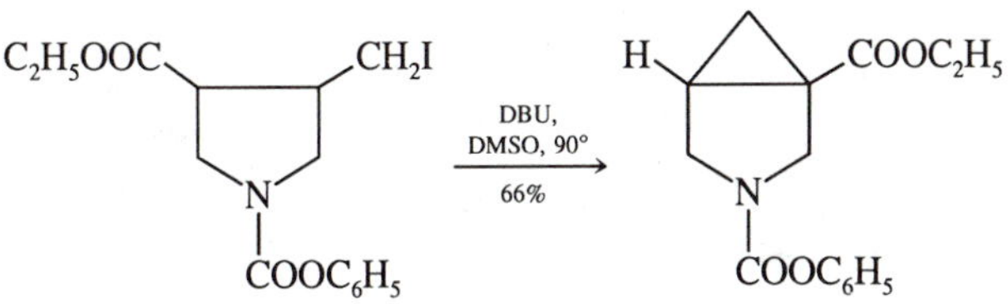

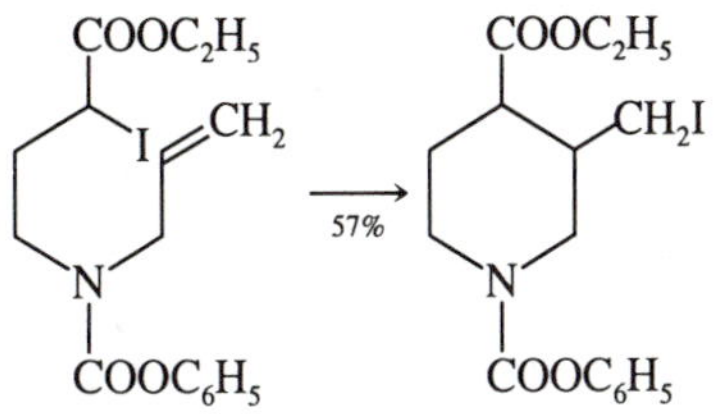

Unsymmetrical biaryls.[14]   The Pd(0)- or Ni(0)-catalyzed coupling of arylzinc derivatives with aryl halides appears to be the method of choice for synthesis of unsymmetrical biaryls (equation I). Both $Pd[P(C_6H_5)_3]_4$ and $Ni[P(C_6H_5)_3]_4$ are usually equally effective with aryl iodides, but only activated aryl bromides can be coupled with Pd(0).

$$(I) \quad Ar^1I \xrightarrow[\substack{ZnCl_2}]{t\text{-BuLi,}} Ar^1ZnCl \xrightarrow[70\text{--}95\%]{\substack{Ar^2Br, \\ Pd(0)}} Ar^1\text{---}Ar^2$$

Primary allylic amines.[15]   In the presence of this catalyst, allylic chlorides or acetates react with sodium p-toluenesulfonamide in THF/DMSO (80:20) to form the corresponding allylic sulfonamide in 60–85% yield. The products are converted into primary allylic amines on reductive cleavage (sodium naphthalenide).

Example:

[1] R. Tamura, K. Hayashi, M. Kakihana, M. Tsuji, and D. Oda, *Tetrahedron Letters*, **26**, 851 (1985).

[2] N. Ono, I. Hamamoto, T. Yanai, and A. Kaji, *J.C.S. Chem. Comm.*, 523 (1985).

[3] R. Tamura, K. Hayashi, M. Kakihana, M. Tsuji, and D. Oda, *Chem. Letters*, 229 (1985).

[4] N. Miyaura, K. Yamada, H. Suginome, and A. Suzuki, *Am. Soc.*, **107**, 972 (1985).

[5] A. Suzuki, *Acc. Chem. Res.*, **15**, 178 (1982).

[6] S. Chatterjee and E. Negishi, *J. Org.*, **50**, 3406 (1985).

[7] M. Mori, G.-E. Purvaneckas, M. Ishikura, and Y. Ban, *Chem. Pharm. Bull.*, **32**, 3840 (1984); M. Mori, Y. Uozumi, and Y. Ban, *J.C.S. Chem. Comm.*, 841 (1986).

[8] M. Kadokura, T. Mitsudo, and Y. Watanabe, *J.C.S. Chem. Comm.*, 252 (1986).

[9] S.-I. Murahashi, T. Naota, and N. Nakajima, *J. Org.*, **51**, 898 (1986); *idem*, *Tetrahedron Letters*, **26**, 925 (1985).

[10] F.-T. Luo and E. Negishi, *Tetrahedron Letters*, **26**, 2177 (1985).

[11] K. Hiroi, K. Suya, and S. Sato, *J.C.S. Chem. Comm.*, 469 (1986).

[12] K. P. C. Vollhardt and L. S. Winn, *Tetrahedron Letters*, **26**, 709 (1985).

[13] M. Mori, Y. Kubo, and Y. Ban, *ibid.*, **26**, 1519 (1985).

[14] E. Negishi, *Acc. Chem. Res.*, **15**, 340 (1982); E. Negishi, T. Takahashi, and A. O. King, *Org. Syn.*, submitted (1985).

[15] S. E. Byström, R. Aslanian, J.-E. Bäckvall, *Tetrahedron Letters*, **26**, 1749 (1985).

Thallium(III) acetate.

Tetrahydrofurans.   Homoallylic alcohols cyclize to tetrahydrofurans on reaction with $Tl(OAc)_3$.[1]

Example:

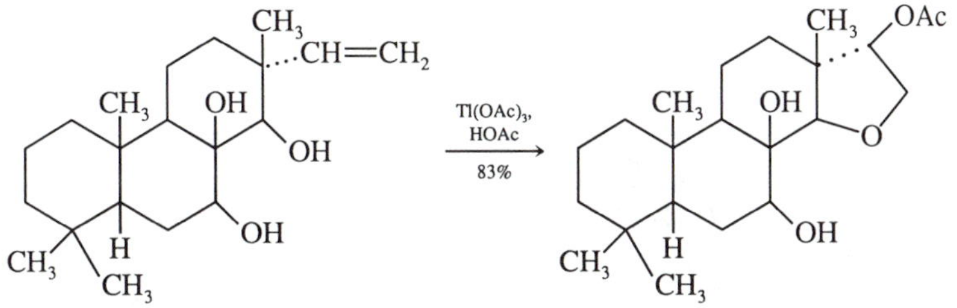

Substituted 4-alkenols (**1**) are converted by Tl(OAc)$_3$ at 0–25° into *trans*-2,5-disubstituted tetrahydrofurans (**2**) in 60–80% yield with only traces of the *cis*-isomer or a tetrahydropyranyl regioisomer.[2] The same reaction using Pb(OAc)$_4$ gives **2** in only 31% yield.

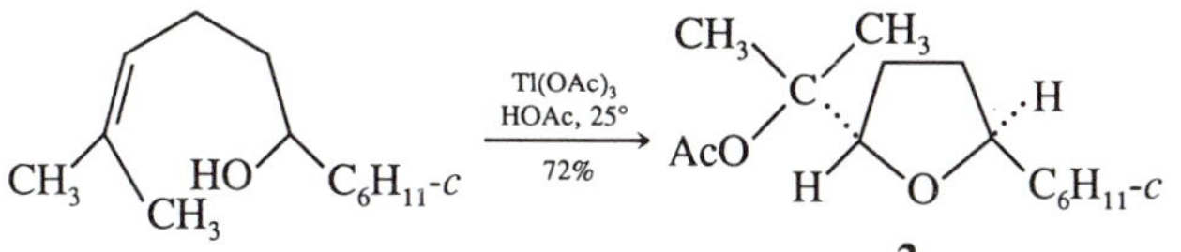

[1] H. M. C. Ferraz, T. J. Brocksom, A. C. Pinto, M. A. Abla, and D. H. T. Zocher, *Tetrahedron Letters*, **27**, 811 (1986).

[2] J. P. Michael, P. C. Ting, and P. A. Bartlett, *J. Org.*, **50**, 2416 (1985).

Thallium(III) trifluoroacetate (TTFA).

Allyl ethers and acetates.[1] Allyltrimethylsilanes and allyltributyltins on treatment with TTFA (1.2 equiv.) in alcohols or acetic acid give the corresponding ethers or acetates in moderate to high yield.

Examples:

$$CH_2{=}CHCH_2M \xrightarrow[\text{ROH}]{\text{TTFA,}} CH_2{=}CHCH_2OR$$

M = Si(CH$_3$)$_3$ R = C$_2$H$_5$,   86%

M = SnBu$_3$ R = C$_2$H$_5$,   78%

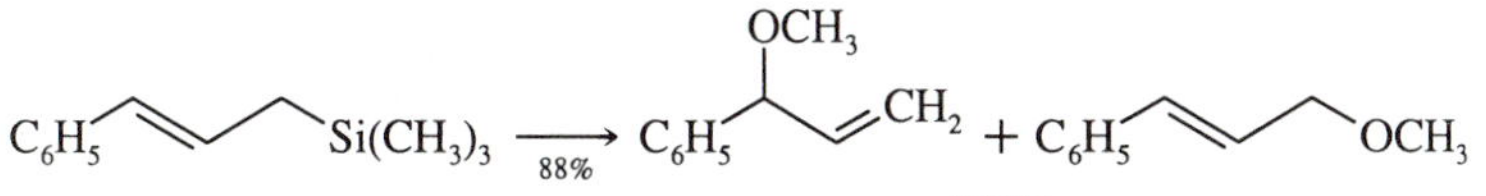

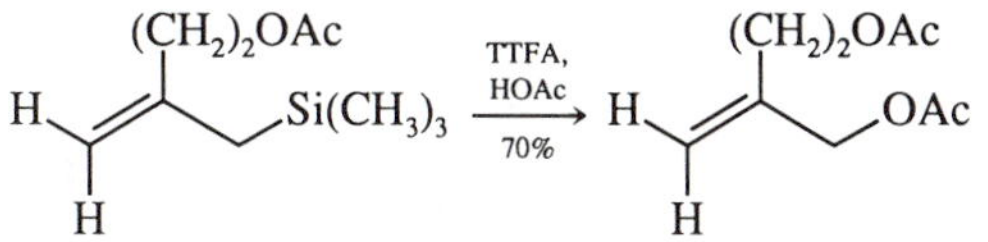

[1] M. Ochiai, E. Fujita, M. Arimoto, and H. Yamaguchi, *Chem. Pharm. Bull.*, **32**, 5027 (1984).

Thallium zeolites, 4A and 13X. The catalyst is prepared from a mixture of Tl_2CO_3 and the crushed molecular sieves in water at 45°, and then washed with acetone and dried at 190°. Co or Cd zeolites are prepared similarly with $CoCO_3$ or $CdCO_3$.[1]

O-Glycosylation. Traditional glycosylation catalysts are silver or mercury salts. Recently silver zeolite[2] has been recommended as the catalyst for preparation of 1,2-*cis*-glycosides. The thallium zeolite is useful when the glycosyl bromide is unstable in the presence of silver catalysts.[1]

Example:

| | Ag(I) | 10% | $\alpha/\beta = 0:100$ |
| | Tl(I) | 60% | $\alpha/\beta = 0:100$ |

[1] D. M. Whitfield, R. N. Shah, J. P. Carver, and J. J. Krepinsky, *Syn. Comm.*, **15**, 737 (1985).
[2] P. J. Garegg, C. Henrichson, T. Norberg, and P. Ossowski, *Carbohyd. Res.*, **119**, 95 (1983).

Thioethoxycarbonylmethylenetriphenylphosphorane, $(C_6H_5)_3P{=}CHCOSC_2H_5$ (1).
Preparation:

$$BrCH_2COOH + HSC_2H_5 \xrightarrow[\text{DMAP}]{\text{DCC,}} BrCH_2COSC_2H_5 \xrightarrow[\text{C}_6\text{H}_6]{P(C_6H_5)_3,}$$

$$(C_6H_5)_3\overset{+}{P}CH_2COSC_2H_5Br^- \xrightarrow[\underset{\text{overall}}{78\%}]{\text{Na}_2\text{CO}_3} 1$$

(E)-α,β-Unsaturated thiol esters.[1] The reagent converts aldehydes mainly into (E)-α,β-unsaturated thiol esters. The *trans*-selectivity can be increased by addition of DMAP (E/Z 90–97:10–3).
Example:

(E/Z = 91:9)

[1] G. E. Keck, E. P. Boden, and S. A. Mabury, *J. Org.*, **50**, 709 (1985).

Thionyl chloride.

Acyl chlorides.[1]   Acyl chlorides are formed rapidly by reaction of carboxylic acids with $SOCl_2$ and pyridine in CH_2Cl_2 at 25°. The dicyclohexylammonium salts of carboxylic acids react particularly rapidly (*ca.* 1 minute). The acid chlorides prepared *in situ* in this way react with amines in the presence of DMAP or DBU to form amides in >85% yield. This $SOCl_2$–Py method is also useful for peptide synthesis with slight racemization.

[1] F. Matsuda, S. Itoh, N. Hattori, M. Yanagiya, and T. Matsumoto, *Tetrahedron*, **41**, 3625 (1985).

Thionyl chloride–Silica gel.

Selective thioacetalization.[1]   On addition of $SOCl_2$ to an equal weight of SiO_2 in CH_2Cl_2, HCl and SO_2 are evolved. The solvent is removed and the dried residue can be stored in sealed vessels for several months. This material is an effective catalyst for conversion of aldehydes to thioacetals with 1,2-ethanedithiol or 1,3-propanedithiol. The reaction proceeds in dry benzene at 20° and is complete after 5 hours. The same reactions with ketones proceed very slowly, but can be effected in 24 hours in refluxing toluene. The difference in rates permits highly selective protection of aldehydes in the presence of ketones. Moist SO_2Cl_2/SiO_2 can also be used to effect dethioacetalization and dethioketalization.

[1] Y. Kamitori, M. Hojo, R. Masuda, T. Kimura, and T. Yoshida, *J. Org.*, **51**, 1427 (1986).

Thiophenol.

α,β-Butenolides.[1]   Reaction of (E)-γ-hydroxy-α,β-unsaturated esters with thiophenol results in cyclization to 3-phenylthiobutyrolactones, which can be converted into 3-phenylthiobutenolides or butenolides.

Example:

[1] R. Tanikaga, H. Yamashita, and A. Kaji, *Synthesis*, 416 (1986).

Threophos,

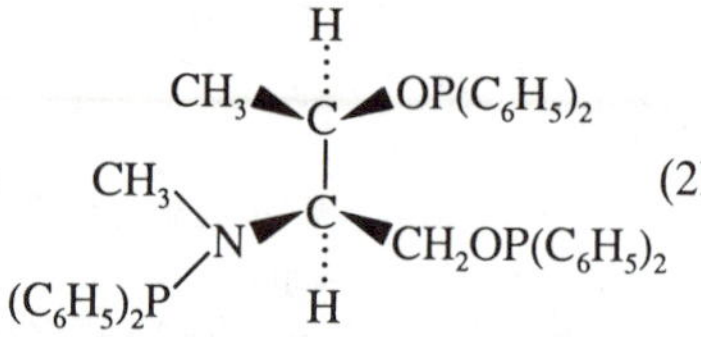

(2R, 3R)-**1**, α_D + 250°. This chiral ami-

nophosphine phosphinite (**1**) is prepared from threonine by formylation, reduction, and phosphinylation.

Asymmetric hydrovinylation.[1]   The reaction of ethylene with 1,3-cyclohexadiene catalyzed by bis(1,5-cyclooctadiene)nickel, diethylaluminum chloride, and **1** gives (+)-(S)-3-vinyl-1-cyclohexene (**2**) in quantitative yield and 93% ee. Related ligands prepared from (S)-proline and D-ephedrine are less effective for asymmetric hydrovinylation.

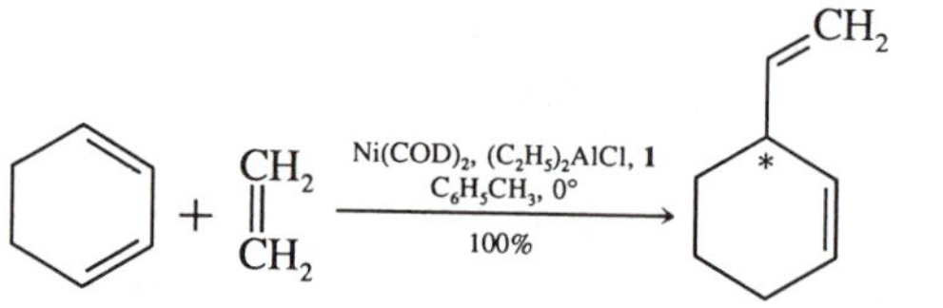

(+)-(S)-**2** (93% ee)

[1] G. Buono, C. Siv, G. Peiffer, C. Triantaphylides, P. Denis, A. Mortreux, and F. Petit, *J. Org.*, **50**, 1781 (1985).

Tin.

Allylation of aldehydes.   Aldehydes undergo selective allylation on reaction with allylic bromides and tin (1.2 equiv.) in H_2O/THF (5:1) in a sonicator.[1] Similar results can be obtained with zinc powder in saturated aqueous NH_4Cl–THF (5:1).[2]
    Example:

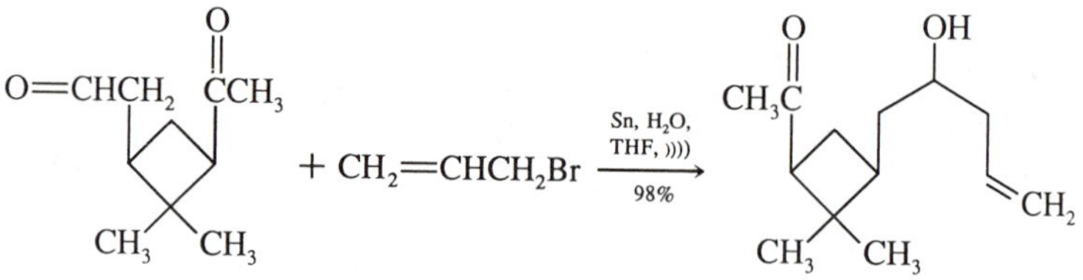

[1] C. Pétrier, J. Einhorn, and J. L. Luche, *Tetrahedron Letters*, **26**, 1449 (1985).
[2] C. Pétrier and J.-L. Luche, *J. Org.*, **50**, 910 (1985).

Tin(II) chloride.

(E)-1,3-Dienes and 1,3,5-trienes.   Tin(II) mediates a reaction of aldehydes with 1-bromo-3-iodopropene to form (E)-1,3-dienes in 40–70% yield.[1]

Example:

$$CH_3(CH_2)_7CHO + BrCH=CHCH_2I \xrightarrow[\substack{45\%}]{\substack{2\ SnCl_2, \\ DMF,\ C_6H_5CH_3}} \begin{array}{c} CH_3(CH_2)_7 \\ \diagdown \\ H \end{array} C=C \begin{array}{c} H \\ \diagup \\ CH=CH_2 \end{array}$$

Aldol condensation.[2]   The yields of the aldol condensation of aldehydes with the trimethylsilyl enol ethers of aldehydes catalyzed by $TiCl_4$ are considerably improved by workup with $SnCl_2$. Formation of polymers and β-elimination are inhibited.

Example:

$$CH_3(CH_2)_2CHO + CH_2=CHOSi(CH_3)_3 \xrightarrow[\substack{68\%}]{\substack{TiCl_4}} CH_3(CH_2)_2\overset{\overset{\displaystyle OH}{|}}{CH}-CH_2CHO$$

$$+ SnCl_2 \quad 81\%$$

Reduction of nitroalkenes.[3]   1-Nitroalkenes are reduced in an alcohol medium by $SnCl_2·2H_2O$ to α-alkoxy oximes in 70–90% yield.

Example:

$$C_6H_5CH=C \begin{array}{c} CH_3 \\ \diagdown \\ NO_2 \end{array} + C_2H_5OH \xrightarrow[\substack{90\%}]{\substack{SnCl_2·2H_2O}} C_6H_5\overset{\overset{\displaystyle OC_2H_5}{|}}{CH}-\underset{\underset{\displaystyle CH_3}{|}}{C}=NOH$$

Reduction of ArNO$_2$.[4]   Aromatic nitro compounds are reduced by $SnCl_2·2H_2O$ in ethanol or ethyl acetate at 70° to amines in 90–99% yield. Under these conditions, carbonyl, cyano, halo, and benzyl groups are not reduced.

[1] J. Augé, *Tetrahedron Letters*, **26**, 753 (1985).
[2] B. A. B. Kohler, *Syn. Comm.*, **15**, 39 (1985).
[3] R. S. Varma and G. W. Kabalka, *Chem. Letters*, 243 (1985).
[4] F. D. Ballamy and K. Ou, *Tetrahedron Letters*, **25**, 839 (1984).

Tin(II) chloride–Aluminum.

Homoallylic alcohols.[1]   This combination generates a Sn(0) species, which is more effective than Sn–Al (**12**, 486) for diastereoselective reaction of cinnamyl chloride with aldehydes to form *anti*-homoallylic alcohols.

Examples:

(99:1)

(90:10)

[1] K. Uneyama, H. Nanbu, and S. Torii, *Tetrahedron Letters*, **27**, 2395 (1986).

Tin(IV) chloride.

Medium-ring cyclic ethers.[1]   Unsaturated eight- and nine-membered cyclic ethers can be obtained by cyclization of unsaturated acetals with $SnCl_4$ (2 equiv.) in CH_2Cl_2.

Examples:

Cyclization of vinylsilanes.[2]   In the presence of Lewis acids, particularly $SnCl_4$, $TiCl_4$, or $Cl_3Ti(O\text{-}i\text{-}Pr)$, vinylsilanes substituted by a mixed acetal group, particularly by MEM ethers, can undergo cyclization to unsaturated five-, six-, and seven-membered oxacyclics. The Si group controls the regio- and stereochemistry of the resulting double bond.

Examples:

$MEM = CH_2OCH_2CH_2OCH_3$

α,β-Dehydro-β-amino acid derivatives.   These amino acid derivatives can be obtained by $SnCl_4$-catalyzed reaction of aliphatic or aromatic nitriles with dialkyl malonates.[3]

Example:

$$(CH_3OOC)_2CH_2 + CH_3CH_2CN \xrightarrow[\text{2) Na}_2\text{CO}_3]{\text{1) SnCl}_4, \text{ ClCH}_2\text{CH}_2\text{Cl}} \left[(CH_3OOC)_2CHCC_2H_5 \atop \overset{\|}{NH} \right]$$

$$55\% \downarrow$$

$$(CH_3OOC)_2C = C \begin{smallmatrix} C_2H_5 \\ \\ NH_2 \end{smallmatrix}$$

[1] L. E. Overman, T. A. Blumenkopf, A. Castañeda, and A. S. Thompson, *Am. Soc.*, **108**, 3516 (1986).
[2] L. E. Overman, A. Castañeda, and T. A. Blumenkopf, *ibid.*, **108**, 1303 (1986).
[3] F. Scavo and P. Helquist, *Tetrahedron Letters*, **26**, 2603 (1985).

Tin(II) trifluoromethanesulfonate.

Michael addition of tin(II) dienolates.[1] The tin(II) dienolates of β,γ-enones undergo γ-selective addition to acyclic α,β-enones to give *trans*-1,7-enediones. If the reaction is carried out at high dilution, the intermediate tin(II) enolate can undergo an intramolecular aldol cyclization to give a cyclohexenol.

Example:

$$(CH_3)_3\overset{O}{\overset{\|}{C}}CCH_2CH=CH_2 \xrightarrow[C_2H_5N]{Sn(OTf)_2,} \left[(CH_3)_3\overset{OSnOTf}{\overset{|}{C}}C=CHCH=CH_2 \right]$$

$$\xrightarrow[\text{high dilution}]{C_6H_5\overset{O}{\overset{\|}{C}}CH=CHCH_3 \quad -45°}$$

73%

83%

$$(CH_3)_3C \underset{\cdots CH_3}{\overset{OH}{\diagup}} COC_6H_5$$

$$(CH_3)_3\overset{O}{\overset{\|}{C}}CCH\underset{(E)}{=\!=}CHCH_2\overset{CH_3}{\overset{|}{C}}HCH_2\overset{O}{\overset{\|}{C}}C_6H_5$$

γ-Keto sulfides.[2] In the presence of trityl tetrafluoroborate, silyl enol ethers react with thioacetals or -ketals to give γ-keto sulfides in 75–95% yield (equation I). The same products can be obtained directly from ketones and thioacetals by *in*

(I)

$$\overset{OSi(CH_3)_3}{\bigcirc} + C_6H_5CH(SC_2H_5)_2 \xrightarrow[95\%]{(C_6H_5)_3CBF_4 \atop CH_2Cl_2} \overset{O}{\overset{\|}{\bigcirc}}\overset{SC_2H_5}{\underset{C_6H_5}{\diagup}}$$

1 (94:6)

$$(II) \xrightarrow{Sn(OTf)_2} \left[\begin{array}{c} OSnOTf \\ \end{array} \right] \xrightarrow[78\%]{\begin{array}{c} C_6H_5CH(SC_2H_5)_2 \\ (C_6H_5)_3CBF_4 \end{array}} \mathbf{1} \ (23{:}77)$$

situ preparation of the tin(II) enolate (equation II). However, the yield is usually lower, and the relative configuration of the γ-keto sulfide is reversed.

Asymmetric allylation of aldehydes.[3]   In the presence of (S)-1-[1-methyl-2-pyrrolidino]methylpiperidine (**1**, 2 equiv.) and Sn(OTf)$_2$ (2 equiv.) prochiral al-

(S)-**1**

dehydes react with allyldialkylaluminums to give chiral secondary homoallyl alcohols. Allyldiisobutylaluminum is more effective than other aluminum reagents. It is prepared by reaction of diisobutylaluminum chloride with allylmagnesium chloride.

Example:

$$C_6H_5CHO + CH_2{=}CHCH_2Al(i\text{-}Bu)_2 \xrightarrow[91\%]{\begin{array}{c} Sn(OTf)_2, \ \mathbf{1} \\ CH_2Cl_2, \ -78° \end{array}} \underset{*}{C_6H_5\overset{\overset{\displaystyle OH}{|}}{C}HCH_2CH{=}CH_2}$$

(S, 83% ee)

Glycosidation.[4]   β-D-Glucosides are obtained stereospecifically by reaction of acetobromoglucose with a protected sugar in CH$_2$Cl$_2$ in the presence of Sn(OTf)$_2$ (1 equiv.) and a base (1 equiv. of collidine, diisopropylethylamine, or 1,1,3,3-tetramethylurea). Addition of 4 Å molecular sieves ensures anhydrous conditions. Yields are 30–60%.

[1] R. W. Stevens and T. Mukaiyama, *Chem. Letters*, 851 (1985); *Idem, ibid.*, 855 (1985).
[2] M. Ohshima, M. Murakami, and T. Mukaiyama, *ibid.*, 1871 (1985).
[3] T. Mukaiyama, N. Minowa, T. Oriyama, and K. Narasaka, *ibid.*, 97 (1986).
[4] A. Lubineau and A. Malleron, *Tetrahedron Letters*, **26**, 1713 (1985).

Titanium(III) chloride.

Pinacol reduction.[1]   In a strongly basic medium (pH 11–12), TiCl$_3$ effects pinacol reduction of aromatic ketones, ArCOCH$_3$ or C$_6$H$_5$COR, previously effected with a Ti(II) species (**7**, 373–374). Both the *dl*- and *meso*-pinacols are formed with marked preference for the former isomer.

Both benzil and benzoin are reduced by TiCl₃ in an acidic medium to hydro-benzoin (95% yield) but in this case the *meso*-isomer is strongly favored over the *dl*-isomer (~80:20).

[1] A. Clerici and O. Porta, *J. Org.*, **50**, 76 (1985).

Titanium(III) chloride–Diisobutylaluminum hydride. A black, solid lower-valent titanium reagent (**1**) is obtained on reduction of TiCl₃·3THF with DIBAH in toluene.[1]

Deoximation.[2] A new prostaglandin synthesis is based on the ability of the O-methyloxime **2**, when complexed with BF₃, to react with the mixed cyanocuprate formed from **3**, and then with **4** to give the prostanoid **5** in 75% yield.

Conversion of **5** to PGE₂ (**6**) requires Lindlar reduction, saponification to the acid group, desilylation (HF), and deoximation. Usual methods for the last step are ineffective, but deoximation was achieved with the solid Ti reagent (**1**) in 73% yield.

[1] L. E. Manzer, *Inorg. Syn.*, **21**, 135 (1982).
[2] E. J. Corey, K. Niimura, Y. Konishi, S. Hashimoto, and Y. Hamada, *Tetrahedron Letters*, **27**, 2199 (1986).

Titanium(III) chloride–Zinc/copper couple.

Intramolecular cyclization of keto aldehydes.[1] The first synthesis of the sesquiterpene bicyclogermacrene (**2**) has been achieved by cyclization of **1** with TiCl₃–Zn/Cu to give **2** and the isomer **3**.

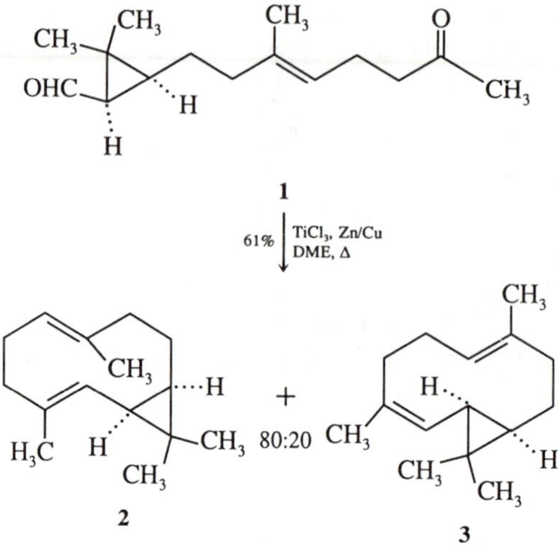

1

61% | TiCl₃, Zn/Cu
DME, Δ

2   80:20   **3**

This cyclization also effects a synthesis of helminthogermacrene (**5**) and β-elemene (**6**) from **4**.[2]

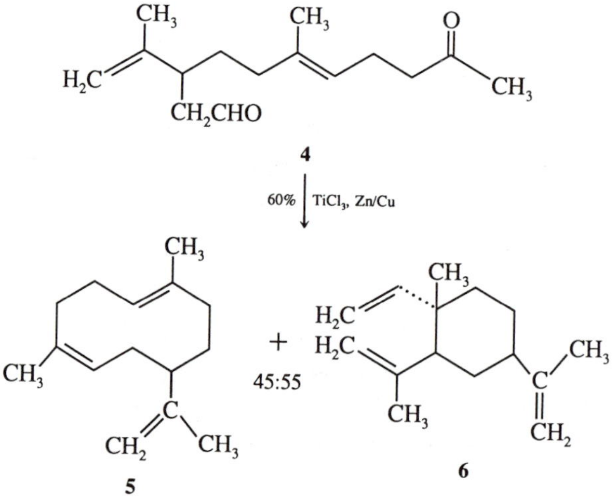

4

60% | TiCl₃, Zn/Cu

5   45:55   **6**

[1] J. E. McMurry and G. K. Bosch, *Tetrahedron Letters*, **26**, 2167 (1985).
[2] J. E. McMurry and P. Kočovský, *ibid.*, **26**, 2171 (1985).

Titanium(IV) chloride.

Conjugate addition of allylsilanes to α,β-unsaturated acylsilanes.[1]   In the presence of TiCl₄ (1–1.5 equiv.) allylsilanes undergo 1,4-addition to α,β-unsaturated acylsilanes. The products are convertible into carboxylic acids or aldehydes.[1]

Example:

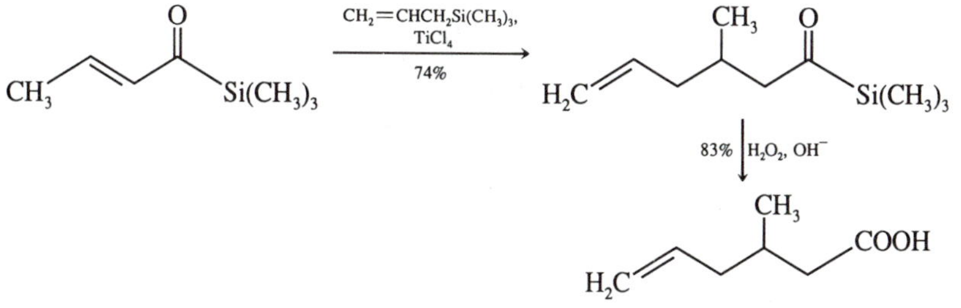

Allenylsilanes react with α,β-unsaturated acylsilanes to give trimethylsilyl-substituted cyclopentenes (equation I); the reaction of allenylsilanes with α-methyl-α,β-unsaturated acylsilanes results in silyl-substituted cyclohexenones.[2]

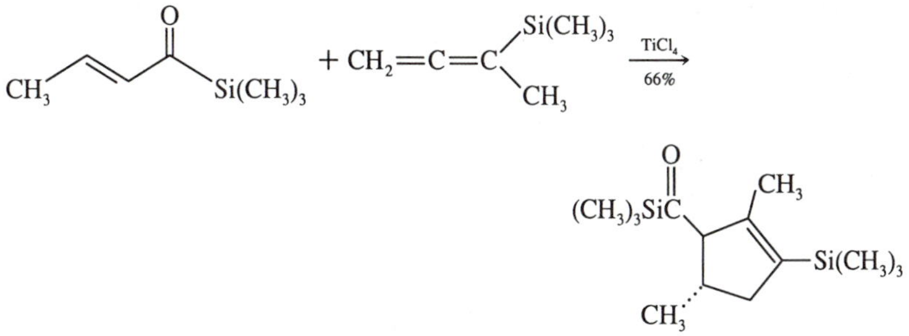

Alkynylation of aldehydes.[3] The reaction of alkynylmetals with chiral aldehydes shows no or slight diastereoselectivity, but in the presence of TiCl$_4$, alkynyltributyltin reagents react with the steroid aldehyde **1** with about 9:1 diastereoselectivity. TiCl$_4$-catalyzed reaction of **1** with allyltrimethylsilane or allyltributyltin also shows similar and comparable diastereoselectivity.

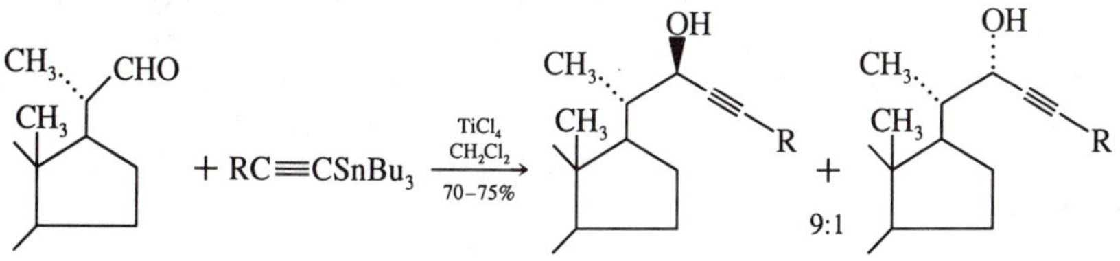

Intramolecular cyclization of allylsilanes.[4] β-Keto esters (or amides) substituted in the α-position by an allylsilane undergo diastereospecific cyclization in the presence of TiCl$_4$ (chelation controlled). Cyclizations induced by F$^-$ or BF$_3$ etherate give mixtures of diastereomers.

Example:

$$CH_3C-C-COC_2H_5$$

TiCl$_4$, CH$_2$Cl$_2$, $-78°$
Bu$_4$NF, THF
BF$_3$·O(C$_2$H$_5$)$_2$, CH$_2$Cl$_2$

88%

2:1
1:4.8

Intramolecular conjugate addition of allylsilanes to enones (**12**, 25).   In an extension to the earlier study, Majetich *et al.*[5] have examined the intramolecular addition of allylsilanes to cyclic enones. The course of cyclization depends markedly on the choice of catalyst and on the size of the ring that is formed.
    Example:

Bu$_4$NF   70%
TiCl$_4$ 78%
C$_2$H$_5$AlCl$_2$ —

Stereoselective Mukaiyama-Michael reactions.[6]   Heathcock *et al.*[7] have investigated the *syn/anti* stereoselectivity in the reaction of twelve silyl enol ethers with a variety of acyclic and cyclic enones catalyzed by TiCl$_4$ or SnCl$_4$. Preliminary results suggest that the stereoselectivity is independent of the geometry of the silyl enol ether, and that silyl enol ethers derived from aliphatic ketones show a preference for *anti*-addition ranging from 1.5:1 to 10:1. The preference for *anti*-addition is even higher in the case of (Z)-silyl enol ethers of aromatic ketones (10:1 to >20:1). However, high *syn*-selectivity is observed with acyclic *t*-butyl enones.
    Example:

(Z)
(E)

87%
76%

$$syn/anti = 96:4$$
$$= 98:2$$

Mukaiyama aldol condensation (**6**, 590–591).[8] This reaction can be effected in the absence of a Lewis acid catalyst under high pressure (10 kbar). Surprisingly the stereoselectivity is the reverse of that of the $TiCl_4$-catalyzed reaction (equation I). The reaction can also be effected in water with the same stereoselectivity, but the yield is low because of hydrolysis of the silyl enol ether. Yields are improved by use of water–oxolane (1:1) and by sonication.[9]

		(*syn*)	(*anti*)
$TiCl_4$, CH_2Cl_2	82%	25:75	
10 kbar, CH_2Cl_2	90%	75:25	
H_2O	23%	85:15	
H_2O/oxolane (1:1)	45%	74:26	
H_2O/oxolane, ((((	68%	73:27	

Aldol condensation of α-amino silyl ketene acetals (**1**).[10]   2-Dibenzylaminoketene trimethylsilyl acetals (**1**) react with aldehydes premixed with $TiCl_4$ to give α-amino-β-hydroxy carboxylic esters (**2**) with moderate to high *syn*-selectivity. Surprisingly, $TiCl_4$-catalyzed reaction of **1** with a chiral α-alkoxy aldehyde proceeds with low asymmetric induction.

$$70-90:30-10$$

***anti*-Selective aldol condensation.**[11]   The reaction of silyl ketene acetals with aldehydes catalyzed by TiCl₄ shows some *anti*-selectivity but is dependent on the structure of the aldehyde. The *anti*-stereoselectivity can be markedly improved by addition of triphenylphosphine (equation I).

(I)

+ Bu₃P	71%	4.5:1
(C₆H₅)₂P(CH₂)₃P(C₆H₅)₂	72%	10.3:1
(C₆H₅)₃P	79%	10.5:1

This variation was used for an enantioselective synthesis of *anti*-α-methyl-β-hydroxy esters using the silylketene acetal derived from (1R, 2S)-N-methylephedrine-O-propionate (equation II).[12]

(II)

anti/syn ≥ 30:1, 94% de

(R)-(−)-3-Benzyloxy-2-methylpropionaldehyde (**2**), obtained from N-methylephedrine, reacts with the ketene *t*-butylthio trialkylsilyl acetal **3** in CH₂Cl₂ in the presence of TiCl₄ to give **4** in >99% de. Use of BF₃ etherate as the Lewis acid catalyst results in three stereoisomers.[13]

¹ R. L. Danheiser and D. M. Fink, *Tetrahedron Letters*, **26**, 2509 (1985).
² *Idem, ibid.*, **26**, 2513 (1985).
³ Y. Yamamoto, S. Nishii, and K. Maruyama, *J.C.S. Chem. Comm.*, 102 (1986).
⁴ G. A. Molander and S. W. Andrews, *Tetrahedron Letters*, **27**, 3115 (1986).
⁵ G. Majetich, K. Hull, J. Defauw, and T. Shawe, *ibid.*, **26**, 2755 (1985).
⁶ T. Mukaiyama, *Org. Reactions*, **28**, 203 (1982).
⁷ C. H. Heathcock, M. H. Norman, and D. E. Uehling, *Am. Soc.*, **107**, 2797 (1985).

[8] Y. Yamamoto, K. Maruyama, and K. Matsumoto, *Am. Soc.*, **105**, 6963 (1983).
[9] A. Lubineau, *J. Org.*, **51**, 2142 (1986).
[10] G. Guanti, L. Banfi, E. Narisano, and C. Scolastico, *Tetrahedron Letters*, **26**, 3517 (1985).
[11] C. Palazzi, L. Colombo, and C. Gennari, *ibid.*, **27**, 1735 (1986).
[12] C. Gennari, A. Bernardi, L. Colombo, and C. Scolastico, *Am. Soc.*, **107**, 5812 (1985).
[13] C. Gennari, A. Bernardi, C. Scolastico, and D. Potenza, *Tetrahedron Letters*, **26**, 4129 (1985).

Titanium(IV) chloride–1,8-Diazabicyclo[5.4.0]undecene-7.

Chlorohydrins.[1] Epoxides are converted into chlorohydrins by $TiCl_4$ and DBU in CH_2Cl_2 at 25°. The conditions allow survival of acid-sensitive groups such as acetals. Yields are lower when DBU is replaced by DMAP.

Example:

[1] C.-L. Spawn, G. J. Drtina, and D. F. Wiemer, *Synthesis*, 315 (1986).

Titanium(IV) chloride–Diethylaluminum chloride.

Diels-Alder reactions of bis(trimethylsilyl)acetylene.[1] A catalyst obtained from $TiCl_4$ and $(C_2H_5)_2AlCl$ (1:20) effects Diels-Alder reactions of this acetylene with butadiene and methyl-substituted derivatives to form 1,2-bis(trimethylsilyl)-cyclohexa-1,4-dienes in 70–78% yield (equation I). The yield is low (15%) only when $R^1,R^4 = CH_3,R^2,R^3 = H$ because of polymerization of the diene. The products undergo thermal dehydrogenation at 240° to form 1,2-bis(trimethylsilyl)benzenes in almost quantitative yield. This cycloaddition has been effected in low yield with an iron-based catalyst.

[1] K. Mach, H. Antropiusova, L. Petrusova, F. Turecek, V. Hanus, P. Sedmera, and J. Schraml, *J. Organomet.*, **289**, 331 (1985).

Titanium(IV) chloride–Lithium aluminum hydride.

$\rangle C{=}O \rightarrow \rangle C{=}CH_2$.[1]   In a synthesis of *trans*-γ-irone (**4**) the methylenation of **1** was carried out in two steps by conversion to the oxide **2** with dimethylsulfonium methylide followed by deoxygenation with $TiCl_4$–$LiAlH_4$ (1:1) and $N(C_2H_5)_3$.

[1] O. Takazawa, K. Kogami, and K. Hayashi, *Bull. Chem. Soc. Japan*, **58**, 389 (1985).

Titanium(IV) chloride–N-Methylaniline.

[3 + 2]*Cycloaddition*.[1]   In the presence of $TiCl_4$–$C_6H_5NHCH_3$ (1–2:1), 2-trimethylsilylmethyl allylic alcohols undergo intramolecular addition to δ,ε-double bonds to provide bicyclo[3.3.0]octanes.

Example:

[1] J. Ipaktschi and G. Lauterbach, *Angew. Chem. Int. Ed.*, **25**, 354 (1986).

Titanium(IV) chloride–Zinc, 8, 487–488.

2,5-*Dihydrothiophenes*.[1]   These thiophenes are available by reductive coupling of diketo sulfides.

Example:

When this reaction is carried out at 0°, 3,4-dihydroxythiolanes are formed in 70–95% yield. These products undergo dehydration to thiophenes when heated in toluene with a catalytic amount of TsOH (85–95% yield).

Porphycene (2).[2] This isomer (2) of porphin is obtained in 2–3% yield by McMurry coupling of the diformylbipyrrole **1** in THF/Py with TiCl$_4$–Zn. The reaction presumably involves a dihydro derivative (**a**).

[1] J. Nakayama, H. Machida, and M. Hoshino, *Tetrahedron Letters*, **26**, 1981 (1985); J. Nakayama, H. Machida, R. Saito, and M. Hoshino, *Tetrahedron Letters*, **26**, 1983 (1985).
[2] E. Vogel, M. Kocher, H. Schmickler, and J. Lex, *Angew. Chem. Int. Ed.*, **25**, 257 (1986).

Titanium(IV) isopropoxide.

Cleavage of 2,3-epoxy alcohols.[1] Titanium(IV) isopropoxide (1–1.5 equiv.) not only increases the rate of reaction of nucleophiles with these alcohols but controls the regioselectivity resulting in highly selective attack at C$_3$. Some other metal alkoxides are also effective, including Zr(O-*i*-Pr)$_4$–*i*-PrOH and Lu(O-*i*-Pr)$_3$.[2]

$(C_2H_5)_2NH$	90%	20:1	$R = N(C_2H_5)_2$
$(CH_3)_2CHOH$	88%	100:1	$R = OCH(CH_3)_2$
C_6H_5SH, C_6H_6	95%	6.4:1	$R = SC_6H_5$
C_6H_5COOH, CH_2Cl_2	74%	100:1	$R = C_6H_5COO$

Typical results are shown for the reaction of some nucleophiles with the oxirane derived from 2-hexen-1-ol.

Cleavage of 2,3-epoxy acids and amides.[3]   These glycidic acids are easily prepared from 2,3-epoxy alcohols by reaction with periodic acid (H_5IO_6) and 2 mole% $RuCl_3$ in CCl_4–CH_3CN–H_2O. This system is superior to the $NaIO_4$–$RuCl_3$ system used previously (**11**, 462–463). Glycidic acids and amides react with dialkylamines or thiophenol preferentially at C_2, but in the presence of at least 1 equiv. of Ti(O-i-Pr)$_4$ the reaction occurs with high regioselectivity at C_3.

Examples:

High regioselectivity also obtains in reaction with LiN_3 and $NaCN$.

α,β-Epoxy alcohols.[4]   In the presence of Ti(O-i-Pr)$_4$, allylic hydroperoxides are converted into α,β-epoxy alcohols. The precursors are readily available by the ene reaction of alkenes with singlet oxygen. The oxygen transfer involves the corresponding allylic alcohol.

Example:

(95:5)

[1] M. Caron and K. B. Sharpless, *J. Org.*, **50**, 1557 (1985).
[2] K. S. Kirshenbaum, *Nouv. J. Chim.*, **7**, 699 (1983).

2

3, α-CH$_3$ (9%)
4, β-CH$_3$ (18%)

Lactonization of alkenoyloxymethyl selenides (or iodides).[4] Bu$_3$GeH is generally more effective than Bu$_3$SnH for cyclization of these substrates to γ- and δ-lactones. The cyclization (exclusively *exo*) generally affords *cis*-fused products.
Example:

Bu$_3$SnH	38%	7%
Bu$_3$GeH	62%	26%

[1] P. J. Stang and M. R. White, *Am. Soc.*, **103**, 5429 (1981).
[2] L. J. Johnston, J. Lusztyk, D. D. M. Wayner, A. N. Abeywickreyma, A. L. J. Beckwith, J. C. Scaiano, and K. U. Ingold, *ibid.*, **107**, 4594 (1985).
[3] A. L. J. Beckwith, D. H. Roberts, C. H. Schiesser, and A. Wallner, *Tetrahedron Letters*, **26**, 3349 (1985).
[4] A. L. J. Beckwith and P. E. Pigou, *J.C.S. Chem. Comm.*, 85 (1986).

Tributyl(iodomethyl)tin, ICH$_2$SnBu$_3$ (**1**).
Methylenation of sulfones.[1] This reagent reacts rapidly with the anion (LDA) of even moderately hindered sulfones to form β-stannyl sulfones, which undergo desulfonylstannylation on treatment with Bu$_4$NF to form alkenes.
Example:

[3] J. M. Chong and K. B. Sharpless, *J. Org.*, **50**, 1560 (1985).
[4] W. Adam, A. Griesbeck, and E. Staab, *Angew. Chem. Int. Ed.*, **25**, 269 (1986).

p-Toluenesulfonyl chloride.

Tosylation of alcohols.[1]   A detailed study of the tosylation of a typical pri
alcohol finds that the highest yield (98%) is obtained by use of a 1:1.5:2 ra
alcohol/tosyl chloride/pyridine in chloroform and at a temperature of 0°.
reaction requires about 2.5 hours. Secondary alcohols under the same cond
react more slowly. The rate can be increased by use of a 1:2:3 ratio of alc
tosyl chloride/pyridine without a significant effect on the yield (85–98%).

[1] G. W. Kabalka, M. Varma, R. S. Varma, P. C. Srivastava, and F. F. Knapp, Jr., *J.*
51, 2386 (1986).

Tosylmethyl isocyanide (TosMIC).

1,5- and 1,6-*Diketones*.[1]   These ketones can be prepared in useful yiel
reaction of 1,3-dibromopropane or 1,4-dibromobutane with 2 equiv. of a mono
derivative of tosylmethyl isocyanide (equation I).

$$
\text{(I)}\quad Br(CH_2)_nBr + 2TosCHN{=}C\underset{CH_3}{|}\xrightarrow{\text{NaH}} \underset{\substack{|\\N{=}C}}{\overset{\substack{Tos\\|}}{CH_3C}}{-}(CH_2)_n{-}\underset{\substack{|\\N{=}C}}{\overset{\substack{Tos\\|}}{CCH_3}}\xrightarrow[82\%]{H_3O^+} CH_3\overset{\substack{O\\||}}{C}(CH_2)
$$

n = 3   91%
n = 4   89%

Unfortunately yields are low when this reaction is extended to 1,2-dibro
ethane, possibly because of steric problems and/or elimination of HBr.

[1] A. M. van Leusen, R. Oosterwijk, E. van Echten, and D. van Leusen, *Rec. Trav.*, **104**
(1985).

Tributylgermane, Bu_3GeH (**1**), b.p. 61–63°/0.3 mm. The reagent is prepared
reduction of Bu_3GeCl with $LiAlH_4$ (91% yield).[1] Tributylgermane is a considerab
poorer donor of hydrogen to alkyl radicals than tributyltin hydride, but is as
fective for abstraction of halogen. Consequently it should be more useful th
Bu_3SnH for slow intramolecular radical cyclizations.[2]

Linear triquinanes.[3]   Reaction of the trienyl bromide **2** with **1** (AIBN, 80
results in at least eight products. The two major ones (27% yield) are the epimer
triquinanes **3** and **4**.

Iodomethyltrimethylsilane has been used previously for this reaction (**9**, 444–445), but the tin reagent (**1**) is at least 600 times more reactive than the silicon reagent. Furthermore, fluoride-induced desulfonylstannylation is much faster than desulfonylsilylation, probably because a C–Sn bond is weaker than a C–Si bond.

Methylenation of nitriles.[2]   Nitriles are alkylated more readily by **1** than sulfones are. But decyanostannylation requires a strong base (CH$_3$Li, −20°). Even so, this two-step reaction is useful for conversion of nitriles into terminal alkenes. Example:

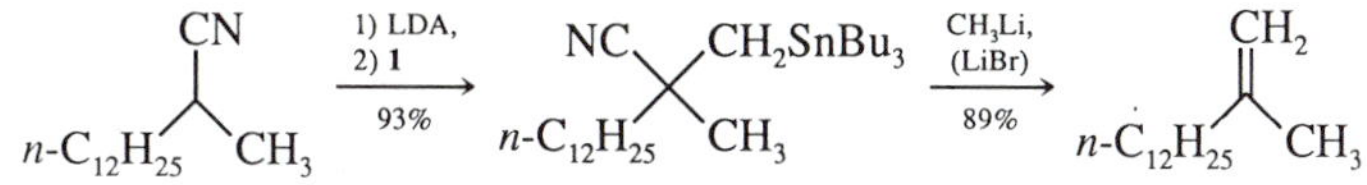

[1] B. A. Pearlman, S. R. Putt, and J. A. Fleming, *J. Org.*, **50**, 3622 (1985).
[2] *Idem, ibid.*, **50**, 3625 (1985).

Tributyltin chloride.

vic-*Dialkylation of cyclopentenones.*   Conjugate addition of trimethylsilyllithium to cyclopentenone generates the lithium enolate of 3-trimethylsilylcyclopentanone. Transmetallation of this enolate with Bu$_3$SnCl provides the stannyl enolate, which undergoes α-alkylation in moderate to high yield (equation I).[1] A complicated mixture of products is obtained in the absence of the tin reagent.

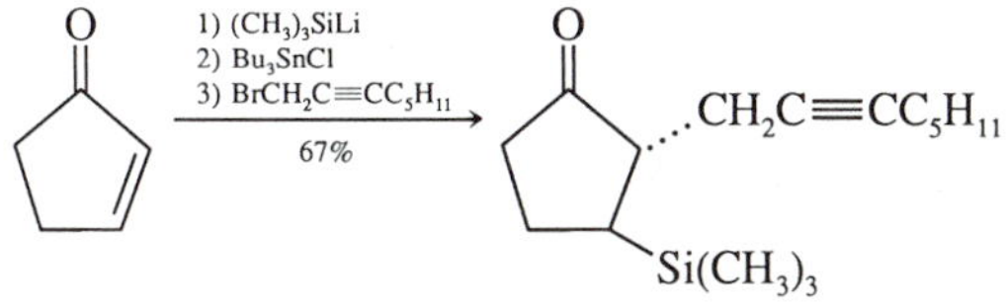

This sequence provides a short synthesis of prostaglandins (equation II).[2]

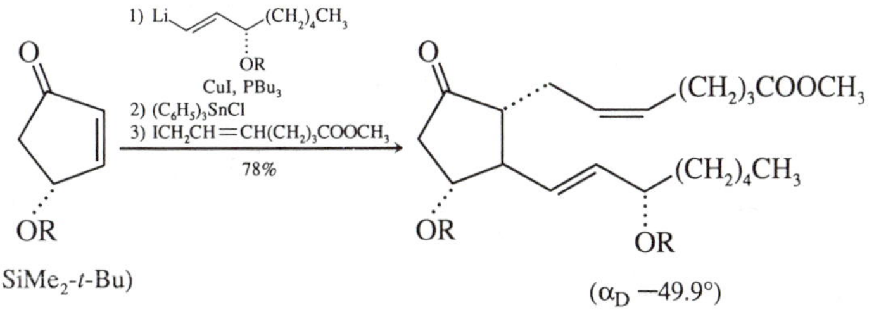

[1] H. Nishiyama, K. Sakuta, and K. Itoh, *Tetrahedron Letters*, **25**, 223 (1984); *idem, ibid.*, **25**, 2487 (1984).
[2] M. Suzuki, A. Yanagisawa, and R. Noyori, *Am. Soc.*, **107**, 3348 (1985).

Tributyltin hydride.

Stereocontrolled addition of an alkane to cyclic allylic alcohols.[1]   The silyl ether **2**, obtained by reaction of allylic alcohol **1** with (bromomethyl)-chlorodimethylsilane, on reaction with Bu$_3$SnH (AIBN) is converted into the cyclic siloxane **3**. H$_2$O$_2$ oxidation followed by oxidation of the resulting diol gives the ketol **4**.[1] The net effect is addition of —CH$_2$OH and —H to the double bond of **1** in such a way that the hydrogen at the ring juncture is *trans* to the original hydroxyl group.

Radical macrocylization.   Although radical cyclization is usually used only for construction of five- and six-membered rings, Porter *et al.*[2] find that this method, although not useful for cyclization to medium-sized ketones, can give macrocyclic ketones in yields as high as 75–80% in favorable cases (absence of steric effects). Bu$_3$SnH is superior to (C$_6$H$_5$)$_3$SnH or Bu$_3$GeH; formation of acyclic products decreases as the concentration of reactants decreases.

Examples:

$$I(CH_2)_7CH\underset{(E)}{=\!\!=\!\!=}CH(CH_2)_2\overset{O}{\overset{\|}{C}}CH=\!\!=\!\!CH_2 \xrightarrow[78\%]{}$$

I_2-*transfer cyclization.*[3]   Free-radical chain reactions are usually terminated by abstraction of a hydrogen atom. In contrast, standard radical cyclization of the 6-iodohexyne (1) induces isomerization to the (iodomethylene)cyclopentane (2). The formation of minor amounts of the reduced product 3 can be suppressed by use of $Bu_3Sn—SnBu_3$ and irradiation (95% yield of 2).

$$\begin{array}{ccc} \textbf{1} & \textbf{2} & \textbf{3 (2–5\%)} \\ & [84\% \text{ (E/Z = 15:1)}] & \end{array}$$

Another example:

$$(E/Z = 6:1)$$

C-Glycosides (**12**, 518).   The radical formed by reduction of a tertiary nitro sugar such as 1 with Bu_3SnH (AIBN) reacts with acrylonitrile to form the C-glycoside 2 in 45% yield.[4]

$$\begin{array}{ccc} \textbf{1} & & \textbf{2} \end{array}$$

C-Disaccharides.[5]   These disaccharides linked by a CH_2 group can be prepared by an extension of the Giese reaction. Thus the radical generated from a glycosyl halide (**1**) couples with an α-methylene lactone (**2**) to give a C-disaccharide (**3**) as a mixture of α/β-isomers.

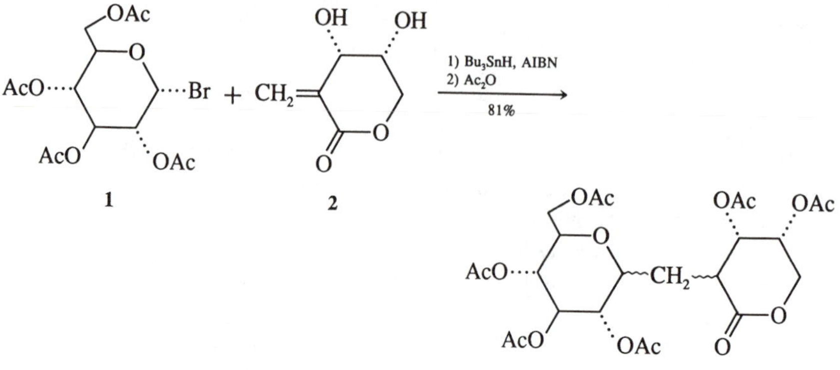

3 (α/β =10:1)

Modified Giese reaction.[6]   The radical obtained by denitration of tertiary nitro compounds undergoes inter- and intramolecular reaction with activated double bonds.

Examples:

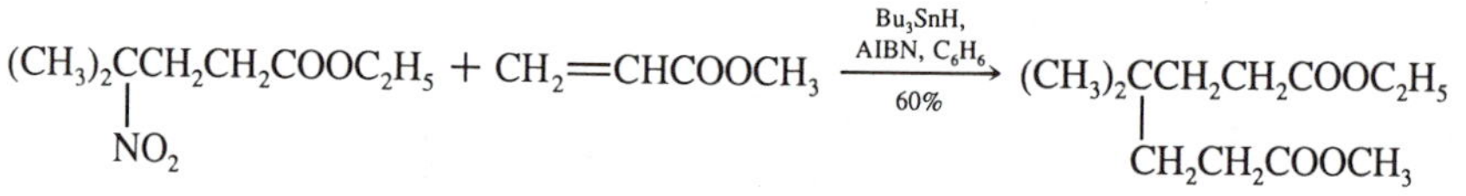

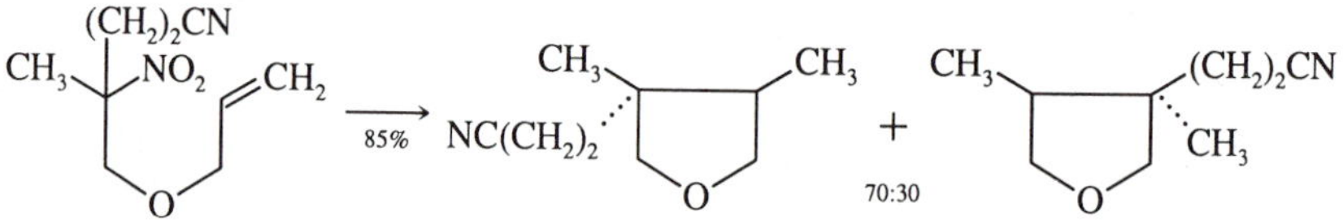

Deoxygenation of sec-alcohols.[7]   2'-Deoxynucleosides (**1**) can be deoxygenated efficiently to 2',3'-dideoxynucleosides (**3**) by reaction with N,N'-thiocarbonyl-diimidazole in DMF (80°) to form imidazolides, which on reaction with methanol are converted into methylthionocarbonates (**2**). These crystalline derivatives are reduced by Bu₃SnH to **3**.

Example:

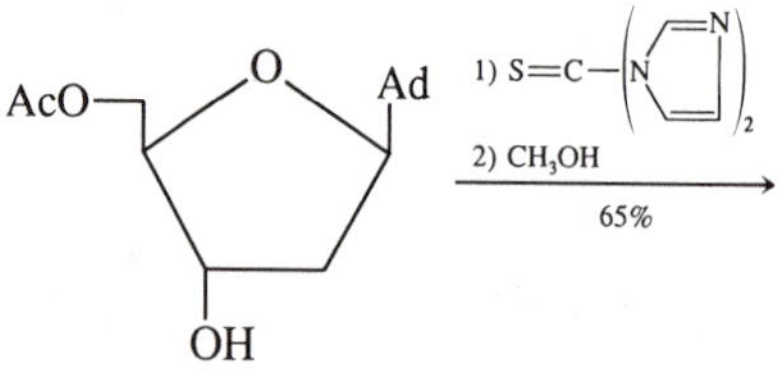

The method is a modification of one used by Barton and McCombie.[8]

Reduction of ketones.[9]   Ketones can be reduced to alcohols by Bu$_3$SnH in the presence of either AIBN or a Lewis acid, but this reaction is limited to unhindered ketones. However, even sterically hindered ketones, such as *t*-butyl methyl ketone, can be reduced under high pressure (10 kbar) in the absence of a catalyst. This method is particularly useful in the case of cyclopropyl and α,β-epoxy ketones, which are reduced to the corresponding alcohols. Reduction of these ketones with Bu$_3$SnH under radical conditions results in ring-opened products.

[1] G. Stork and M. Kahn, *Am. Soc.*, **107**, 500 (1985).
[2] N. A. Porter, D. R. Magnin, and B. T. Wright, *ibid.*, **108**, 2787 (1986).
[3] D. P. Curran, M.-H. Chen, and D. Kim, *ibid.*, **108**, 2489 (1986).
[4] J. Dupuis, B. Giese, J. Hartung, and M. Leising, *ibid.*, **107**, 4332 (1985).
[5] B. Giese and T. Witzel, *Angew. Chem. Int. Ed.*, **25**, 450 (1986).
[6] N. Ono, H. Miyake, and A. Kaji, *Chem. Letters*, 635 (1985).
[7] E. J. Prisbe and J. C. Martin, *Syn. Comm.*, **15**, 401 (1985).
[8] D. H. R. Barton and S. W. McCombie, *J.C.S. Perkin I*, 1574 (1985).
[9] M. Degueil-Castaing, A. Rahm, and N. Dahan, *J. Org.*, **51**, 1672 (1986).

Tributyltin hydride–Dichlorobis(triphenylphosphine)palladium(II).

Deprotection of allyl or allyloxycarbonyl derivatives of amino acids.[1]   This combination of reagents effects rapid reductive cleavage of allyl or allyloxycarbonyl derivatives of amino acids in CH$_2$Cl$_2$ containing a proton donor (*p*-NO$_2$C$_6$H$_5$OH, acetic acid, H$_2$O). The actual catalyst is probably bis(triphenylphosphine)palladium(0). Benzyl, Boc, and Cbo groups are stable to these conditions. This cleavage does not induce racemization.

[1] F. Guibe, O. Dangles, and G. Balavoine, *Tetrahedron Letters*, **27**, 2365 (1986).

Tributyltinlithium (1).

α- or β-C-Glycosides.[1]   Reaction of the protected α-hexapyranosyl chloride **3**, prepared by hydrochlorination of the D-glucal **2**, with **1** produces selectively the equatorial β-D-tributyltin-glycopyranoside **4**. Transmetallation of **4** (BuLi) followed by reaction with an electrophile results in β-*C*-glycosides with overall retention.

The axial α-D-tribuyltin-glycopyranoside can be obtained by reaction of **3** with lithium naphthalenide and then with Bu_3SnCl. Transmetallation (BuLi) and reaction with an electrophile results in α-C-glycosides.

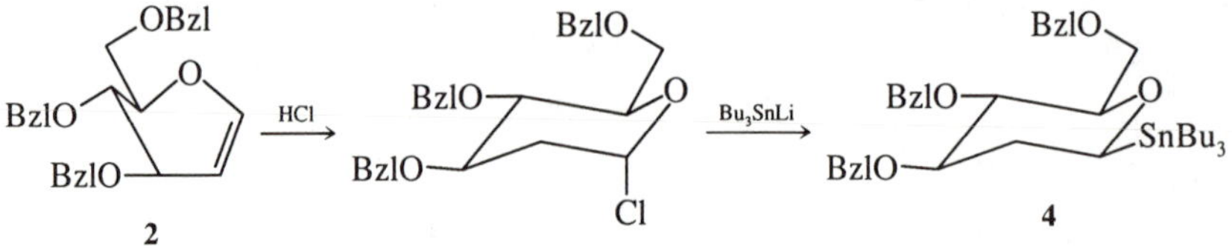

[1] P. Lesimple, J.-M. Beau, and P. Sinaÿ, *J.C.S. Chem. Comm.*, 894 (1985).

Tributyltinlithium–Trimethylaluminum.

Stereoselective deoxygenation of epoxides.[1] The ate complex $[Bu_3SnAl-(CH_3)_3]^-Li^+$ (**1**) formed from Bu_3SnLi and $Al(CH_3)_3$ converts epoxides to alkenes with overall retention of stereochemistry. The results can be explained by inversion in the epoxide cleavage and *anti*-elimination of the Bu_3Sn and $OAl(CH_3)_3$ groups.

Example:

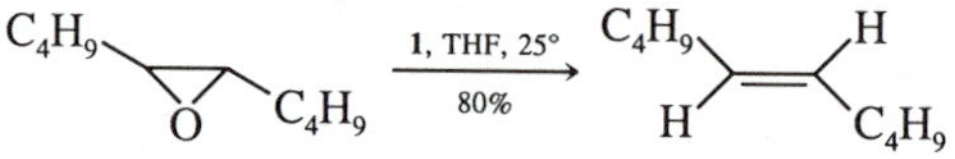

[1] S. Matsubara, T. Nonaka, Y. Okuda, S. Kanemoto, K. Oshima, and H. Nozaki, *Bull. Chem. Soc. Japan*, **58**, 1480 (1985).

Tri-μ-carbonylhexacarbonyldiiron, $Fe_2(CO)_9$.

Deoxygenation of furan-aryne adducts. At temperatures of 50–60°, these adducts on treatment with $Fe_2(CO)_9$ in benzene form a complex with $Fe(CO)_4$, which decomposes gradually to an aromatic hydrocarbon.[1]

Example:

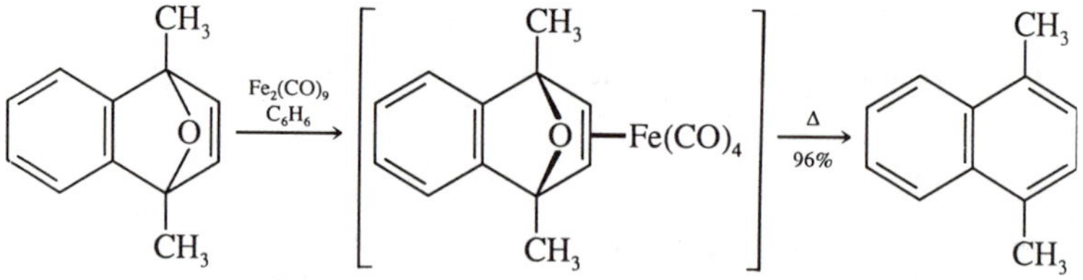

This method is recommended for deoxygenation of isobenzofuran-aryne adducts.[2]

Example:

[1] W. M. Best, P. A. Collins, R. K. McCulloch, and D. Wege, *Aust. J. Chem.*, **35**, 843 (1982).
[2] S. L. Crump, J. Netka, and B. Rickborn, *J. Org.*, **50**, 2746 (1985).

Trichloroacetonitrile (1).

Glycosides. The reaction of tetrabenzylglucose (**2**) with trichloroacetonitrile (**1**) in the presence of K_2CO_3 results in rapid formation of the β-trichloroacetimidate (β-**3**) in 78% yield. In contrast, use of NaH results in slow formation of the more stable α-trichloroacetimidate (α-**3**) in almost quantitative yield. The α-isomer of **3**, in the absence of a neighboring group effect and in the presence of a Lewis acid reacts with an alcohol preferentially with inversion to give β-glycosides. The β-isomer of **3** in the presence of trimethylsilyl triflate reacts to give mainly the more stable α-glycoside.[1]

Preliminary studies indicate that trichloroacetimidates undergo substitution with N-, S-, and C-nucleophiles.

[1] R. R. Schmidt, *Angew. Chem. Int. Ed.*, **25**, 212 (1986).

Trichloroisocyanuric acid (Chloreal).

α-Chloro ketones.[1] Trichloroisocyanuric acid, activated by BF_3 etherate, is an efficient reagent for α-chlorination of ketones at the more substituted position. It can be used to convert acetophenone into α-chloro-, α,α-dichloro-, and α,α,α-trichloroacetophenone (81% yield).

[1] G. A. Hiegel and K. B. Peyton, *Syn. Comm.*, **15**, 385 (1985).

Trichlorosilane–*t*-Amines. A general synthesis of allyltrimethylsilanes is illustrated by a synthesis of (E)-crotyltrimethylsilane (equation I).[1]

$$\text{(I)} \quad CH_3CH{=}CHCH_2Br \xrightarrow[84\%]{\substack{\text{1) } HSiCl_3, Bu_3N \\ \text{2) } 3CH_3MgBr}} CH_3CH{=}CHCH_2Si(CH_3)_3$$
$$\text{(Z/E = 2.5:1)} \qquad\qquad\qquad \text{(E/Z = 97:3)}$$

[1] S. A. Carr and W. P. Weber, *J. Org.*, **50**, 2782 (1985).

Triethyl phosphonoacetate.

Wittig-Horner reaction. Strong bases [BuLi, NaH, $KOC(CH_3)_3$] are commonly used in the Wittig-Horner reaction with phosphonates. Actually, the weak base triethylamine can be used if metal salts, particularly LiBr or $MgBr_2$, are also present. Under these conditions reactions of triethyl phosphonoacetate proceed in high yield with various aldehydes or ketones, with the exception of methyl ketones.[1] Reactions using K_2CO_3 as base in H_2O at 20° or in refluxing THF or toluene result in the (E)-alkene. The mild conditions permit use of sensitive aldehydes without protection of substituent groups.[2]

Ethyl α-(hydroxymethyl)acrylate.[3] This useful acrylate (2) can be prepared in ~75% yield by reaction of triethyl phosphonoacetate with a 30% aqueous solution of formaldehyde. It is converted to ethyl α-(bromomethyl)acrylate by reaction with PBr_3 in ether (~85% yield).

$$(C_2H_5O)_2\overset{\overset{\displaystyle O}{\|}}{P}CH_2COOC_2H_5 + HCHO \xrightarrow[75\%]{\substack{K_2CO_3, \\ H_2O}} CH_2{=}\underset{\underset{\displaystyle CH_2OH}{|}}{C}{-}COOC_2H_5$$

$$\qquad\qquad\quad \mathbf{1} \qquad\qquad\qquad\qquad\qquad\qquad\qquad\qquad \mathbf{2}$$

[1] M. W. Rathke and M. Nowak, *J. Org.*, **50**, 2624 (1985).
[2] J. Villieras, M. Rambaud, and M. Graff, *Tetrahedron Letters*, **26**, 53 (1985).
[3] J. Villieras and M. Rambaud, *Org. Syn.*, submitted (1985).

$$CH_3$$
$$|$$

(α-Triethylsilyl)propionaldehyde *t*-butylimine, $(C_2H_5)_3SiCHCH=NC(CH_3)_3$ **(1).**
The imine is prepared as an oil in 75% yield by triethylsilylation of the *t*-butylimine
of propionaldehyde [LDA, $(C_2H_5)_3SiCl$] in 73% yield.

α-Methyl-α,β-unsaturated aldehydes.[1] The anion of **1** (*sec*-BuLi) converts
the aldehyde **2** to the α,β-unsaturated aldehyde **3** in 77% yield. No other reagent
was found to be useful for this olefination.

[1] R. H. Schlessinger, M. A. Poss, S. Richardson, and P. Lin, *Tetrahedron Letters*, **26**, 2391
(1985).

1,1,1-Trifluoroacetone, CF_3COCH_3 **(1),** b.p. 22°. Flammable, lachrymator. Supplier: Aldrich.

Aldol condensations.[1] Under usual conditions, **1** is not useful for crossed-aldol condensation because of predominant self-condensation. However in the
presence of pyridine and acetic acid **1** undergoes aldol condensation with aromatic
and α,β-unsaturated aldehydes. Yields are moderate to high if the concentration
of **1** is kept low (inverse addition). This reaction can be used to obtain all-*trans*-
19,19,19- and 20,20,20-trifluororetinal (**2**).

[1] D. Mead, R. Loh, A. E. Asato, and R. S. H. Liu, *Tetrahedron Letters*, **26**, 2873 (1985).

Trifluoroacetyl nitrate, CF_3COONO_2 (1).

1-*Nitro*-1,3-*dienes*.[1] The reagent, prepared *in situ* from ammonium nitrate and TFAA in CH_2Cl_2, adds to 1,3-dienes to give a mixture of 1,2- and 1,4-adducts. The mixture on treatment with KOAc in ether is converted into a single 1-nitro-1,3-diene.

Examples:

$$CH_2\!\!=\!\!CHCH\!\!=\!\!CH_2 \xrightarrow{\ 1\ } [CH_2\!\!=\!\!CHCHCH_2NO_2 + CF_3OCOCH_2CH\!\!=\!\!CHCH_2NO_2]$$

with $OCOCF_3$ on the first fragment

$$89\% \downarrow KOAc,\ O(C_2H_5)_2$$

70%

[1] A. J. Bloom and J. M. Mellor, *Tetrahedron Letters*, **27**, 873 (1986).

Trifluoromethanesulfonic anhydride.

Intramolecular [2 + 2] cycloadditions of keteniminium triflates. Two laboratories[1,2] report that unsaturated keteniminium salts, generated from an unsaturated amide with 1 equiv. each of triflic anhydride and collidine, undergo intramolecular [2 + 2] cycloadditions.

Examples:

[1] I. Marko, B. Ronsmans, A.-M. Hesbain-Frisque, S. Dumas, L. Ghosez, B. Ernst, and H. Greuter, *Am. Soc.*, **107**, 2192 (1985).
[2] B. B. Snider, R. A. H. F. Hui, and Y. S. Kulkarni, *ibid.*, **107**, 2134 (1985).

N-(p-Trifluoromethylbenzyl)cinchoninium bromide (1). The phase-transfer catalyst is available from Chemical Dynamics Corp., South Plainfield, N.J.

Enantioselective Robinson annelation.[1]   Alkylation of indanone **2** with 1,3-dichloro-2-butene (E/Z = 4:1) catalyzed by **1** gives (S)-**3** in 92% ee. The enantiomer, (R)-**3**, is obtained by the same alkylation but catalyzed by N-(p-trifluoromethylbenzyl)cinchonidinium bromide in 78% ee and 99% yield. Optically pure **3** undergoes hydrolysis and cyclization in high yield. Demethylation and alkylation provides the desired tricyclic enone **4** in 83% overall yield from (R)-**3**.

(S)-3 (92% ee)

(R)-3   **4**

See Phase-transfer catalysts (this volume) for another example of a chiral quaternary ammonium salt.

[1] A. Bhattacharya, U.-H. Dolling, E. J. J. Grabowski, S. Karady, K. M. Ryan, and L. M. Weinstock, *Angew. Chem. Int. Ed.*, **25**, 476 (1986).

Trimethylamine N-oxide (1).

Oxidation of organoboranes (**6**, 624; **8**, 507; **10**, 423).[1]   The oxidation of tributylborane with **1** in $CHCl_3$ proceeds stepwise to borinate, then boronate, and finally borate esters (equation I). The oxidation of unsymmetrical trialkylboranes shows some selectivity; thus tertiary groups are oxidized more readily than secondary groups, which in turn are more easily oxidized than primary groups. In contrast oxidation of B-alkyl-9-BBN compounds with **1** (1 equiv.) results in oxaborabicyclo[3.3.2]decanes (equation II).

$$\text{(I) Bu}_3\text{B} \xrightarrow{\ 1\ } \text{Bu}_2\text{BOBu} \xrightarrow{\ 1\ } \text{BuB(OBu)}_2 \xrightarrow{\ 1\ } \text{B(OBu)}_3$$

(II)

Deprotonation of t-amine oxides.[2] LDA (excess) converts trimethylamine oxide into a reactive ylide (**a**) that dimerizes to the dimethylpiperazine (**2**). If **a** is generated in the presence of an alkene, pyrrolidines (**3**) are formed via a 1,3-dipolar cycloaddition. The ylide reacts with cyclic alkenes to form bicyclic pyrrolidines in 40–90% yield.[2]

Similar deprotonation of N-methylpiperidine N-oxide in the presence of alkenes affords a useful route to *trans*-octahydroindolizines.[3]

[1] J. A. Soderquist and M. R. Najafi, *J. Org.*, **51**, 1330 (1986).
[2] R. Beugelmans, L. Benadjila-Iguertsira, J. Chastanet, G. Negron, and G. Roussi, *Can. J. Chem.*, **63**, 725 (1985).
[3] J. Chastanet and G. Roussi, *J. Org.*, **50**, 2910 (1985).

Trimethyl orthoformate.

Esterification of sulfonic acids.[1]   Sulfonic acids are converted into methyl or ethyl esters on reaction with trimethyl or triethyl orthoformate. Yields are >80% in esterification of arenesulfonic acids, but are somewhat lower in the reaction of alkanesulfonic acids because of volatility of the products.

[1] A. A. Padmapriya, G. Just, and N. G. Lewis, *Syn. Comm.*, **15**, 1057 (1985).

Trimethyloxonium tetrafluoroborate.

Nitrones.[1]   O-Trimethylsilyloximes undergo N-methylation on reaction with Meerwein's reagent or methyl trifluoromethanesulfonate. The products are converted into nitrones on addition of KF or Bu$_4$NF. Although both (E)- and (Z)-nitrones are formed initially, purification results in isolation of the more stable (Z)-nitrones.

Example:

[1] N. A. LeBel and N. Balasubramanian, *Tetrahedron Letters*, **26**, 4331 (1985).

Trimethylsilyl chlorochromate, $ClCrO_2OSi(CH_3)_3$ **(1).** The orange-red reagent is prepared *in situ* from equimolar amounts of $ClSi(CH_3)_3$ and CrO_3 in CH_2Cl_2.[1] Attempted isolation resulted in a violent explosion.

Oxidations.   The reagent oxidizes benzylic alcohols to the corresponding aldehydes and secondary alcohols to the ketones in satisfactory yield. Of more significance it oxidizes thiols to disulfides in CH_2Cl_2 in high yield (~95%). Oximes are rapidly cleaved by **1** in good yield. The reagent (4 equiv.) effects oxidation of methyl- and ethyl-substituted benzenes to benzaldehydes in 50–80% yield. It is particularly useful for oxidative cleavage of benzylic esters to carboxylic acids.[1]

The combination of **1** and iodine (2:1.2–1.5) effects iodination of arenes and trimethylsilyl enol ethers.

t-Butyldimethylsilyl ethers are stable to Jones reagent and pyridinium dichromate (PDC), but they are oxidized by **1** or by PDC in combination with chlorotrimethylsilane in methylene chloride solution at 25°. Yields are high and comparable with both oxidants.[2]

[1] J. M. Aizpurua, M. Juaristi, B. Lecea, and C. Palomo, *Tetrahedron*, **41**, 2903 (1985).
[2] F. P. Cossio, J. M. Aizpurua, and C. Palomo, *Can. J. Chem.*, **69**, 225 (1986).

Trimethylsilyldiazomethane.

Acylsilanes.[1]   The anion (**1**, prepared with BuLi) of trimethylsilyldiazomethane reacts with alkyl halides to form an α-trimethylsilyldiazoalkane (**2**), which is oxidized to an acylsilane by m-chloroperbenzoic acid in C_6H_6/H_2O buffered to a pH of 7.6 (equation I).

$$\text{(I)}\quad RX + (CH_3)_3SiCN_2\ \underset{Li}{|}\ \xrightarrow[60-85\%]{THF} \overset{N_2}{\overset{||}{R\overset{}{C}Si(CH_3)_3}}\ \xrightarrow[55-70\%]{ClC_6H_4CO_3H,\ H_2O,\ C_6H_6} R\overset{O}{\overset{||}{C}}Si(CH_3)_3$$

$$\qquad\qquad\qquad\qquad 1 \qquad\qquad\qquad\qquad 2 \qquad\qquad\qquad\qquad 3$$

[1] T. Aoyama and T. Shioiri, *Tetrahedron Letters*, **27**, 2005 (1986).

β-Trimethylsilylethanesulfonyl chloride, $(CH_3)_3Si(CH_2)_2SO_2Cl$ (1).
Preparation:

$$(CH_3)_3SiCH{=}CH_2\ \xrightarrow[70\%]{NaHSO_3\ \ C_6H_5CO_3C(CH_3)_3}\ (CH_3)_3SiCH_2CH_2SO_3Na\ \xrightarrow[60\%]{PCl_5,\ CCl_4}\ 1$$

Protection of primary and secondary amines.[1]   The reagent reacts with amines in the presence $N(C_2H_5)_3$ (aliphatic amines) or NaH (aromatic amines) in DMF to form stable sulfonamides, which can be cleaved by CsF or Bu_4NF.
Example:

$$C_6H_5CH_2NH_2 + 1\ \xrightarrow[92\%]{}\ (CH_3)_3Si(CH_2)_2\overset{O}{\underset{}{\overset{\diagdown}{S}}}\overset{O}{\underset{}{\diagup}}{-}NHCH_2C_6H_5\ \xrightarrow[86\%]{F^-}$$

$$C_6H_5CH_2NH_2 + CH_2{=}CH_2 + SO_2 + FSi(CH_3)_3$$

[1] S. M. Weinreb, D. M. Demko, T. A. Lessen, and J. P. Demers, *Tetrahedron Letters*, **27**, 2099 (1986).

Trimethylsilyllithium–Hexamethyldisilane.
Deoxygenation of pyridine N-oxide.   Pyridine N-oxide is converted into pyridine by reaction with $[(CH_3)_3Si]_2$ (1 equiv.) and a catalytic amount of $(CH_3)_3SiLi$ in HMPT at $0 \rightarrow 20°$ in 82% yield. N-Deoxygenation is also effected in comparable yield by reaction with excess $(CH_3)_3SiLi$ in HMPT. The reaction is considered to involve formation of α-trimethylsilylpyridine N-oxide, which then undergoes a 1,2-elimination of $(CH_3)_3SiO^-$.[1] This deoxygenation has also been effected with a catalytic amount of Bu_4NF in THF.[2]

[1] J. R. Hwu and J. M. Wetzel, *J. Org.*, **50**, 400 (1985).
[2] H. Vorbrüggen and K. Krolikiewicz, *Tetrahedron Letters*, **24**, 5337 (1983).

3-Trimethylsilyl-1-propyne, $HC\equiv CCH_2Si(CH_3)_3$ **(1).**

Preparation.[1]

Allenylation of acetals.[2] $TiCl_4$ catalyzes a reaction of propargylsilanes with acetals to provide α-allenyl ethers in good yield.

Examples:

$$HC\equiv CCH_2Si(CH_3)_3 \; + \; (CH_3)_2CHCH_2CH(OC_2H_5)_2 \xrightarrow[70\%]{TiCl_4} CH_2=C=CHCHCH_2CH(CH_3)_2$$
$$\underset{\textbf{1}}{} \qquad\qquad\qquad\qquad\qquad\qquad\qquad\qquad\qquad \underset{OC_2H_5}{|}$$

$$CH_3C\equiv CCH_2Si(CH_3)_3 \; + \; (CH_3)_2CHCH_2CH(OC_2H_5)_2 \xrightarrow{92\%} CH_2=C=\overset{\overset{\displaystyle OC_2H_5}{|}}{\underset{\underset{\displaystyle CH_3}{|}}{C}}CHCH_2CH(CH_3)_2$$

[1] J. Pornet, K. Jaworski, N. B. Kolani, D. Mesnard, and L. Miginiac, *J. Organomet. Chem.,* **236**, 177 (1982).
[2] J. Pornet, L. Miginiac, K. Jaworski, and B. Randrianoelina, *Organometallics,* **4**, 333 (1985).

Trimethylsilyl trifluoromethanesulfonate.

Preparation: A new method involves reaction of trifluoromethanesulfonic anhydride with hexamethyldisiloxane, $[(CH_3)_3Si]_2O$ (88% yield).[1]

Silyl enol ethers.[2] Trimethylsilyl enol ethers can be obtained from α-silyl ketones by thermal rearrangement or by catalysis with $HRh(CO)[P(C_6H_5)_3]_3$, $(CH_3)_3SiOTf$, or $ISi(CH_3)_3$. The first two methods are (E)-selective in the case of unsymmetrical ketones, whereas the latter two are (Z)-selective.

Example:

175°	96%	E/Z = 93/7
Rh catalyst	97%	E/Z = 75/25
$(CH_3)_3SiOTf$	95%	E/Z = 14/86
$ISi(CH_3)_3$	90%	E/Z = 25/75

Intramolecular cyclization of α,β-*enamide esters.*[3] Reaction of the α,β-enamide (**1**) with either trimethylsilyl triflate (1 equiv.) or *t*-butyldimethylsilyl triflate (1 equiv.) and $N(C_2H_5)_3$ at 15° results in the benzo[*a*]quinolizidine **2** in high yield.

$(CH_3)_3SiOTf$	75%
t-BuSi$(CH_3)_2$OTf	83%

The reaction provides a key step in a synthesis of the antineoplastic alkaloid tylophorine (**3**).

3

Aldol cyclization.[4]   The β-keto ester **1** does not undergo aldol cyclization with a wide variety of acids or bases, but is converted to the desired product **2** in 42%

yield when refluxed in benzene with $(CH_3)_3SiOTf$ (3 equiv.) and $N(C_2H_5)_3$ (2 equiv.) with elimination of $[(CH_3)_3Si]_2O$. The product was used for the first total synthesis of $\Delta^{9(12)}$-capnellene-8β,10α-diol (3), a tricyclic sesquiterpene, isolated from the soft coral *Capnella imbricata*, with an unusual bisallylic alcohol group.

Deoxygenation of sulfoxides.[5] α-Cyano and carboalkoxy sulfoxides undergo eliminative deoxygenation with trimethylsilyl triflate in the presence of the hindered base hexamethyldisilazane to give vinyl sulfides.

Example:

$$X{=}CN,\ COOC_2H_5$$

1,1-Dialkoxy-2-trimethylsilylalkenes.[6] Ketene acetals on reaction with trimethylsilyl triflate and triethylamine in ether are converted into trimethylsilyl ketene acetals (equation I).

$$(I)\ \ RCH{=}C(OR^1)_2\ +\ (CH_3)_3SiOTf\ \xrightarrow[\text{ether, 0 }-25°]{\substack{N(C_2H_5)_3,\\ 55-85\%}}\ \begin{array}{c}(CH_3)_3Si\\ R\end{array}\!\!\!\!C{=}C(OR^1)_2$$

[1] J. M. Aizpurua and C. Palomo, *Synthesis*, 206 (1985).
[2] I. Matsuda, S. Sato, M. Hattori, and Y. Izumi, *Tetrahedron Letters*, **26**, 3215 (1985).
[3] M. Ihara, M. Tsuruta, K. Fukumoto, and T. Kametani, *J.C.S. Chem. Comm.*, 1159 (1985).
[4] M. Shibasaki, T. Mase, and S. Ikegami, *Am. Soc.*, **108**, 2090 (1986).
[5] R. D. Miller and R. Hässig, *Tetrahedron Letters*, **26**, 2395 (1985).
[6] D. Schulz and G. Simchen, *Synthesis*, 928 (1984).

Trimethylstannylmethyllithium (1).

β,γ-Unsaturated ketones.[1] The reaction of acyclic α,β-epoxy ketones with two equivalents of **1** results in the cyclopropanols **2**, which are converted into β,γ-enones on treatment with BF_3 etherate or HCl.

[1] T. Sato, T. Kikuchi, N. Sootome, and E. Murayama, *Tetrahedron Letters*, **26**, 2205 (1985).

Triphenylphosphine–Carbon tetraiodide. CI_4 is prepared by reaction of CCl_4 and C_2H_5I in the presence of $AlCl_3$. It is obtained as bright red crystals in about 60% yield.[1]

1,1-*Diiodoalkenes* (*cf.*, **4**, 550).[2]   These alkenes are obtained in 60–85% yield by reaction of $P(C_6H_5)_3$ and CI_4 with an aldehyde in CH_2Cl_2 for 2 hours, after which Zn is added to facilitate the work-up. The reaction fails with ketones.

[1] R. E. McArthur and J. H. Simons, *Inorg. Syn.*, **3**, 37 (1950).
[2] F. Gaviña, S. V. Luis, P. Ferrer, A. M. Costero, and J. A. Marco, *J.C.S. Chem. Comm.*, 296 (1985).

Triphenylphosphine–Diethyl azodicarboxylate.

2-*Oxazolines*.[1]   Hydroxy amides, prepared from β-hydroxy amines and acid chlorides, are converted into 2-oxazolines with the Mitsunobu reagent at $0 \rightarrow 25°$ (equation I).

(I) $R^1CHCH_2NHCR^2$

Aziridines.[2]   2-Aminoethanols can be converted into aziridines by the Mitsunobu reagent if the two carbon atoms between oxygen and nitrogen bear at least one substituent.
    Example:

[1] D. M. Roush and M. M. Patel, *Syn. Comm.*, **15**, 675 (1985).
[2] J. R. Pfister, *Synthesis*, 969 (1984).

Triphenylphosphine–Diethyl azodicarboxylate–Lithium halides.

Alkyl halides or nitriles.[1]   Alcohols are converted conveniently into alkyl halides or nitriles by consecutive addition of a lithium halide or LiCN and the alcohol to a preformed complex of $P(C_6H_5)_3$ and DEAD in anhydrous THF at 0° to 25°. Yields range from 60% to 97%. Reaction of secondary alcohols is accompanied by complete inversion.

[1] S. Manna, J. R. Falck, and C. Mioskowski, *Syn. Comm.*, **15**, 663 (1985).

Triphenylphosphine–2,2'-Dipyridyl disulfide (1).

Penams. The key step in a synthesis of 6-methyl-4-thia-1-azabicyclo-[3.2.0]heptane-7-one (**3**) is formation of the β-lactam ring by reaction of **2**, prepared in several steps from ethyl α-cyanopropionate, with Mukaiyama's reagent (**1**). Even though **2** is a mixture of two diastereomers, a single stereoisomer (**3**) is obtained in high yield. The methyl group at C_6 is essential for a satisfactory yield; in its absence the yield of the penam is 8%.[1]

2 (1:1)

3

[1] T. Chiba, T. Takahashi, J. Sakaki, and C. Kaneko, *Chem. Pharm. Bull.*, **33**, 3046 (1985).

Triphenylphosphine dibromide.

β-Lactams.[1] β-Lactams are obtained by addition of an acetic acid, triethylamine (excess), and an imine to a suspension of $(C_6H_5)_3PBr_2$ in CH_2Cl_2 at 25° (equation I). When R^1 is a phthalimide or a *p*-methoxyphenyl group, *trans*-β-

lactams are formed; in the other reported cases, *cis*-β-lactams are obtained. Dimethyl disulfide dibromide can replace triphenylphosphine dibromide, but yields are lower. Use of carbodiimides in place of imines results in 4-imino-β-lactams (equation II).

[1] F. P. Cossio, I. Ganboa, and C. Palomo, *Tetrahedron Letters*, **26**, 3041 (1985).

Triphenylphosphine–Thiocyanogen.

Review.[1] Use of this reagent in organic synthesis has been reviewed. The report includes cyanation of indoles, pyrroles, and enamines.

[1] R. A. Cherkasov, G. A. Kutyrev, and A. N. Pudovik, *Tetrahedron*, **41**, 2567 (1985).

Triphenyl phosphite ozonide (TPPO). Preparation.[1]

α'-Hydroxylation of α,β-enones. This hydroxylation can be carried out by conversion of an α,β-enone to the trimethylsilyl enol ether [LDA, ClSi(CH₃)₃], which is then allowed to react with triphenyl phosphite ozonide at −78° followed by triphenylphosphine.[2]

Examples:

trans/cis = 2:1

trans/cis = 2.2:1

m-Chloroperbenzoic acid can be used instead of TPPO (**8**, 100–101); in this case, the yields are similar, but the stereoselectivity is the opposite of that observed with TPPO.[3]

[1] P. D. Bartlett and H.-K. Chu, *J. Org.*, **45**, 3000 (1980).
[2] C. Iwata, Y. Takemoto, A. Nakamura, and T. Imanishi, *Tetrahedron Letters*, **26**, 3227 (1985).
[3] C. Iwata, Y. Takemoto, H. Kubota, T. Kuroda, and T. Imanishi, *ibid.*, **26**, 3231 (1985).

Triphenylsilane, (C₆H₅)₃SiH.

Deoxygenation of esters. Esters can be reduced to hydrocarbons by (C₆H₅)₃SiH in the presence of a radical generator, di-*t*-butyl peroxide, at 140°. Highest yields are obtained with acetates; yields based on the alcohol decrease in the order: secondary > primary > tertiary. Other silanes are much less effective than triphenylsilane, which is required in excess for high yields. Radical initiators such as AIBN or benzoyl peroxide are not useful.[1]

Examples:

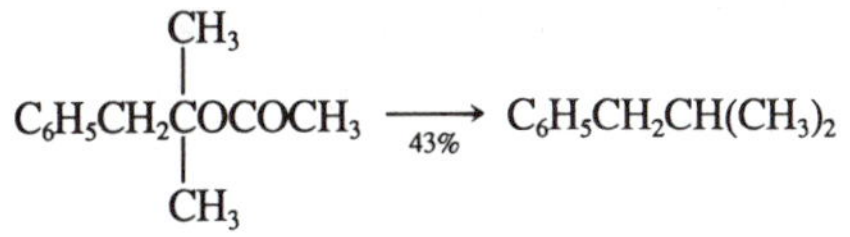

$$\text{C}_6\text{H}_5\text{CH}_2\overset{\overset{\text{CH}_3}{|}}{\underset{\underset{\text{CH}_3}{|}}{\text{C}}}\text{COCOCH}_3 \xrightarrow[43\%]{} \text{C}_6\text{H}_5\text{CH}_2\text{CH(CH}_3)_2$$

[1] H. Sano, M. Ogata, and T. Migita, *Chem. Letters*, 77 (1986).

Triphenyltin hydride.

Spiro compounds.[1]   Diphenyl diselenoketals (**6**, 361–362) on transmetallation react with aldehydes to form hydroxy selenides. On treatment with $(\text{C}_6\text{H}_5)_3\text{SnH}$ and AIBN, these products undergo radical cyclization to spiro compounds.

Example:

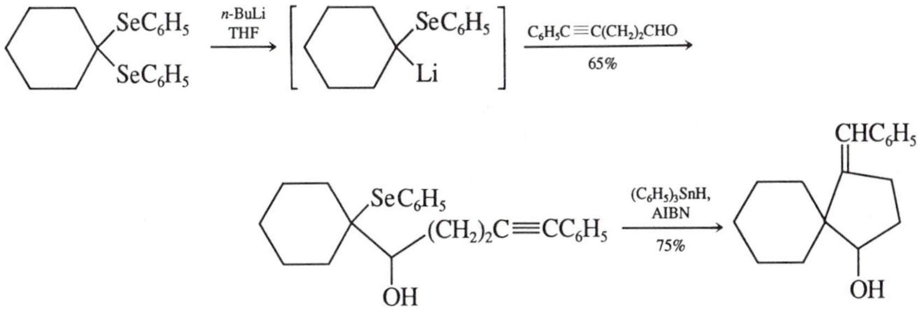

Radical cyclization of β-bromoalkynes.[2]   Addition of $(\text{C}_6\text{H}_5)_3\text{SnH}$ and AIBN to benzene solutions of a β-bromoalkyne and a Michael acceptor results in annelated cyclopentanes in moderate yield.

Example:

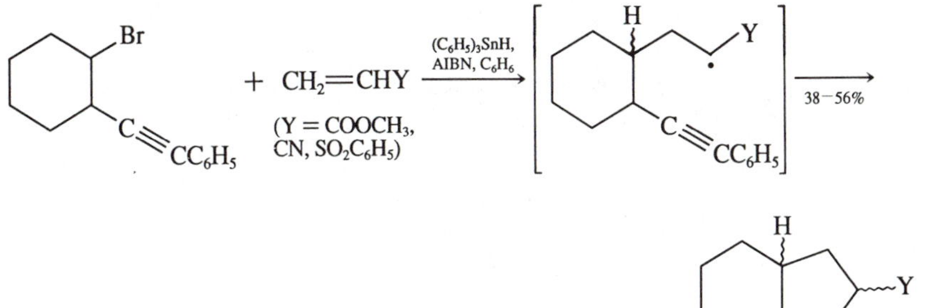

[1] L. Set, D. R. Cheshire, and D. L. J. Clive, *J.C.S. Chem. Comm.*, 1205 (1985).
[2] A. G. Angoh and D. L. J. Clive, *ibid.*, 980 (1985).

Tris(dimethylamino)sulfonium difluorotrimethylsilicate (TASF, **1**).

Reaction of α-halocarbanions with aldehydes.[1]   The reaction of α- or β-poly-halosilanes with **1** at 25° generates halocarbanions which are sufficiently stable to undergo addition to aldehydes in moderate to high yield.

Example:

Polyhalovinylsilanes undergo a similar reaction with aldehydes in the presence of TASF.

Example:

$$n\text{-}C_{10}H_{21}CHO + (C_2H_5)_3SiCF{=}CF_2 \xrightarrow[59\%]{1} n\text{-}C_{10}H_{21}\overset{OSi(C_2H_5)_3}{\underset{}{C}}HCF{=}CF_2$$

Addition of silyl enol ethers to nitroarenes.[2]   In the presence of 1 equiv. of TASF, silyl enol ethers add to nitroalkenes to form unstable *ortho* and/or *para* nitronates, which are oxidized *in situ* by Br_2 or DDQ to nitroaryl carbonyl compounds. The position of substitution depends on the substitution pattern of the arene and the size of the silicon reagent. With less hindered silyl derivatives *ortho* addition is strongly favored.

Example:

Deoxyfluoro sugars.[3]   Carbohydrate triflates are converted into deoxyfluorides by reaction with TASF (3 equiv.) in CH_2Cl_2 (0 → 20°, 10 min.) in 65–70% yield.

[1] M. Fujita and T. Hiyama, *Am. Soc.*, **107**, 4085 (1985).
[2] T. V. RajanBabu, G. S. Reddy, and T. Fukunaga, *ibid.*, **107**, 5473 (1985).
[3] W. A. Szarek, G. W. Hay, and B. Doboszewski, *J.C.S. Chem. Comm.*, 663 (1985).

Tris[2-(2-methoxyethoxy)ethyl]amine (**1**, TDA-1). Supplier: Aldrich. The reagent is obtained by reaction of the monomethyl ether of bis(ethyleneglycol), CH_3O-$(CH_2)_2O(CH_2)_2OH$, with ammonia at 150° catalyzed by Raney nickel.

1 (TDA-1)

Solid-liquid phase-transfer catalyst.[1] The reagent represents a new class of catalysts, acyclic cryptands or tridents. It is singled out of a group as the best compromise of efficiency/price/toxicity. It solubilizes salts of alkali metals as well as of transition metals such as $RuCl_3$ and $PdCl_2$, probably because of the flexibility of the molecule. In addition the trident is sensitive to the nature of the anion, but anionic activation is less than that obtained with cryptands.

It is comparable to crown ethers or R_4NBr for phase-transfer catalyzed aliphatic nucleophilic substitutions; in addition it catalyzes aromatic nucleophilic substitutions such as the Ullmann synthesis.

Examples:

[1] G. Soula, *J. Org.*, **50**, 3717 (1985).

Tris(phenylthio)antimony, $(C_6H_5S)_3Sb$.

RCOOH → ROH.[1] This transformation can be carried out by reaction of the ester (**2**), prepared from a carboxylic acid and the thiohydroxamic acid **1**, with

$(C_6H_5S)_3Sb$ in ether or chlorobenzene at 25° for 4–12 hours. The alcohols are obtained in 80–90% yield. The reaction is considered to involve species such as **a**, which should be easily oxidized by air to species such as **b** and **c**.

[1] D. H. R. Barton, D. Bridon, and S. Z. Zard, *J.C.S. Chem. Comm.*, 1066 (1985).

Tris(tetrabutyl)ammonium pyrophosphate,

$$\underset{Bu_4NO}{\overset{O}{\underset{\|}{HOP}}}-O-\underset{ONBu_4}{\overset{O}{\underset{\|}{P}}}-ONBu_4 \qquad (1).$$ The

reagent is prepared by reaction of disodium dihydrogenpyrophosphate and tetra-butylammonium hydroxide. It is extremely hygroscopic but can be stored in a dessicator over P_2O_5.

Allylic pyrophosphate esters.[1] Allylic chlorides react with **1** in acetonitrile to provide allylic pyrophosphate esters typically in yields of 60–90% after ion-pair chromatographic purification.

Example:

[1] V. M. Dixit, F. M. Laskovics, W. I. Noall, and C. D. Poulter, *J. Org.*, **46**, 1967 (1981); A. B. Woodside, Z. Huang, and C. D. Poulter, *Org. Syn.*, submitted (1985).

Trityllithium, $(C_6H_5)_3CLi$ **(1).** The reagent is prepared by reaction of triphenyl-methane with BuLi in THF at 0–10°.

Diastereoselective aldol reaction. Alkyl trityl ketones **(2)** are readily pre-pared by reaction of **1** with an aldehyde followed by CrO_3 oxidation (50–70% overall yield). Because of steric effects these ketones undergo highly diastereo-selective aldol condensation to give *syn*-adducts (95–99% *syn*). After protection of the hydroxyl group, the adduct is cleaved by lithium triethylborohydride.[1]

Example:

syn:anti = 99:1

Alkenyl trityl ketones, prepared by reaction of **1** with a ketone followed by dehydration, undergo exclusive conjugate addition on reaction with organolithium compounds. The adducts are cleaved to primary alcohols on reduction with $LiB(C_2H_5)_3H$.

[1] D. Seebach, M. Ertas, R. Locher, and W. B. Schweizer, *Helv.*, **68**, 264 (1985).

Trityl perchlorate.

Michael reaction.[1]   In the presence of $(C_6H_5)_3CClO_4$, silyl enol ethers undergo Michael addition to α,β-enones. The adducts can be isolated or rearranged to 1,5-diketones by base. The intermediates cannot be isolated from reactions catalyzed by $TiCl_4$ or CsF.

Example:

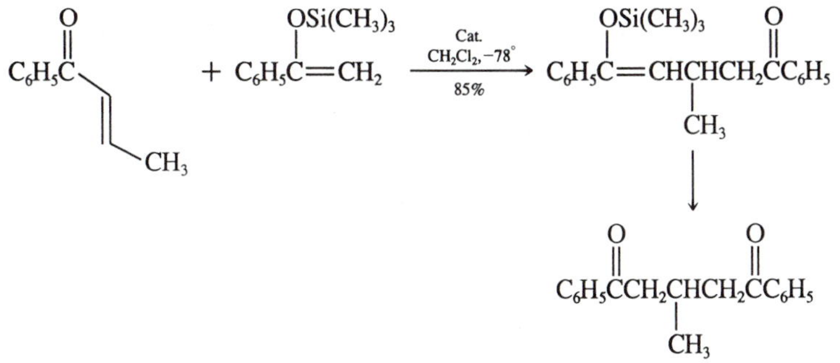

Michael-aldol reactions.[2]   In the presence of $(C_6H_5)_3CClO_4$, the conjugate adducts of silyl enol ethers to enones can undergo aldol reactions to provide γ-acyl-δ-hydroxy ketones.

Example:

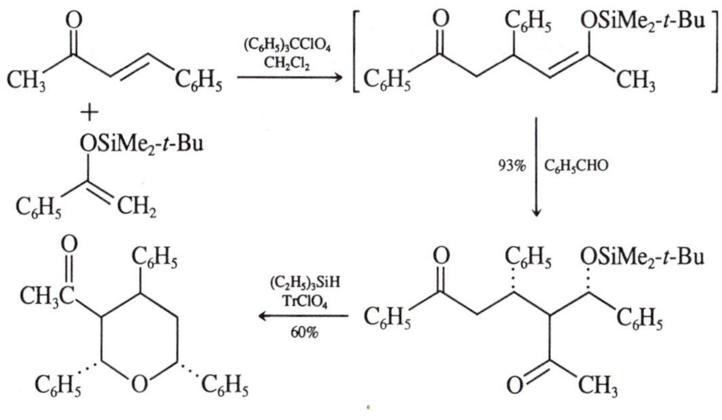

***anti*-Crossed aldols.**[3]   The reaction of trimethylsilyl enol ethers with aldehydes catalyzed by $(C_6H_5)_3CClO_4$ results in 75–90% yields of β-hydroxy ketones, but shows low diastereoselectivity. However, *anti*-aldols are obtained predominantly when *t*-butyldimethylsilyl enol ethers are used. This diastereoselectivity is independent of the geometry of the enol.

Example:

$$\underset{\substack{\text{E/Z} = 76:24 \\ = 6:94}}{\overset{\overset{\displaystyle OSiMe_2\text{-}t\text{-}Bu}{\underset{|}{}}}{C_2H_5C}}\!\!=\!\!CHCH_3 \; + \; \underset{\substack{89\% \\ 87\%}}{C_6H_5CHO} \;\; \xrightarrow[\text{CH}_2\text{Cl}_2]{(C_6H_5)_3CClO_4} \;\; \underset{\substack{| \\ CH_3}}{\overset{\overset{\displaystyle O \quad OSiMe_2\text{-}t\text{-}Bu}{\| \quad\;\; |}}{C_2H_5CCHCHC_6H_5}}$$

$$\substack{syn/anti = 16:84 \\ = 27:73}$$

Homoallyl methyl ethers.[4]   Trityl perchlorate catalyzes a reaction of allyltrimethylsilanes with dimethyl acetals or ketals to form homoallyl methyl ethers in 60–90% yield. Diphenylboryl triflate is a somewhat less efficient catalyst.

Example:

$$C_6H_5CH(OCH_3)_2 \; + \; CH_3CH\!\!=\!\!CHCH_2Si(CH_3)_3 \;\; \xrightarrow[\text{CH}_2\text{Cl}_2]{(C_6H_5)_3CClO_4} \;\; \underset{\substack{| \\ CH_3}}{\overset{\overset{\displaystyle OCH_3}{|}}{C_6H_5CH}}\!\!-\!\!CHCH\!\!=\!\!CH_2$$

$$(anti/syn = 71:29)$$

Ether synthesis.[5]   Unsymmetrical ethers can be obtained from an aldehyde or ketone and an alkoxytrimethylsilane with trityl perchlorate (equation I).

$$\text{(I)} \quad R^1COR^2 \; + \; R^3OSi(CH_3)_3 \;\; \xrightarrow[65-88\%]{\substack{1)\,(C_6H_5)_3CClO_4 \\ 2)\,(C_2H_5)_3SiH}} \;\; \underset{R^2}{\overset{R^1}{>}}\!\!CHOR^3 \; + \; (C_2H_5)_3SiOSi(CH_3)_3$$

[1] S. Kobayashi, M. Murakami, and T. Mukaiyama, *Chem. Letters*, 953 (1985).
[2] S. Kobayashi and T. Mukaiyama, *ibid.*, 221 (1986).
[3] T. Mukaiyama, S. Kobayashi, and M. Murakami, *ibid.*, 447 (1985).
[4] T. Mukaiyama, H. Nagaoka, M. Murakami, and M. Ohshima, *ibid.*, 977 (1985).
[5] J. Kato, N. Iwasawa, and T. Mukaiyama, *ibid.*, 743 (1985).

V

L-Valinol, $(CH_3)_2CHCH(NH_2)CH_2OH(\mathbf{1})$, b.p. 81°/8 mm., α_D + 14.6°.

Chiral 4,4-dialkyl-2-cyclopentenones.[1] The chiral bicyclic lactam **2**, derived from levulinic acid and **1**, on monoalkylation exhibits slight if any selectivity regardless of the electrophile. However, a second alkylation exhibits high *endo*-selectivity. This product (**3**), after reductive cleavage, furnishes a keto aldehyde that is cyclized by base to a chiral 4,4-disubstituted-2-cyclopentenone (**4**). Either antipode of **4** can be prepared by the sequence of alkylation.

$$
\mathbf{2} \xrightarrow[\text{2) } sec\text{-BuLi, C}_6\text{H}_5\text{CH}_2\text{Br}]{\text{1) } sec\text{-BuLi, C}_2\text{H}_5\text{Br}} \mathbf{3}\,(endo/exo = 95:5) \longrightarrow \mathbf{4}
$$

Diastereoselective aza-Claisen rearrangement.[2] The oxazole **1** prepared by reaction of L-valinol with propionic acid is readily convertible into an N-allylketene acetal (**2**), which rearranges at 150° to **3** in 94% de and 80% overall chemical yield. Acid catalyzed hydrolysis of **3** gives (R)-(−)-2-methyl-4-pentenoic acid (**4**) (85% yield) with recovery of L-valinol.

$$
\mathbf{1} \xrightarrow[\text{2) } n\text{-BuLi}]{\text{1) } CH_2=CHCH_2OTs} \mathbf{2} \xrightarrow[\substack{80\% \\ \text{overall}}]{150°} \mathbf{3}\,(94\%\ de) \xrightarrow[85\%]{H_3O^+} \mathbf{4}\,(93\%\ ee)
$$

[1] A. I. Meyers and K. T. Wanner, *Tetrahedron Letters*, **26**, 2047 (1985).
[2] M. J. Kurth, O. H. W. Decker, H. Hope, and M. D. Yanuck, *Am. Soc.*, **107**, 443 (1985).

Vilsmeier reagent.

Formylation of indoles.[1] 3-Arylindoles are formylated at C_2 by the Vilsmeier reagent ($POCl_3$/DMF) in 40–86% yield or by 1,1-dichloromethyl methyl ether in the presence of a slight excess of $TiCl_4$ or $SnCl_4$ (35–52% yield).

Example:

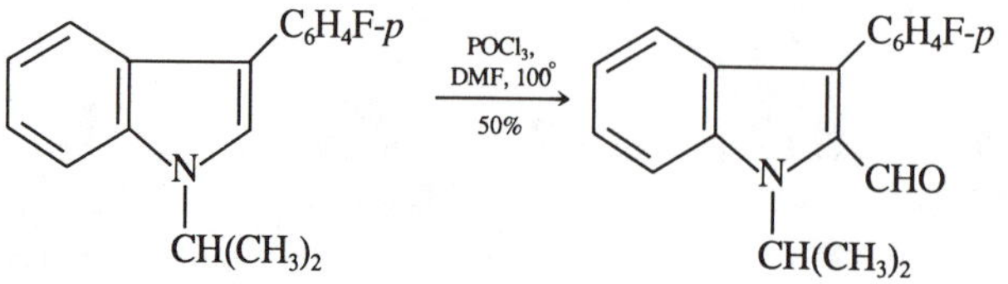

Formylation of alcohols; formate esters.[2] The Vilsmeier adduct **1** of DMF and benzoyl chloride reacts with alcohols to give stable imidate ester chlorides (**2**). These are hydrolyzed by dilute acid to formate esters (**3**).

$$\underset{\mathbf{1}}{\underset{Cl^-}{(CH_3)_2\overset{+}{N}{=}CHO\overset{\overset{O}{\|}}{C}C_6H_5}} \xrightarrow[60-75\%]{ROH} \underset{\mathbf{2}}{\underset{Cl^-}{(CH_3)_2\overset{+}{N}{=}CHOR}} \xrightarrow{H_3O^+} \underset{\mathbf{3}}{ROCHO}$$

RCOOH → RCHO. The iminium salt (**1**) formed on reaction of a carboxylic acid with the Vilsmeier reagent (formed from DMF and oxalyl chloride) is reduced by lithium tri-*t*-butoxyaluminum hydride (1 equiv.) to an aldehyde. The chemoselectivity is noteworthy; ester, nitrile, keto, and halide groups are not affected.[3]

$$RCOOH \xrightarrow{ClCH{=}\overset{+}{N}(CH_3)_2Cl^-} \underset{\mathbf{1}}{\left[\underset{Cl^-}{RCOOCH{=}\overset{+}{N}(CH_3)_2}\right]} \xrightarrow[55-93\%]{LiAlH(O\text{-}t\text{-}Bu)_3} \underset{\mathbf{2}}{RCHO}$$

[1] R. E. Walkup and J. Linder, *Tetrahedron Letters*, **26**, 2155 (1985).
[2] J. Barluenga, P. J. Campos, E. Gonzalez-Nuñez, and G. Asensio, *Synthesis*, 426 (1985).
[3] T. Fujisawa, T. Mori, S. Tsuge, and T. Sato, *Tetrahedron Letters*, **24**, 1543 (1983); T. Fujisawa and T. Sato, *Org. Syn.*, submitted (1985).

Vinylene carbonate, 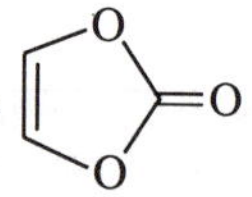

(**1**). Mol. wt. 86.05, m.p. 22°. Supplier: Aldrich.

Diels-Alder reaction.[1] The first step in a synthesis of citreoviral (**5**), a metabolite of *Penicillium citreoviride*, from 2,4-dimethylfuran (**2**) is a Diels-Alder reaction with **1**, to give a 7:5 mixture of *endo*- and *exo*-adducts (**3**). Both are converted in 3 steps to the diol (**4**), a precursor to **5**.

[1] Y. Shizuri, S. Nishiyama, H. Shigemori, and S. Yamamura, *J.C.S. Chem. Comm.*, 292 (1985).

Vinyltrimethylsilane.

Allylic amines.[1]   The trimethylsilylisoxazolidines formed on cycloaddition of vinyltrimethylsilanes with nitrones (**12**, 566) are converted on reductive cleavage of the N—O bond and concomitant Peterson elimination of $(CH_3)_3SiOH$ into allylic amines.

Example:

This sequence can be used to convert allyltrimethylsilanes into homoallylic amines.

[1] P. DeShong, J. M. Leginus, and S. W. Lander, Jr., *J. Org.*, **51**, 574 (1986).

Vinyl(triphenyl)phosphonium bromide (1).

[2 + 2 + 2]Cyclohexene annelation.[1]   Reaction of ketone enolates with 2 equiv. of this salt results in two sequential Michael reactions to give a cyclohexenyl(triphenyl)phosphonium bromide, which can be hydrolyzed to a cyclohexenyl(diphenyl)phosphine oxide.[1]

Example:

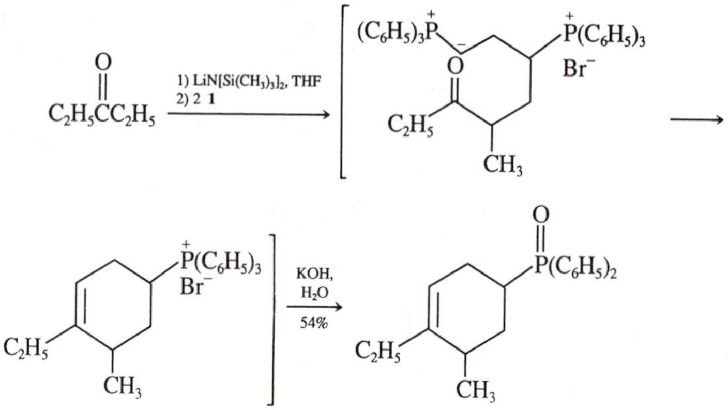

1,4-*Dienes*.[2]   Cuprates react with vinyltriphenylphosphonium bromide **(1)** to form phosphoranes, which react with aldehydes to form alkenes. Addition of HMPT to the cuprate favors formation of (Z)-alkenes.

$$Bu_2CuLi \; + \; CH_2{=}CH\overset{+}{P}(C_6H_5)_3 \underset{Br^-}{} \xrightarrow[\text{HMPT}]{\text{THF}} [BuCH_2CH{=}P(C_6H_5)_3] \xrightarrow[80\%]{C_6H_5CHO} BuCH_2CH{=}CHC_6H_5$$

$$\textbf{1} \qquad\qquad\qquad (Z/E = 97{:}3)$$

This reaction can be extended to the synthesis of (Z,Z)-1,5-disubstituted 1,4-pentadienes (equation I).

(I)

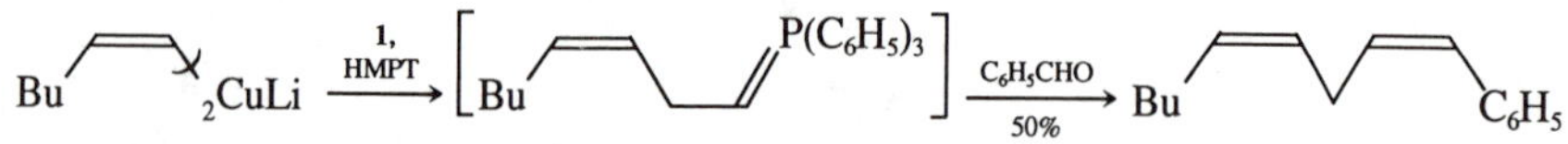

[1] G. H. Posner and S.-B. Lu, *Am. Soc.*, **107**, 1424 (1985).
[2] G. Just and B. O'Connor, *Tetrahedron Letters*, **26**, 1799 (1985).

X

Xenon(II) fluoride.

Fluorodecarboxylation.[1]   Reaction of alkanoic acids with XeF_2 in CH_2Cl_2 can result in alkyl fluorides by replacement of the COOH group by F. Yields are highest with primary, tertiary, and benzylic acids. Aryl or vinylic acids do not undergo this reaction.

[1] T. B. Patrick, K. K. Johri, D. H. White, W. S. Bertrand, R. Mokhtar, M. R. Kilbourn, and M. J. Welch, *Can. J. Chem.*, **64**, 138 (1986).

Z

Zinc.

Coupling of allylic acetates.[1]   Allylic acetates are coupled to 1,5-dienes with zinc dust in the presence of $Pd[P(C_6H_5)_3]_4$ in THF. Addition of methanol or 1,2-ethanediol increases the preservation of the stereochemistry and the rate of reaction.

Examples:

$$(C_6H_5)_2C = CHCH_2OAc \xrightarrow[83\%]{THF, CH_3OH} (C_6H_5)_2C = CH(CH_2)_2CH = C(C_6H_5)_2 \ +$$

70:30

$$\underset{(C_6H_5)_2C = CHCH_2\overset{|}{C}(C_6H_5)_2}{\overset{CH=CH_2}{\overset{|}{}}}$$

Carboxylation.[2]   Ultrasonically dispersed zinc powder in DMF promotes direct carboxylation (CO_2 or dry ice) of perfluoroalkyl iodides.

Example:

$$C_8F_{17}I \xrightarrow[72\%]{CO_2, Zn, (((, DMF}} C_8F_{17}COOH$$

β-Keto-γ-butyrolactones; tetronic acids.[3]   Zinc, activated with $Cu(OAc)_2$,[4] promotes a Reformatsky-type reaction of α-bromo esters with O-silylated cyanohydrins to provide β-keto-γ-butyrolactones or tetronic acids.

Reformatsky reaction.[5]   The Reformatsky reaction of γ-bromocrotonic acid with aldehydes or ketones in THF gives rise to a mixture of branched and linear unsaturated hydroxy acids. The former are the kinetic products and, on equilibration, rearrange to the latter products.

346

Example:

$$Br \diagdown \diagup \diagdown CO_2ZnBr \ + \ C_6H_5CHO \xrightarrow{\substack{1)\ Zn,\ THF \\ 2)H_3O^+}}$$

0 , 10 days	70%
4 hrs, reflux	72%
100 hrs, reflux	66%

$$\underset{C_6H_5}{\overset{H_2C}{}} \diagup \overset{COOH}{} \diagdown OH \quad + \quad \underset{OH}{C_6H_5} \diagdown \diagup \diagdown \overset{COOH}{}$$

87:13
77:23
0:100

[1] S. Sasaoka, T. Yamamoto, H. Kinoshita, K. Inomata, and H. Kotake, *Chem. Letters*, 315 (1985).
[2] T. Kitazume and N. Ishikawa, *ibid.*, 137, 1453 (1982); *idem*, *Org. Syn.*, submitted (1985).
[3] L. R. Krepski, L. E. Lynch, S. M. Heilmann, and J. K. Rasmussen, *Tetrahedron Letters*, **26**, 981 (1985).
[4] E. LeGoff, *J. Org.*, **29**, 2048 (1964).
[5] M. Bellassoued, F. Habbachi, and M. Gaudemar, *Tetrahedron*, **41**, 1299 (1985).

Zinc amalgam.

Intramolecular reductive amino cyclization.[1]   The key step in a synthesis of the ergot alkaloid aurantioclavine (**3**) from 3-formylindole (**1**) involves treatment of the nitro olefin **2** with amalgamated zinc in refluxing methanolic 2 *N* hydrochloric acid. Zn–HOAc, Fe–HOAc, SnCl$_2$–HCl are not useful for this purpose. Cyclization is believed to involve an intermediate hydroxylamine, which can be isolated from short-term reactions.

1   4 steps →   **2**   Zn(Hg), HCl, CH$_3$OH / 67% (31% overall) →   **3**

[1] F. Yamada, Y. Makita, T. Suzuki, and M. Somei, *Chem. Pharm. Bull.*, **33**, 2162 (1985).

Zinc/copper couple.

Conjugate additions to enals and enones.[1]   In the presence of zinc (or tin) allyl halides undergo 1,2-addition to carbonyl compounds. The reaction is facilitated by sonication and proceeds in highest yield in saturated aqueous ammonium chloride/THF (5:1).[1]

The same group[2] now finds that a zinc/copper couple prepared by sonication of Zn and CuI in ethanol/H_2O (9:1) permits conjugate addition of alkyl halides to enones and enals. The order of reactivity is RI > RBr and *tert* > *sec* >> primary. THF/H_2O or Py/H_2O or even pure water can be used as solvent. This reaction can hardly involve a classical organometallic reagent, but probably involves an alkyl radical.

Examples:

[1] C. Pétrier and J. L. Luche, *J. Org.*, **50**, 910 (1985); C. Petrier, J. Einhorn, and J. L. Luche, *Tetrahedron Letters*, **26**, 1449 (1985).
[2] C. Petrier, C. Dupuy, and J. L. Luche, *ibid.*, **27**, 3149 (1986).

Zinc/silver–graphite.
A laminated Zn/Ag–C_8 can be prepared by reaction of C_8K with $ZnCl_2$/AgOAc in THF.

Reformatsky reaction.[1]   This form of zinc is highly effective for Reformatsky reactions with both α-bromo- and α-chloro esters and with both cyclic and acyclic ketones. Reactions proceed in THF at −78 to 0° in 10–30 minutes. It even effects diastereospecific Reformatsky reactions with chiral carbohydrate ketones.

Example:

[1] R. Csuk, A. Furstner, and H. Weidmann, *J.C.S. Chem. Comm.*, 775 (1986).

Zinc bromide.

Intramolecular ene reaction of **1,7-dienes.**[1]   The ZnBr$_2$-catalyzed reaction of the activated 1,7-diene **1** gives the *trans*-disubstituted cyclohexane **2** (>98% selectivity.) Under the same conditions, the chiral 1,7-diene **3** [from (R)-citronellal] is converted into two diastereomeric *trans*-disubstituted cyclohexanes, **4** and **5** in the ratio 96.5:3.5. Slightly higher diastereoselectivity obtains with (C$_2$H$_5$)$_2$AlCl (96% de), but the thermal ene reaction gives **4** and **5** in the ratio 86:14.

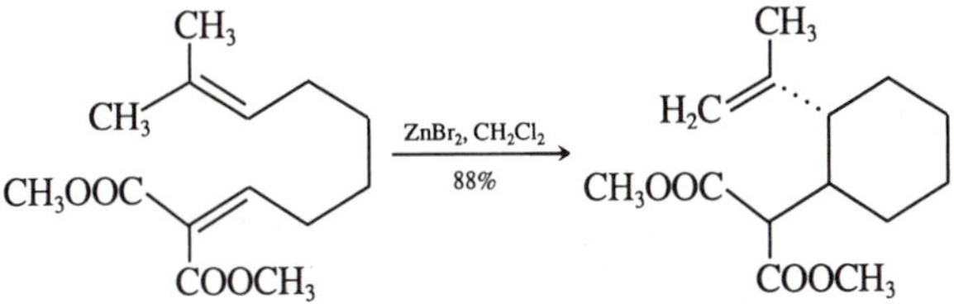

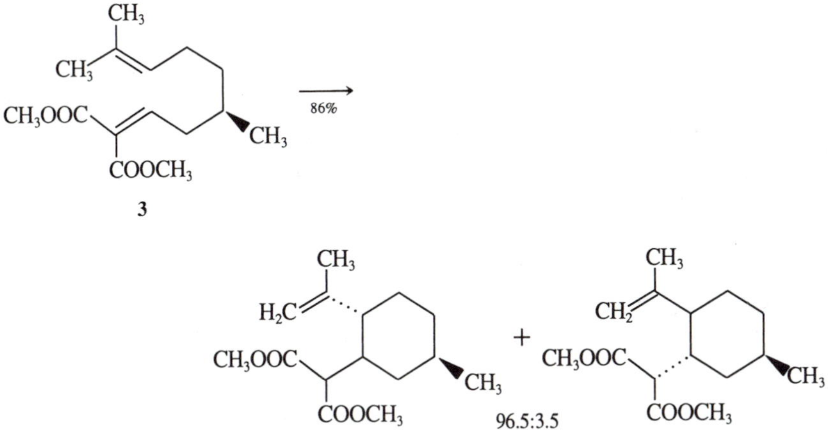

[1] L. F. Tietze and U. Biefuss, *Angew. Chem. Int. Ed.*, **24**, 1042 (1985); *Idem, Tetrahedron Letters*, **27**, 1767 (1986).

Zinc chloride.

Homo-Reformatsky reaction.[1]   The reaction of 1-ethoxy-1-trimethylsilyloxy-cyclopropane (**1**) with an aldehyde in the presence of ZnCl$_2$ results in γ-silyloxy esters via a zinc homoenolate (**a**) of ethyl propionate (equation I). ZnI$_2$ is the preferred catalyst in the case of reactions with acetophenone and benzaldehyde dimethyl acetal and in reactions of 1-isopropoxy-1-(*t*-butyldimethylsilyl-oxy)cyclopropane with aromatic aldehydes.

(I) with $\mathrm{OSi(CH_3)_3}$ and $\mathrm{OC_2H_5}$ (**1**) $\xrightarrow{\mathrm{ZnCl_2}}$ $\mathrm{C_2H_5OCCH_2CH_2ZnCl}$ (**a**) $\xrightarrow[50-90\%]{\mathrm{RCHO}}$ $\mathrm{RCH(CH_2)_2COOC_2H_5}$ with $\mathrm{OSi(CH_3)_3}$

The zinc homoenolate **a** reacts with acid chlorides in $\mathrm{CH_2Cl_2}$ to form 1-(acyl-oxy)cyclopropanes.

Example:

$$\mathbf{a} + \mathrm{C_6H_5(CH_2)_2COCl} \xrightarrow[47\%]{\mathrm{CH_2Cl_2}}$$

79% $\left|\begin{array}{l}\mathrm{(CH_3)_3SiCl}\\ \mathrm{HMPT,\ ether}\end{array}\right.$

$$\mathrm{C_6H_5(CH_2)_2\overset{O}{\overset{\|}{C}}(CH_2)_2COOC_2H_5}$$

Aldol condensations of enones with α-ketols.[2] $\mathrm{ZnCl_2}$ promotes the crossed-aldol condensation of enones with protected α-ketols. This reaction has been used for a short synthesis of frontalin (**1**).

[1] H. Oshino, E. Nakamura, and I. Kuwajima, *J. Org.*, **50**, 2802 (1985).
[2] H. Hagiwara and H. Uda, *J.C.S. Perkin I*, 91 (1984); *idem, ibid.*, 283 (1985).

Zinc iodide.

β-*Lactams*.[1] In the presence of $\mathrm{ZnI_2}$ and *t*-butyl alcohol as a proton source, N-trimethylsilyl imines (**1**), prepared as shown, react with silyl ketene acetals (**2**) to form N-unsubstituted azetidinones (**3**) in 60–80% yield. Other Lewis acids and Bu₄NF or KF are not useful catalysts.

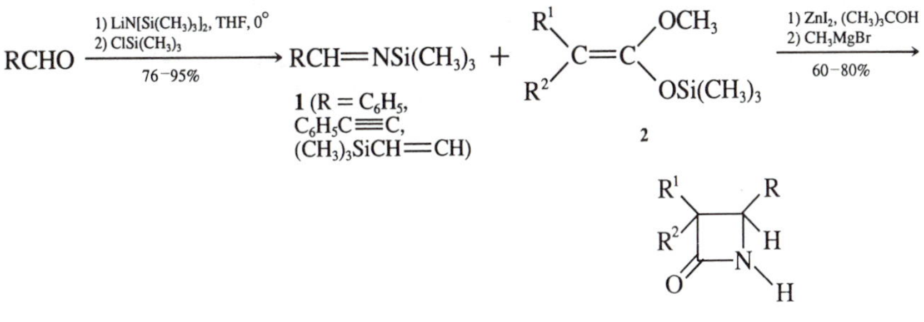

$$RCHO \xrightarrow[\substack{76-95\%}]{\substack{1)\ LiN[Si(CH_3)_3]_2,\ THF,\ 0° \\ 2)\ ClSi(CH_3)_3}} RCH=NSi(CH_3)_3 +$$

1 (R = C₆H₅,
C₆H₅C≡C,
(CH₃)₃SiCH=CH)

2

3

---

¹ E. W. Colvin and D. G. McGarry, *J.C.S. Chem. Comm.*, 539 (1985).

Zirconium(IV) acetylacetonate, Zr(acac)₄. Supplier: Alfa.

*Regioselective acylation of diols.*¹ 3-Acyl-2-oxazolones² are useful for acylation of amines and thiols to give peptides and thioesters. They also acylate alcohols in high yield in the presence of this catalyst. With this catalyst, the reactivity for acylation of hydroxyl groups in primary >> phenolic > secondary. However, for selective acylation of primary hydroxyl groups of 1,2-diols, the Cp₂ZrHCl-Mg couple is superior to Zr(acac)₄.

Examples:

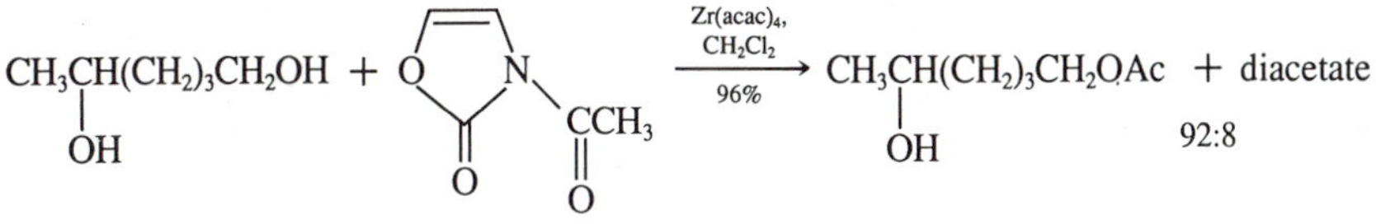

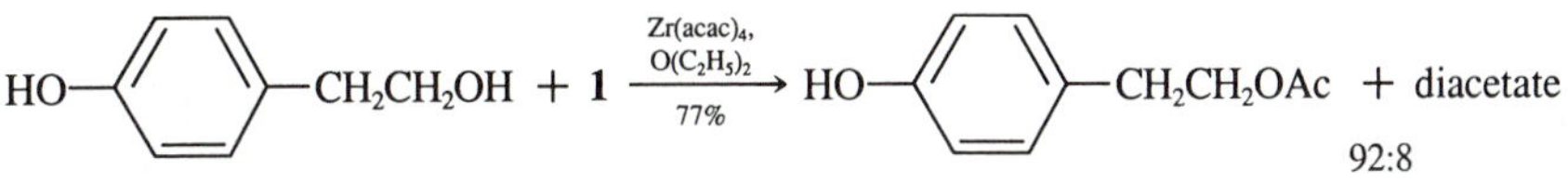

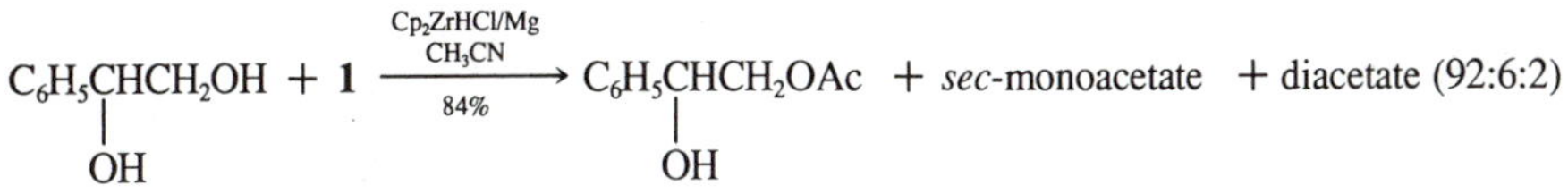

¹ T. Kunieda, T. Mori, T. Higuchi, and M. Hirobe, *Tetrahedron Letters*, **26**, 1977 (1985).
² T. Kunieda, T. Higuchi, Y. Abe, and M. Hirobe, *Tetrahedron*, **39**, 3253 (1983).

Zirconium(IV) isopropoxide, $Zr[OCH(CH_3)_2]_4$ (1).

Reduction of pyrimidine-2(1H)-ones.[1]   The pyrimidinone **2** is reduced by metal hydrides such as lithium tri-*t*-butoxyaluminum hydride to a mixture of the 3,6- and 3,4-dihydro derivatives in the ratio 9:1. In contrast, Meerwein-Ponndorf-Verley reduction with **1** in isopropanol results only in the 3,4-dihydro derivative (**3a**) but the reaction is slow and stops after two days to provide only a 25% yield. However, introduction of a halo substituent at C_5 results in enhanced yields of the 3,4-dihydro derivatives, with the highest yields obtained with the 5-chloro derivative.

The traditional reagent for this type of reduction, triisopropoxyaluminum, is less efficient than **1**.

2a, X = H
2b, X = Cl
2c, X = Br

3a 25%
b 91%
c 63%

Intramolecular Michael-aldol cyclization.[2]   One route to a hydrindane involves base-catalyzed cyclization of the keto aldehyde **1** to the hydrindene **2**. Although $Zr(O\text{-}i\text{-}Pr)_4$ is useful, $Zr(O\text{-}n\text{-}Pr)_4$ is the most satisfactory base, both in respect to yield and selectivity.

$Zr(O\text{-}n\text{-}Pr)_4$	80%	10:1:1
$Mg(OCH_3)_2$	51%	10:1.5:1
$NaOCH_3$	80%	3.4:1.2:1

In this case, an intramolecular Diels-Alder reaction is less stereoselective.

[1] T. Høseggen, F. Rise, and K. Undheim, *J.C.S. Perkin I*, 849 (1986).
[2] S. K. Attah-Poku, F. Chau, V. K. Yadav, and A. G. Fallis, *J. Org.*, **50**, 3418 (1985).

TYPE OF REACTION INDEX

SYNTHESIS INDEX

TYPE OF COMPOUND INDEX

401

INDEX OF SYNTHETIC TARGETS

AUTHOR INDEX

REAGENT INDEX